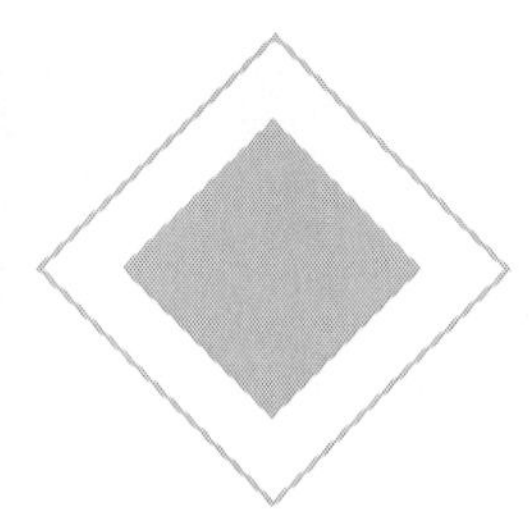

Construction Contract Claims

3rd Edition

Reg Thomas, BSc (Hons), FCIOB, ACIArb

Formerly Executive Director of James R. Knowles

Mark Wright, FCIArb, Barrister

Managing Director, Wright Contractual Services Limited

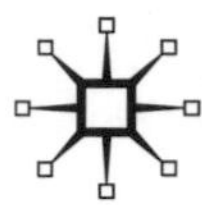

First edition 1993
Second edition 2001
This edition first published 2011 by
PALGRAVE MACMILLAN

Palgrave Macmillan in the UK is an imprint of Macmillan Publishers Limited, registered in England, company number 785998, of Houndmills, Basingstoke, Hampshire RG21 6XS.

Palgrave Macmillan in the US is a division of St Martin's Press LLC, 175 Fifth Avenue, New York, NY 10010.

Palgrave Macmillan is the global academic imprint of the above companies and has companies and representatives throughout the world.

Palgrave® and Macmillan® are registered trademarks in the United States, the United Kingdom, Europe and other countries.

ISBN 978–0–230–24285–2

This book is printed on paper suitable for recycling and made from fully managed and sustained forest sources. Logging, pulping and manufacturing processes are expected to conform to the environmental regulations of the country of origin.

A catalogue record for this book is available from the British Library.

A catalog record for this book is available from the Library of Congress.

Contents

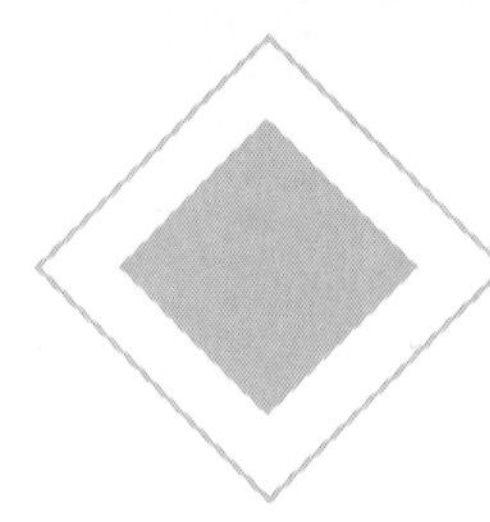

Foreword to the Third Edition

Reg Thomas, an old friend and former colleague, wrote the first two editions of this book and passed on the mantle to Mark Wright, another old friend and former colleague, to produce the Third Edition. It is therefore a great pleasure for me to again be asked to write the Foreword.

The government is implementing cuts amounting to £80billion to its expenditure over the five years from 2000 to 2015, a significant part of which will fall on much-needed construction projects. Competition for work will become very tough, with those responsible for procurement, demanding significantly reduced prices with no reduction in the quality. Disputes will, as a result, be on the increase and books such as this will be a godsend. However an appropriate textbook is of little use if it is out of date, in fact its use could be dangerous. Mark Wright has therefore done a first-class job of amending Reg Thomas' book to include all relevant changes to standard contracts and case law.

The drafters of standard contracts are forever introducing changes and this book reflects the latest revisions to those contracts extensively employed in the United Kingdom and on international projects. The JCT family of contracts comprise the standard contracts most commonly used in the United Kingdom. In 2005 these contracts were the subject of a major overhaul, with three further revisions since, the latest of which (2011) has been catered for in this revision. The NEC contract, first published in the early 1990s, with the latest revision published in 2005, has only in the past five years or so risen to prominence. Mark has introduced amendments to reflect the increase in use of this form. One aspect of the NEC contract, the Compensation Events provisions which deal with extension of time and additional payments, has caused a great deal of controversy. It is therefore very reassuring to note that Appendix C includes an 'Example of the Procedure for the Notification and Assessment of a Compensation Event under the NEC 3 Contract'.

The book brings the reader up to date with regard to recent case law. The legal and contractual requirements for written notices to be submitted by Contractors and Subcontractors to support claims for extensions of time and additional payment have for many years been the subject of dispute. Over the years there has been a steady stream of cases which deal with this matter. The cases included in this edition bring the subject up to date. While mentioning extensions of time, it seems appropriate to refer to concurrent delays. Disputes concerning entitlements to extensions of time when concurrent delays are concerned can quickly turn into heated argument. A number of cases have been brought before the courts, where concurrent delays have been involved. Unfortunately the courts have sent out mixed messages as to entitlement, leaving behind a great deal of uncertainty, all of which have been delicately dealt with in this book.

To bring the book right up to date, the provisions of the Local Democracy, Economic Development and Construction Act 1999 which updates the Housing Grants, Construction and Regeneration Act 1996, together with the effect on the Scheme for Construction Contracts have received comprehensive coverage.

Finally, while not what would normally be expected in a book which deals with construction claims, there is included a very helpful review of the Competition Act 1998 and the Enterprise Act 2002. In view of the fines totalling £129.5m levied on hundred-and-three contractors in 2009 for infringements and the possibility of a jail sentence for those who transgress, it is a very useful subject to be included.

I am pleased to be able to record that Mark Wright in writing this Third Edition of Construction Contract Claims has maintained the very high standard set by Reg Thomas.

Roger Knowles
FRICS FCIArb FQSi Barrister

Preface to the Third Edition

Nothing stays the same forever and so it is in the field of construction claims. Since the second edition of this book, there have been a number of significant developments in the way construction contract claims are resolved in the United Kingdom, primarily due to the approach of the courts to decisions of adjudicators. Adjudication, and mandatory payment requirements, were introduced into construction contracts by the Housing Grants, Construction and Regeneration Act 1996 and adjudication has been a great success as a means of dispute resolution. However, both the adjudication and payment provisions of the Act have required refining and amendments to the Housing Grants, Construction and Regeneration Act 1996 have been introduced, most notably by the Local Democracy, Economic Development and Construction Act 2009. As a consequence of the amendments, the Scheme for Construction Contracts has been amended. The relevant provisions of the amended Housing Grants, Construction and Regeneration Act 1996 and the amended Scheme for Construction Contracts are set out in Appendices D and E to this third edition.

Most of the recognised standard forms of contract have been amended since the second edition of this book and employing the conveyor belt principle the oldest of the forms referred to in the second edition have been taken off the conveyor and latest editions added while those which the latest editions have replaced are still included if it is thought that they may still be in use. Although the JCT has tried to do something about it, the Management Form of Contract has not been the success it was hoped and it has been revised. The big success among the JCT stable of contracts is the design and build standard form which employers have recognised as a form which reduces their risk and attendant claims.

The NEC form referred to in the second edition has been revised and it is now in its third edition and goes from strength to strength. The procedure for making a claim under the NEC 3 form of contract is significantly different to that under other standard forms of contract. An example of a claim under the NEC 3 form of contract is set out in Appendix C.

The practice of employing nominated subcontractors has been recognised as having had its day and it is no longer included in many of the standard forms of subcontract in the United Kingdom.

Since the second edition of this book, there have been developments in competition law both within the United Kingdom and within Europe concerning what is and what is not allowable at tender stage. Statutory legislation has had a significant effect by way of the Competition Act 1998 and the Enterprise Act 2002 as many construction companies have recently found to their cost. These developments are to be ignored at peril as there are significant financial penalties for infringements and the possibility of imprisonment for the perpetrators. All involved in the tendering process need to be aware of the relevant law as the United Kingdom construction industry has innocently acted in contravention of the law for many years and as some contractors and individuals have recently found out, ignorance of the law is no defence to contravention of it. The pitfalls are referred to in Chapter 3.

In recent years public procurement has become an important part of the construction industry and as more contractors are prepared to look further afield, contractors should be aware of the relevant regulations and, in particular, remedies that contractors may have if a competitor obtains an unfair advantage. Public procurement is governed throughout Europe by European legislation which has filtered down into domestic legislation. The United Kingdom and European Courts are keen to enforce the relevant legislation. Failure to adhere to the relevant legislation will result in unsuccessful tenderers being able to make claims against the tendering authority. The regulations are identified in Chapter 3.

Mark Wright

Preface to the First Edition

There are a number of excellent text books on construction law, contracts and claims. The author has referred to *Hudson's Building and Engineering Contracts, tenth edition* for a number of early cases, and readers are advised to refer to this invaluable source for a better understanding of many issues discussed in this book. Publications by James R. Knowles listed in the bibliography have also been invaluable in the preparation of this book and are recommended for further reading. Knowles' publications and summaries of the cases cited in References may be purchased from Knowles Publications, Wardle House, King Street, Knutsford, Cheshire WA16 6PD. The contents of this book are intended to present to readers a general view of the practical problems which exist and how they might be avoided or resolved. The views expressed by the author represent several years' experience of looking backwards at projects which have gone wrong. In practice, many projects go well, are completed without major claims, and where they do occur, they are often settled promptly, professionally and amicably. Unfortunately, there is an increasing incidence of claims, most of which are brought about by financial pressures which stretch the resources of consultants, contractors and subcontractors alike. Many firms do not have sufficient allowances built into their fees, or into the contract price, to carry out their obligations properly. Some firms lack sufficient staff with the skills required to manage projects efficiently and to deal with claims in a professional manner. Insufficient attention to training staff, so that they can be better prepared to deal with claims, is another reason for many of the problems which exist in the industry. Whilst many claims are well presented and dealt with professionally by the recipient, some of these failures are evidenced in the presentation and quality of some claims submitted by large and small firms alike and in the response made by some architects, engineers and quantity surveyors.

The chapters which follow attempt to guide readers through the history of developments in law and contracts so that they may understand more fully the reasons for good contracts administration as a means of avoiding or minimising the effects of claims for delay and disruption.

Some of the arguments and methods of quantifying claims in this book should be regarded as possible means of persuasion according to the circumstances and records which are available to support a claim. In some cases, a lack of records may not be fatal to a claim, but it may be an uphill battle to persuade the recipient of a claim to pay out large sums of money on the basis of hypothetical calculations which have no real foundation. Readers should be aware that there is no real substitute for good records when it comes to quantifying a claim for an extension of time or for additional payment. Nevertheless, if the contractor has been delayed at almost every turn, it must be right that he receives some relief and compensation so far as it can be established by applying commonsense according to the circumstances. As a consultant to contractors and subcontractors, a duty is owed to them to use every means available, providing that they are honest and justifiable, to obtain the best possible settlement of their claims. As a consultant to employers (or to contractors defending a claim from subcontractors), a duty is owed to them to defend all claims and to discredit any unmeritorious claims. Nevertheless, employers (and contractors as the case may be) will need to be advised on the possible worth of a claim in order to facilitate a decision as to settlement or arbitration or litigation.

Whilst some practitioners may seek refuge in cases in which claims have been rejected on the grounds that the records and/or the method of quantification were lacking, the author supports the view expressed in *Penvidic Contracting Co. Ltd* v. *International Nickel Co. of Canada Ltd* (1975) 53 DLR (3d) 748 (quoting Davies J in *Wood* v. *Grand Valley Railway Co*) – see *A Building Contract Casebook* by Dr Vincent Powell Smith and Michael Furmston at page 316:

> 'It was clearly impossible under the fact of that case to estimate with anything approaching to mathematical accuracy the damages sustained by the plaintiffs, but it seems to me clearly laid down there by the learned Judges that such an impossibility cannot "relieve the wrongdoer of the necessity of paying damages for his breach of contract" and that on the other hand the tribunal to estimate them, whether jury or Judge, must under such circumstances do "the best it can" and its conclusion will not be set aside even if '*the amount of the verdict is a matter of guess work.*' (emphasis added).

However, the above quotation should not be relied upon to cure all ills. The terms of the contract and other circumstances may require a more robust approach when defending any claim which is clearly deficient in the essential ingredients to justify anything less than total or partial rejection.

It is hoped that this book will provide useful guidance for those responsible for dealing with claims so that they can be resolved with the minimum cost and without any party being seriously disadvantaged.

Reginald W. Thomas
Spring 1992

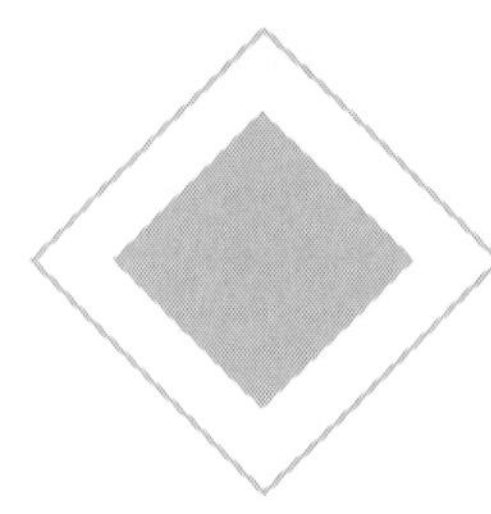

Preface to the Second Edition

Since the first edition of this book, there have been several important changes in contracts and law which are worthy of note. There have also been a number of excellent new publications, in particular, the eleventh edition of *Hudson's Building and Engineering Contracts*, *Delay and Disruption in Construction Contracts* by Keith Pickavance and a number of 'up-to-the minute' regular publications by James R. Knowles to which the author has been fortunate to have had access and which have been invaluable in the preparation of this book. As in the first edition of this book, the author has referred to the tenth edition of '*Hudson*' and its supplements for a number of early cases and to *Construction Contracts: Principles and Policies in Tort and Contract* by the same author, much of which is now reproduced in the eleventh edition of '*Hudson*'. In addition the author has sourced a number of important US cases of interest from *Construction Delay Claims* by Barry B. Bramble and Michael T. Callahan, a publication which ought to be read by those wishing to have an account of many aspects of claims which are seldom covered in detail in the UK.

During the decade since the first edition, the report 'Constructing the Team' published under the chairmanship of Sir Michael Latham (The Latham Report) stimulated constructive discussion about the direction of contracting in the UK. The Government, in its enactment of the *Housing Grants, Construction and Regeneration Act 1966* (*The Construction Act*), took on board many of the recommendations in the Latham Report. The publication of 'The New Engineering Contract' (NEC), now reissued under a new title, 'The Engineering and Construction Contract', helped to promote a new approach to contracting based on good contract administration and minimisation of disputes (as well as quick and effective resolution of such disputes if they arose). The decade ended with the NEC gaining ground in the UK and internationally.

Unfortunately, the good ingredients in the NEC have not been grasped by institutions promoting other forms of contract. Whilst it is true that dispute resolution in the UK has been given new dimensions by requirements imposed by the Construction Act, little has been done to follow some of the better principles found in the NEC. At the end of the decade, Fédération International des Ingénieurs-Conseils (FIDIC) introduced its test editions of four new contracts, three of which were to replace existing contracts and one of which was entirely new (for small works). The test editions illustrated an attitude which promoted adversity by the introduction of stringent notice provisions which could only have increased the incidence of poor relationships and disputes. Fortunately, after consultation with contractors and other interested parties, FIDIC has softened its approach to some extent. It has introduced a Dispute Adjudication Board into all of its standard contracts which fits broadly into the recommendations of the Latham Report. FIDIC has also improved procedures for contract administration by the contractor, but has not seen fit to bring the employer into the team in the same way as the NEC.

This book does not seek to promote any one single contract over another. The criticism of FIDIC when compared to NEC is intended to illustrate the author's view that, in spite of several important changes in contracts and law, the fundamental divide between employers and contractors is still fairly deep rooted. Claims and disputes are unlikely to change in substance and form in the near future unless all sides of the industry recognize that co-operation is more effective than separation. A new wind of change is still needed if the highly experienced and expensive resources currently engaged in the claims and arbitration business are to be better used in designing, managing and constructing exciting projects in the twenty-first century.

> 'An offending brother is more unyielding than a fortified city, and disputes are like the gates of a citadel.' Proverbs 18: 19 (NIV)

Reg Thomas
Spring 2000

Acknowledgements

The authors would like to thank the following for providing assistance in the production of the third edition of the book:

- Phil Walters for reviewing the second edition of the book and providing constructive criticism concerning what should be added, omitted or revised.
- Peter Dale of P W Dale Consulting Limited for revising the JCT claim for extensions of time and reimbursement of loss and expense found in Appendix A.
- Chris Holden for producing the example of a Compensation Event which is to be found in Appendix C.

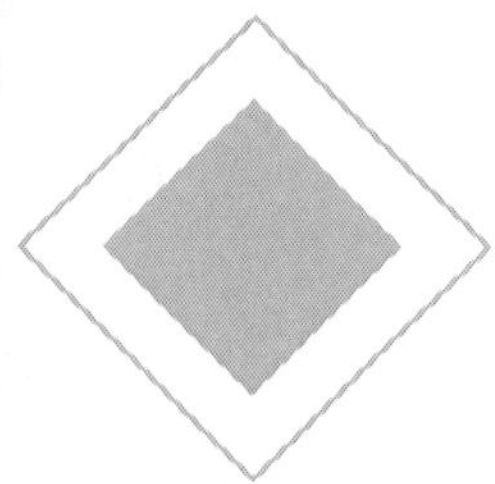

Abbreviations used in Case References

AC	*Law Reports Appeal Cases.*
AD	*South African Supreme Court Appellate Division.*
ALJR	*Australian Law Journal Reports.*
All ER	*All England Law Reports.*
ASBCA	*Army Service Board of Contract Appeals (USA).*
BCA	*Board of Contract Appeals (USA).*
BLR	*Building Law Reports.*
CA	*Court of Appeal (UK).*
CLD	*Construction Law Digest.*
CMLR	*Common Market Law Reports.*
ConLR	*Construction Law Reports.*
DLR	*Dominion Law Reports.*
EG	*Estates Gazette.*
F	*Federal Circular (USA).*
HKLR	*Hong Kong Law Reports.*
HL	*House of Lords.*
JP	*Justice of the Peace and Local Government Review.*
KB	*King's Bench.*
Lloyds LR	*Lloyd's Law Reports.*
LT	*Law Times Report.*
M & W	*Meeson & Welsby.*
MLJ	*Malaysian Law Journal.*
NZLR	*New Zealand Law Reports.*
PC	*Privy Council.*
QBD	*Queen's Bench Division.*
SA	*South African Law Reports.*
SC	*Session Cases.*
SCR	*Supreme Court Reports.*
SLR	*Singapore Law Reports.*
SLT	*Scots Law Times.*
SW	*South Western Reporter (USA).*
TPD	*South African Law Reports Transvaal Provincial Division.*
VR	*Victorian Reports.*
WLR	*Weekly Law Reports.*

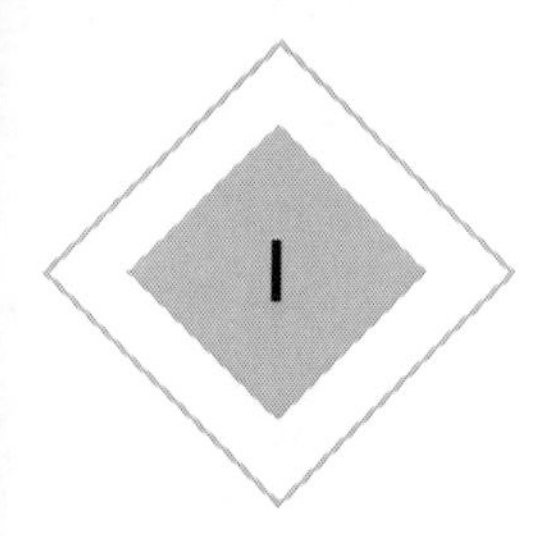

Brief History of Construction Contracts and Case Law

1.1 Introduction

Modern contracts are used in a commercial environment which has encouraged the development of claims in construction contracts for a number of years. Nevertheless, some of the conditions of contract used today are based on documents that were drawn up in the nineteenth century, and much of the construction law that is relied upon in the courts and in arbitration has been made as a result of cases that took place in the industrial revolution.

Civil engineering contracts evolved significantly in the nineteenth century, mainly as a result of the growth in transport, such as canals and railways. Most early contracts had the essential ingredients governing price, time for completion, damages and specification of the work to be done, but it was the construction of the canals and railways which eventually caused entrepreneurs to consider additional provisions such as health, safety and welfare and to make contractual provisions governing the requirements which were necessary to protect the workforce and the community. In his book *The Railway Navvies* (Penguin Books, 1981), Terry Coleman describes how the Chester and Holyhead Railway Company stipulated in contracts that the contractors should provide huts for the men where there was no room for them in the villages along the line, and that the men should be paid on stated days in money, with no part paid in goods.

At the same time as the growth in civil engineering there was an increasing demand for buildings such as mills, factories and hostels for a working population which had flooded into the towns and cities. Building contracts had to take account of new pressures to complete on time, and new standards and specifications had to be drawn up to cope with new materials, such as cast iron, which were becoming available in commercial quantities. It is evident from reported cases throughout the nineteenth century that the roles of architect, or engineer or surveyor included that of an independent certifier when carrying out certain duties under construction contracts.

Gradually the contents of construction contracts became more sophisticated and included a host of new provisions; some brought about by Statute and others by the influence of the new professional institutions and trade associations that were being formed and which were to play an important role in a fast growing industry.

The method of tendering, in the early years of the industrial revolution, is best illustrated by Firbank, quoted by Coleman in *The Railway Navvies* (*supra*):

> 'Firbank himself used to tell a story of one Mr Wythes (probably George Wythes, who undertook, among other lines, that from Dorchester to Maiden Newton) who was thinking of submitting an offer for a contract. He first thought £18000 would be reasonable, but then consulted his wife and agreed it should be £20000. Thinking it over, he decided not to take any risk, so made it £40000. They slept on it and the next morning his wife said she thought he had better

make it £80000. He did; it turned out to be the lowest tender notwithstanding, and he founded his fortune on it.'

Fortunes could be made quickly, but many contractors went broke from underestimating the practical difficulties of constructing the work to strict standards in all weathers and a lack of awareness of the consequences of delay and other serious breaches of contract. It was soon realised that a major area of risk was inherent in the uncertainty of the quantity of work to be done and the variable ground conditions. Civil engineering contracts developed on the basis that all work would be remeasured at rates which were agreed at the outset; a reasonable solution bearing in mind the uncertainty of ground conditions which affected most of the work which was to be carried out. On the other hand, it was thought that building work was capable of quantification with reasonable accuracy (with the exception of changes ordered after the contract was agreed).

Therefore, building contracts were generally not subject to remeasurement and the contractor bore the risk of any mistakes which he may have made when measuring the work to be done from the drawings. The high cost of tendering for building work caused tendering contractors to engage a 'surveyor' who was responsible for measuring all of the work from the drawings and whose fees would be shared by all tenderers. Very soon this practice was overtaken by the employer (or his architect) engaging the surveyor to measure the work and for the 'quantities' to be provided for each tendering contractor for pricing the work. The surveyor's fees for measuring the work was usually required to be shown at the foot of the priced bill of quantities to be submitted with the tender and the successful contractor would then pay the surveyor out of the proceeds of interim certificates. This meant that each tendering contractor started by pricing the work based on the same bills of quantities, thereby reducing the cost of tendering and reducing the risk of error in quantifying the work to be done.

This practice, which survived for many years, caused problems if the building owner decided not to proceed with the work. Some building owners contended that they had no liability to pay the quantity surveyor's fees if the contract did not go ahead: *Moon* v. *Whitney Union* (1837), and *Waghorn* v. *Wimbledon Local Board* (1877); (*Hudson's Building and Engineering Contracts, tenth edition*, at pp 113 and 114 – the eleventh edition, at paragraph 2.069, considers the point to be redundant today). Even as late as the 1920s some standard forms of contract reflected this practice. The form of contract which was known by the short title as *The Model Form of Contract* (one of the RIBA publications referred to hereinafter), contained the following clause 14 prior to 1931:

'(a) The fees for the Bills of Quantities and the Surveyor's expenses (if any) stated therein shall be paid by the Contractor to the Surveyor named therein out of and immediately after receiving the amount of the certificates in which they shall be included. The fees chargeable under clause 13 [Variations] shall be paid by the Contractor before the issue by the Architect of the certificate for final payment. (b) If the Contractor fails or neglects to pay as herein provided, then the Employer shall be at liberty, and is hereby authorised, to do so on the certificate of the Architect, and the amount so paid by the Employer shall be deducted from the amount otherwise due to the Contractor.'

Until 1963 the RIBA standard forms of contract contained optional provisions (clause 10) whereby the contractor could be responsible for paying the quantity surveyor's fees out of monies certified by the architect. However the quantity surveyor generally became engaged by the building owner, or his architect, who were responsible for paying the fees.

While much of the case law which was relevant to construction contracts was shaped in the nineteenth century, there continued to be cases of note during the twentieth century and this century. In parallel, non-standard and standard forms of contract evolved. The first 'standard forms of contract' were probably developed by public corporations. Revisions to many forms of contract were often prompted by decisions in the courts and these revisions (or the interpretation and application of them) sometimes became the subject of later cases which were to have a continuing influence on the draftsmen of new contracts and on the understanding of the law which affects contracts in construction.

Standard forms of contract which came into general use in building contracts were developed by the Royal Institute of British Architects (RIBA). By the early twentieth century the use of the RIBA form of contract was widespread. This form of contract, which was to be the subject of several editions and revisions, was to become the basis of most building contracts and was the forerunner of the Joint Contracts Tribunal (JCT) forms of contract of 1963, 1980, 1998, 2005 and the recently introduced 2011 forms. In civil engineering, the first edition of the Institution of Civil Engineers (ICE) conditions of contract was launched in 1945. The seventh edition (1999) is currently in use. One of the features of these standard forms of contract is that they are approved and accepted by the professional institutions and the contractors' associations. Several other standard forms of contract developed independently, such as GC/Works/1 for use by government departments and forms published by other professional bodies.

Internationally, particularly where there was British influence, standard forms of contract developed on the same lines as in the United Kingdom. Forms of contract which were (almost verbatim) the same as the RIBA/JCT forms of contract came into use in Cyprus, Jamaica, Gibraltar, Bahrain, Hong Kong and Singapore. In Cyprus, one of the first editions of the RIBA form of contract (probably used in the United Kingdom about the time of the First World War) has been used alongside a variant of the 1963 edition of the JCT form of contract.

In Hong Kong a variant of the 1963 edition of the JCT form of contract is widely used and a draft based on the 1980 edition of the JCT form had been awaiting sanction since the early 1980s. In 2005 the Joint Contracts Committee of Hong Kong produced 2 new standard form contracts. Until recently, the form of contract used in Singapore was a variant of the 1963 edition of the JCT form. However, since 1980 the Singapore Institute of Architects has departed from following developments in the United Kingdom and has adopted an entirely new form of contract which bears no resemblance to any other standard form of contract used in the United Kingdom. In civil engineering a standard form of contract for use internationally was developed and agreed by the Fédération Internationale des Ingénieurs-Conseils (FIDIC) using almost entirely the same format and conditions as the ICE conditions of contract. Various editions of FIDIC are currently being used internationally. The fourth edition of FIDIC (published in 1987) is the last to be used, based on the ICE format, and the 1999 editions are likely to be used in the future.

In all forms of international contracting, it is important to be aware that there are significant differences in law in various parts of the world. There are four main categories of law:

- Common law based on the English legal system;
- Civil law based on the French or German codes;
- Local law (such as the *Shari'a* in the Middle East);
- Combinations of various laws and legal systems.

Common law jurisdictions

This type of legal system is found mainly in Commonwealth countries. As in the UK, there are a number of statutory laws and it is here that the main departures from English law can be found. Some examples are given later in this chapter.

Civil law jurisdictions

While this is evident in France and its former colonies, many countries have developed their own Civil Codes using the French Civil Codes as a model. In the Middle East, Egypt was the first country to adopt a codified legal system based on the French Codes. The draftsman of the Egyptian Codes also drafted the Kuwait Civil Codes and, while there have been changes from the original French versions in both adaptations, in many respects Egyptian and Kuwaiti law follows French law. In the Far East, Thailand has its own Codes which are based on the French Codes.

Examples of significant differences between some of the Civil laws and English common law are:

- *Termination* – In some civil law jurisdictions, regardless of the contractual provisions, it is not possible to terminate a contract without obtaining an Order from the courts.
- *Quantum meruit* – In many civil law jurisdictions, *quantum meruit* is not recognised. The contract price must be agreed or determined by an agreed method. In contrast, the Kuwaiti Commercial and Civil Code contains the following provisions:

  > 'If no consideration is mentioned in the contract, the Contractor shall be entitled to be paid at the prevailing rate for similar work at the date of conclusion of the contract.'

- *Consideration* – Under English law, consideration is an essential element of a contract (with certain exceptions). Often, civil law jurisdictions do not require consideration as an essential element of a contract. A contract can be made without any consideration.
- *Time for acceptance of offers* – While, with certain exceptions, an offer can be withdrawn at any time before acceptance under English law, some countries have introduced laws to make it a condition that offers are kept open for a specified period and cannot be withdrawn before the period has expired (for example, in Kuwait). Obligations of honesty and good faith are recognised by the courts in civil law jurisdictions. Therefore the revocation of an offer may be seen as a breach of judge-made law that offers must be kept open for a reasonable time.
- *Letters of intent* – The original purpose of a letter of intent was only a statement to the effect that the employer intended to enter into a contract at some later

stage and the letter imposed no obligations on the parties under English law. In many civil law jurisdictions, a letter of intent is an 'Agreement in Principle'. That is to say that all of the terms may not have been agreed but the principle of an agreement has. The parties are required to negotiate in good faith and conclude a contract in due course.

- *Liquidated damages and penalties* – Under English law, a penalty clause cannot be enforced. Roman Dutch law recognises penalty clauses and they can be enforced. Sometimes the law includes the powers given to the Court to modify a penalty if the amount shall be considered excessive or derisory. In South Africa, there are limited provisions for modifying penalties (see 1.4). In many Middle Eastern countries, the distinction between liquidated damages and penalties is a matter of translation (there is no Arabic word for 'liquidated damages') and 'penalties' are construed as if they were liquidated damages.

Local law

Some countries have developed their own laws and have been almost uninfluenced by the laws of other countries. In some cases, much of the law is based on religious teachings. For example, Saudi Arabia law is almost entirely based on the *Shari'a* (Islamic Law), in which there are four main 'Sunni Schools' (*Hanafi*, *Maliki*, *Shafi'i* and *Hanbali*). The main differences between the four schools is the priority which is given to the *Qur'an*. Some countries have combined *Shari'a* with modern statutes or codes to a greater or lesser extent.

Combinations of various laws and legal systems

Combinations vary widely from country to country. Bahrain and the UAE have traditionally followed the common law legal systems, with an element of Shari'a and written regulations. However, Bahrain and the UAE appear to be moving in the direction of civil law based on the French and Egyptian legal systems. Kuwait have a fully codified legal system and is one of the most advanced of the Gulf States in that it has been a civil law jurisdiction for many years. Its Civil Code is similar to that of Egypt but some local laws still apply.

Since 1907, Japan's laws have been developed on a Civil and Commercial Code (of German influence) followed by a degree of American influence after the Second World War. In spite of this history, complex contractual arrangements are generally avoided, even for some major projects. The traditional Japanese philosophy has remained almost unaffected by recent developments in law.

The law of the contract and the procedural law

The law of the contract

This is the law which governs the interpretation and application of the contract. It is important to establish any impediment to foreign laws which cannot be enforced in the country in which the contract is made.

The procedural law

This is a law which governs litigation, arbitration and adjudication and is normally the law of the country in which the proceedings will take place. It is not necessarily the same as the law of the contract.

Difficulties can arise if the choice of law is ambiguous. In general, the choice of law specified in the contract will be upheld unless:

- it is contrary to public policy of the place where the proceedings are held;
- the choice is not exercised for *bona fide* and legal reasons;
- *dicta* of Denning LG in *Bouissevan* v. *Weil* (1948) 1KB 482 applies – 'I do not believe the parties are free to stipulate by what Law the validity of their contract is to be determined. Their intention is only one of the factors to be taken into account.'

Problems can arise if the choice of the law of contract for main contracts and subcontracts are not the same (see 6.11 *infra*).

1.2 Bills of quantities

Contractors who calculated their own quantities from drawings supplied by the building owner adopted methods of measurement according to their own style. The first quantity surveyors also prepared the bills of quantities in their own style and adopted their own particular methods of measurement. In the beginning this was probably confusing as the tendering contractors must have placed their own interpretation on the method of measurement. No doubt the quantity surveyors gradually developed methods which were fairly consistent and contractors became familiar with each individual quantity surveyor's method of measurement. The courts dealt with many cases involving liability for inaccurate bills of quantities and the decisions appear to be inconsistent. The apparent inconsistency was due in part to the distinguishing features of the various contracts and representations which were made regarding the quantities. However, it was held in *Bolt* v. *Thomas* (1859) (*Hudson's Building and Engineering Contracts, eleventh edition*, at paragraph 2.286, page 407 (but more extensively quoted in the *tenth* edition at page 196)) that where it was stipulated that the builder should pay the architect for the calculation of the quantities, and he had done so, then the builder was entitled to compensation from the architect if the bill was not reasonably accurate.

As late as the 1920s the Model Form of Contract (RIBA) did not incorporate a standard method of measurement, nor did it expressly state that the bills of quantities was a contract document. Nevertheless it was implied that the bills of quantities had contractual status and the contract contained provisions in clause 12a as follows:

> 'Should any error appear in the Bills of Quantities other than in the Contractor's prices and calculations, it shall be rectified, and such rectification shall constitute a variation of the Contract, and shall be dealt with as hereinafter provided.'

The provisions in the above contract have survived to the present day and a similar provision is included at clause 2.14 of the JCT Standard Building Contract With Quantities 2011. A similar provision also appears in the seventh edition of the ICE conditions of contract in clause 55(2).

In the absence of a standard method of measurement, errors in composite descriptions and alleged omissions of items, as opposed to errors in measurement, became a constant source of argument. The first steps to rectify these difficulties probably took place in 1909, when the Quantity Surveyors' Association appointed a committee to prepare and publish pamphlets recommending the method of measurement for three trades. The first edition of the Standard Method of

Measurement (SMM) was published in 1922 with the agreement of representatives of the Surveyors' Institution, the Quantity Surveyors' Association, the National Federation of Building Trades Employers and the Institute of Builders. The situation which existed prior to the publication of the first edition is perhaps best described in the opening paragraph of the preface to this historic document:

> 'For many years the Surveyors' Institution and the Quantity Surveyors' Association (which bodies are now amalgamated) were accepted as the recognised authorities for deciding disputed points in connection with the measurement of building works. The frequency of the demands upon their services for this purpose directed attention to the diversity of practice, varying with local custom, and even with the idiosyncrasies of individual surveyors, which obtained. This lack of uniformity afforded a just ground of complaint on the part of contractors that the estimator was frequently left in doubt as to the true meaning of items in the bills of quantities which he was called upon to price, a circumstance which militated against scientific and accurate tendering.'

As might be expected, it took several years for the quantity surveying profession to become aware of the SMM and to use it in practice. Several years after the publication of the first SMM, in *House and Cottage Construction, Volume IV*, Chapter ll (Caxton Publishing Company Limited), Horace W. Langdon Esq., F.S.I, a practising Chartered Quantity Surveyor, made no reference to a standard method of measurement and he described how the quantity surveyor ought to explain the method of measurement used to prepare the bills of quantities.

The second edition of the SMM was published in 1927, and in 1931 the RIBA published its revised form of contract which (in clause 11) incorporated the SMM, where quantities formed part of the contract. The first test as to the valid incorporation of the SMM into the contract and the application and interpretation of the principles laid down in the standard method of measurement took place in 1938: *Bryant and Sons Ltd* v. *Birmingham Saturday Hospital Fund* [1938] 1 All ER 503. It was held that clause 11 of the contract, and the SMM, had been incorporated into the contract and that the contractor was entitled to extra payment for excavation in rock which ought to be measured separately pursuant to the principles laid down in the SMM.

It is evident that the decision in the Bryant case turned on the special wording in the standard form in clause 11, to the effect that the bills *unless otherwise stated* should be deemed to have been prepared in accordance with the current standard method of measurement. A similar provision appears in clause 2.13 of the JCT Standard Building Contract With Quantities 2011 and is the basis of many claims which persist in the construction industry today. The development of more sophisticated standard methods of measurement, while desirable in many respects, has done little to eliminate this type of claim. The provisions of SMM7 require the quantity surveyor to provide more detailed information than that required by the SMM (where necessary) (A1) and for the employer to provide information on groundwater (D3.1) or to state what information is assumed.

Civil engineering quantities developed along similar lines to building quantities and standard methods of measurement became incorporated into contracts for civil engineering work. Clause 57 of the fifth, sixth and seventh editions of the ICE conditions of contract contains similar provisions regarding the status and application of the Civil Engineering Standard Method of Measurement (CESMM) referred to therein. Any work carried out by the contractor which is not measured separately in accordance with the CESMM may (unless there is a statement to the contrary) be subject to a claim for additional payment: *A.E. Farr Ltd* v. *Ministry of Transport* (1965) 5 BLR 94.

In international contracting, it is unfortunate that the clear advantages resulting from standard methods of measurement which seek to address the problems stated in the preface to the 1922 SMM (*supra*) have not been grasped. Little could be simpler than to select one of the many standard methods of measurement for building or civil engineering work and to specify that the works have been measured accordingly. This would remove the uncertainty in pricing large and complex projects based on bills of quantities. One of the UK SMMs or a local SMM (such as exist in Jamaica and Hong Kong) or the International SMM may suit the purpose.

The situation which prevails all too often is for the contract to say (in this example quoting from the 1999 FIDIC Red Book, clause 12.2 (b)):

> '...the method of measurement shall be in accordance with the Bill of Quantities or other applicable Schedules.'

Such provisions can (unless the contract sets out in considerable detail the methods of measuring each element of work) only lead to estimators being left in doubt as to the true meaning of items in the bills of quantities.

1.3 Variations

Building and civil engineering contracts are of such a nature that it is almost impossible, especially where work has to be carried out in the ground, to design and construct a project so that the final product is identical in every way to the original design which formed the basis of the contractor's tender. Changes to the original design and/or details may come about for technical reasons or because the building owner desires a revision to the plans or details.

Where technical reasons are the cause of a variation (for example, unsuitable ground conditions) the employer, or his architect, or engineer, will have limited control over the scope of the change in the work to be done by the contractor. Where the employer desires a change to the plans or details (for example, for aesthetic, or practical, or financial reasons), the scope of the change is to a large extent within the control of the employer. Without a suitable provision in a contract which allows the works to be varied, such changes would not be permitted (under the terms of the contract) and in the event of unavoidable changes for technical reasons the contractor would no longer be obliged to complete the work. Changes could only be executed with the agreement of the contractor or by way of a separate contract.

The standard forms of contracts used in building and civil engineering forms of contract provide for variations which are necessary or desirable (the latter being the employer's prerogative, but it does not exclude variations initiated by the contractor). The JCT forms of contract expressly provide for the architect to sanction a variation made by the contractor without an instruction issued by the architect.

Sometimes arguments are raised concerning the limit beyond which it may be regarded that the changes were outside the scope of the variation clause. Such arguments, if successful, would enable the contractor to refuse to execute the revised works or to escape from the contract rates and recover on a *quantum meruit* basis (a reasonable valuation in all the circumstances). There are no finite guidelines to assist in this matter. Some early forms of contract expressly stated a percentage of the contract price as the yardstick for determining the extent of variations permitted under the terms of the contract. The international form of contract (FIDIC) provides for a limited revision to the contract price if the sum total of all changes and remeasurement (with some exceptions) exceeds 10 per cent (clause 52(3) of the third edition) or 15 per cent (clause 52.3 of the fourth edition). However, this cannot be construed as being a true valuation on a *quantum meruit* basis. In the absence of stated limits such as a percentage, it is necessary to decide whether or not the scope of the changes went beyond that which was reasonably contemplated by reference to the contract documents and the surrounding circumstances of the case.

In *Bush* v. *Whitehaven Port and Town Trustees* (1888) 52 JP 392, the contractor was to lay pipes and possession of the site was to be given to the contractor for the performance of the work. Owing to delay in giving possession of the site to the contractor, the work had to be done in the winter, whereas it was contemplated that the work would be done in the summer. It was held that the contractor was entitled to payment on a *quantum meruit* basis (a reasonable price for the work in all the circumstances).

Modern contracts contain variation provisions which are so wide that it may appear doubtful that any claim for payment on a *quantum meruit* basis would succeed. However, in *Wegan Construction Pty. Ltd.* v. *Wodonga Sewerage Authority* [1978] VR 67 (Supreme Court of Victoria), the contractor successfully claimed on a *quantum meruit* basis. This case is worthy of further consideration on the grounds that the contractual provisions for variation were very wide (being similar to the ICE fifth and sixth editions and FIDIC fourth edition) and is summarised in Chapter 5.

Another problem which has come before the courts over the years, is the vexed question about omissions when the employer intends to have the work done by others. It is an increasingly common practice, when progress is delayed by the contractor, for the employer (through his architect) to omit work.

Presumably the employer believes that if the work is omitted, the architect does not have to issue any (late) instructions to carry out the work, which would have the effect of defeating the employer's claim to liquidated damages. It is well established in law that the power to omit work, even where the contract provides that no variation should in any way vitiate or invalidate the contract, is limited to genuine omissions, that is, work not required at all. It does not extend to work taken out of the contract for it to be done by another contractor: *Amec Building Limited* v. *Cadmus Investment Co. Limited* (1996) Cons L. R. 105.

1.4 Extensions of time and liquidated damages: penalties

An extension of time provision is inserted in a contract for the benefit of both the contractor and the employer. However, its insertion is primarily for the benefit of the employer. Without such a provision, once the employer had caused delay, the contractor would no longer be bound to complete the works by the contract completion date and the employer would no longer be able to rely on the liquidated damages provisions in the contract. These fundamental points are often not appreciated by employers or their agents who are responsible for making extensions of time, in spite of the fact that decisions in the courts spanning almost two centuries have consistently reflected this view. In *Holme* v. *Guppy* (1838) 3 M & W 387, the contractors were responsible for delay of one week and the employer was responsible for delay of four weeks. There was no extension of time clause. It was held that the employer could not deduct liquidated damages from monies due to the contractor.

Draftsmen of contracts for building and civil engineering work recognised that there were many possible causes of delay to projects which were to be constructed over a period of years, in all weathers, and which were almost certainly going to be subject to delay by events within the control of the employer. Delays which were due to neutral events (such as inclement weather) and events which were generally within the control of the contractor were of no concern to the employer, and if contracts were delayed by such matters, then the contractor would have to take the necessary measures to make up the delay or face the consequences by payment of liquidated damages.

The use of contracts with onerous provisions which held the contractor liable for damages for every type of delay was not commercially satisfactory, as it encouraged cautious contractors to increase their prices and the reckless ones probably went out of business. Neither of these options were in the interests of the employer nor were they in the interests of the industry as a whole. On the other hand, delays on the part of the employer would extinguish the employer's rights to liquidated damages and it was therefore essential that the contract should include suitable provisions to enable the employer, or his agent, to make an extension in the event of delay for any cause which was within the employer's control or for which the employer was responsible (such as obtaining statutory approvals).

The drafting of suitable provisions which would protect the employer in the event of delay caused by him, and which would permit extensions of time for neutral causes and causes of delay which were generally within the control or at the risk of the contractor, proved to be a major problem. Very general provisions such as 'circumstances wholly beyond the control of the builder' proved to be of no effect in circumstances where delay had been caused by the employer. This was held in *Wells* v. *Army and Navy Co-operative Society Ltd* (1902) 86 LT 764, where the extension of time clause contained the words 'or other causes of delay beyond the contractor's control'.

In spite of the decision in the *Wells* case (which was reported in the fourth edition of *Hudson's Building Contracts* in 1914), draftsmen of building and civil engineering contracts continued to use general terms which were almost certainly bound to be ineffective where the employer caused delay. Over fifty years later in *Perini Pacific Ltd* v. *Greater Vancouver Sewerage and Drainage District Council* [1967] SCR 189, delays of ninety-nine days occurred which included forty-six days on the part of the employer. The extension of time clause in the contract contained the provisions to extend time for completion due to 'extras or delays occasioned by strikes, lockouts, *force majeure* or other cause beyond the control of the contractor'. It was held that the extension of time clause did not cover delays caused by the employer and no liquidated damages could be recovered.

The fourth, fifth and sixth editions of the ICE form of contract and the third edition of FIDIC contain the general terms 'other special circumstances of any kind whatsoever'. It is evident, in view of the decisions in the *Wells* and *Perini Pacific* cases, that these standard forms of contract, some of which may still be in use today, do not cover delay by the employer (with the exception of certain specified 'other cause of delay referred to in these Conditions'). It is conceivable that several causes of delay by the employer could occur in a civil engineering contract, which delays are not expressly covered elsewhere in the contract and which would therefore deprive the employer of its rights to deduct liquidated damages.

For many years standard forms of building contract appear to have been drafted in recognition of the difficulties caused by the *Wells* decision. Since the early part of this century the RIBA forms of contract have listed several causes of delay within the control of the employer (and other causes of delay) for which an extension of time could be granted. However, unless such a list is comprehensive, any delay which is not included therein would not qualify for an extension. If the non-qualifying delay was the employer's responsibility, no extension could be granted and the employer's rights to deduct liquidated damages would be extinguished. This point was clearly emphasised in *Peak Construction (Liverpool) Ltd* v. *Mckinney Foundations Ltd* (1970) 1 BLR 111. In this case a subcontractor (Mckinney) was guilty of defective work in the piling for foundations as a result of which there was a suspension of work. The subcontractor submitted design proposals to remedy the defects. The employer (Liverpool Corporation, a local authority) took an unreasonably long time to approve the subcontractor's proposals and the contractor was unable to continue with the works until some fifty-eight weeks later. The employer deducted liquidated damages for the period of delay and the contractor sought to recover the damages from the subcontractor. The contract contained an extension of time clause which set out the causes of delay for which an extension of time could be made, but it did not cover the employer's delay in approving the subcontractor's proposals. It was held that since part of the delay was due to the employer's default, and since there was no applicable extension of time provision, the employer could not deduct liquidated damages and he was left to recover such damages as he could prove flowed from the subcontractor's breach.

More recently in the case of *Rapid Building Group Ltd* v. *Ealing Family Housing Association Ltd* (1984) 29 BLR 5, the contractor was prevented from having full possession of the site on the due date. The contract was the 1963 edition of the JCT standard form of contract. There was delay and the works were completed late. The architect extended time for completion and issued a certificate that the works ought reasonably to have been completed by the extended date for completion. The employer deducted liquidated damages for the period after the extended date for completion until the date when the contractor completed the works. It was held that the 1963 edition of the JCT form of contract did not provide for extensions of time due to the employer's breach of contract in failing to give possession of the site in accordance with the terms of the contract and the employer could not deduct liquidated damages from monies due to the contractor. The JCT Standard Building Contract 2011 includes failure to give possession of the site as a cause of delay (a relevant event – clause 2.29.3) for which an extension of time may be granted.

Recent drafting (such as the fourth edition of FIDIC, GC/Works/1 (1998), the seventh edition of ICE and the Singapore Institute of Architects forms of contract) includes a list of causes of delay for which an extension of time can be made and there is a 'catch-all' provision intended to cover 'any act or default of the employer'. It is unlikely that this type of catch-all provision will enable the employer to cause delay with impunity. Some delays may well be beyond the contemplation of such a clause and the contractor may have grounds to determine his employment.

Even if a contract contains an effective extension of time clause, the employer's rights to deduct liquidated damages may be extinguished if the power to extend time for completion is not exercised within the time contemplated by the contract terms. In *Miller* v. *London County Council* (1934) 151 LT 425, the contract contained the following terms:

> 'it shall be lawful for the engineer, if he thinks fit, to grant from time to time, and at any time or times, by writing under his hand such extension of time for completion of the work and that either prospectively or retrospectively, and to assign such other time or times for completion as to him may seem reasonable.'

The contractor completed the works on 25 July 1932 and, on 17 November 1932, the engineer extended time for completion to 7 February 1932 and certified that liquidated damages were payable for the period from 7 February to 25 July 1932. It was held that the extension of time clause empowered the engineer to look back (retrospectively) at the delay as soon as the cause of the delay had ceased to operate and to fix a new completion date 'within a reasonable time after the delay has come to an end' (Du Parcq, J, quoting from *Hudson on Building Contracts, sixth edition* at page 360). The power to grant an extension of time had been exercised too late and the employer could not rely on the liquidated damages provision in the contract.

In another case, *Amalgamated Building Contractors* v. *Waltham Holy Cross UDC* [1952] 2 All ER 452, the contract was an RIBA form of contract which contained the following provisions in clause 18:

> 'If in the opinion of the architect the works be delayed … (ix) by reason of labour and materials not being available as required...then in any such case the architect shall make a fair and reasonable extension of time for completion of the works …'.

In this case the contractor was delayed owing to non-availability of labour and during the month prior to the contract completion date he made two applications for an extension of time which the architect formally acknowledged. The date for completion was 7 February 1949 and the contractor completed the works in August 1950. In December 1950 the architect made an extension of time to May 1949. The contractor argued that an extension of time cannot be made to a date which has passed and therefore the extension was given too late. It was held, distinguishing *Miller* v. *London County Council*, that the extension of time could be made retrospectively and the extension was valid.

The different decisions in the *Miller* and *Amalgamated Building Contractors* cases are due to several distinguishing matters which are relevant. In *Miller* the engineer's decision on extensions of time was final and the wording in the two contracts were not the same. Perhaps more importantly, the cause of delay in *Miller* was within the control of the employer, whereas in *Amalgamated Building Contractors*, the cause of delay was beyond the control of the employer. In the latter case the delay was continuous, over a period of several months, thereby making it difficult, if not impossible, to estimate the length of the delay until the works had been completed. A detailed explanation of the law as it applies to this subject is given in the judgement in *Fernbrook Trading Co. Ltd* v. *Taggart* [1979] 1 NZLR 556. (For an excellent summary of this case, refer to *A Building Contract Casebook* by Dr Vincent Powell-Smith and Michael Furmston at page 355.)

Contractors seeking to argue that the contract does not provide for extensions of time (for delay by the employer), or that an extension of time was made too late, thereby being invalid, may not necessarily be in a better position than they might have been by accepting a reasonable extension of time, valid or otherwise. If the contractor's arguments are successful and the contract completion date is no longer applicable, the contractor's obligation is to complete within a reasonable time (time is at large) and the employer cannot rely on the liquidated damages provision to deduct the sums stated in the contract. In these circumstances the contractor does not have all the time in the world to complete the works, nor does he escape liability for general damages which the employer may suffer as a result of delay within the control of the contractor. Nevertheless, contractors may find it attractive to escape from the contractual period and the potential liability for delay at the rate stated as liquidated damages in the contract on the basis that the burden of proof shifts from the contractor to the employer. In *Wells* v. *Army and Navy Co-operative Society* (*supra*), Wright, J, the trial judge said:

> 'The defaults were, in my opinion, sufficiently substantial to cast upon the defendants [the employer] the burden of showing that the defaults did not excuse the delay.'

And in *Peak Construction (Liverpool) Ltd* v. *Mckinney Foundations Ltd*, (*supra*) Salmon, LJ said:

> 'If the failure to complete on time is due to fault of both the employer and the contractor, in my view, the clause does not bite. I cannot see how, in the ordinary course, the employer can insist on compliance with a condition if it is partly his own fault that it cannot be fulfilled ... I consider that unless the contract expresses a contrary intention, the employer, in the circumstances postulated, is left to his ordinary remedy; that is to say, to recover such damages as he can prove flow from the contractor's breach.'

The term *time at large* is a principle of English law which may be inappropriate in some countries. In many civil law jurisdictions and, for example, in South Africa, the principle is recognised but not by that title, namely that a debtor is excused from performing an obligation on time if his creditor wrongfully prevents him from doing so.

In *Group 5 Building Limited* v. *The Minister of Community Development* 1993(3) SA 629 (A), the Plaintiff, Group 5, was delayed arising from delays in giving variation orders and instructions and unauthorised suspension orders which constituted, so it was alleged, breaches of the contract by the Defendant.

The extension of time clause, clause 17(ii), contained the standard provision in regard to delays occasioned 'by any other causes beyond the contractor's control'. Group 5 contended that the delays arising out of the alleged breaches of contract fell outside the ambit of clause 17(ii) and as a consequence time was at large (following strictly English principles of law and indeed the South African law at that time as provided in the judgement in *Kelly & Hingles Trustees* v. *Union Government (Minister of Public Works)* 1928 TPD 272).

His Lordship Mr Justice Nienaber said:

> 'In my opinion the words "or by any other causes beyond the contractor's control" in clause 17(ii) are wide enough to embrace a wrongful conduct by the employer or his agent. Such conduct would entitle the contractor to apply for an extension of time and, if the application is refused, to have the matter tested

> in a court of law. In addition, the contractor can recover any losses he may have suffered as a result of the owner's wrongful conduct by means of an action for damages. The express terms of the contract accordingly provide for the very eventuality which the Plaintiff (Group 5) alleges occurred in this instance.'

The essential difference between the South African approach and that adopted by the English courts is that the latter are prepared to give a much narrower interpretation to the provisions of the extension of time clauses. In the Group 5 decision the phrase 'or any other causes beyond the contractor's control' was given a much broader interpretation to include for breaches by the employer.

The Group 5 decision may also have been influenced by the fact that 'the other causes' complained of by Group 5 were in fact expressly covered elsewhere and qualified for extensions of time in any event. No doubt Group 5 used this argument to overcome its failure to give notice.

Recent legal decisions indicate that there may be a wind of change. It may be that the contractor's argument that 'time is at large' if the engineer or architect fails to grant an extension at the appropriate time is losing favour (see 5.3 and 7.2, *infra*).

It is often argued that the employer cannot recover more in general damages than he would have been able to recover by way of liquidated damages. It appears from *Rapid Building Group Ltd* v. *Ealing Family Housing Association Ltd* (*supra*), that if the employer has lost his rights to liquidated damages, his claim for general damages may not be limited by the amount specified in the contract for liquidated damages. This point was not decided in the *Rapid Building* case but it must be at least arguable that this may be the case in certain circumstances.

In *Temloc Ltd* v. *Erril Properties Ltd* (1987) 39 BLR 31, the sum specified for liquidated damages was '£nil' and the employer sought to recover unliquidated damages arising out of delay in completion by the contractor. The Court of Appeal decided that by inserting a £nil rate for liquidated damages (to be calculated pursuant to clause 24.2.1 of a 1980 edition of the JCT form of contract), the parties had agreed that there should be no damages for late completion. However, in this case the Court of Appeal took the view that an extension of time which had been made by the architect after the twelve-week period required by clause 25.3.3 of the contract did not invalidate the liquidated damages provision and general damages could not be recovered as an alternative. Accordingly, the matter of the employer's rights in the event of the liquidated damages provisions being inapplicable did not have to be considered.

Nevertheless, notwithstanding the *Temloc* case, it appears likely that in the event of the contractor successfully arguing that the liquidated damages provisions are no longer applicable, then he may run the risk of being liable for general damages in excess of the liquidated damages. On the other hand, an employer who caused the liquidated damages provision to be invalidated, for any reason, for the purposes of claiming a higher amount of general damages than he might have recovered under the contractual provisions would be unlikely to find favour in the courts (see further commentary on the *Temloc* case in Chapter 7). This practice would surely fall foul of the rule of law which prevents a party from taking advantage of his own wrong, *Alghussein Establishment* v. *Eton College* [1988] 1 WLR 587.

The law relating to liquidated damages is substantially different in countries where the law is based on Indian law, for example Cyprus and Malaysia. The precise differences vary from country to country and perhaps the situation in Malaysia is most at odds with the established principles of English law.

The Contracts Act of Malaysia, Section 75, provides for the actual loss (as a result of delay or other default) to be proved and the right of recovery is limited by the amount of liquidated damages stipulated in the contract, that is the stipulated liquidated damages is a ceiling on the amount recoverable. There is no room to argue that a plaintiff may recover a genuine pre-estimate of loss without proof of actual loss: *Larut Matang Supermarket Sdn Bhd* v. *Liew Fook Yung* [1995] 1 MLJ 375; *Song Toh Chu* v. *Chan Kiat Neo* [1973] 2 MLJ 206; *Woon Hoe Kan & Sons Sdn Bhd* v. *Bandar Raya Development Bhd* [1973] 1 MLJ 60.

Penalties are not enforceable in English law. Statutory enforcement of penalties is the exception rather than the rule in international systems of law. In the absence of such statutory enforcement, various attempts have been made by contract draftsmen to avoid the general principle that a penalty is not enforceable by referring to the deduction as 'pre-ascertained sums by way of liquidated damages'. The difficulty with this remains that in the event that the amount is out of proportion to the loss actually suffered by the employer then, irrespective of the description, the sum will be seen as a penalty and will be unenforceable.

Roman Dutch law embodies the maxim *pacta sunt servanda*. Contracts are made to be enforced. The Conventional Penalties Act in South Africa, Act 15 of 1962, provides that where parties agree upon a sum to be deducted for each period of delay (day, week or month) for which the contract overran the contractual completion date then, irrespective of whether the sum was a penalty or otherwise, the parties should be bound by the terms of their agreement.

The Conventional Penalties Act thereby created statutory enforcement of penalties prescribed by the contract. The employer is entitled to enforce the application of penalties through the dispute mechanism or through the courts.

Another vexed question arises in contracts where the employer intends to have phased completion and where the form of contract (usually a standard form) does not deal properly with this issue. In *Bramall and Ogden* v. *Sheffield City Council* (1983) 29 BLR 73, the contract incorporated the 1963 JCT conditions with liquidated damages 'at the rate of £20 per week for each uncompleted dwelling'. Extensions of time were granted but the contractor contended that further extensions were due and he disputed the employer's rights to deduct liquidated damages. The arbitrator awarded £26150 as liquidated damages. On appeal it was held that the contract did not provide for sectional completion and the employer could not deduct liquidated damages.

In the case of *Philips Hong Kong Ltd* v. *The Attorney General of Hong Kong* (1990) 50 BLR 122, the plaintiffs followed a similar argument to the one put forward in *Bramall and Ogden* v. *Sheffield City Council*. It was argued that a minimum figure for liquidated damages together with a provision for a reduction in liquidated damages in the case of sectional completion amounted to a penalty. The argument succeeded in the High Court of Hong Kong, but was overturned on appeal in *Philips Hong Kong* v. *The Attorney General of Hong Kong* (1993) 61 BLR 41 (P.C.). It is now unlikely that liquidated damages provisions will be construed as penalties merely on arithmetical grounds.

It will be seen from the cases referred to that extensions of time and liquidated damages provisions in contracts merit careful drafting and that the interpretation placed on many provisions is open to dispute at almost every turn. The courts have generally taken a very strict view and the *contra proferentem* rule has usually been applied (that is, the clause is usually construed against the interests of the party putting forward the clause and seeking to rely on it): *Peak Construction (Liverpool) Ltd* v. *Mckinney Foundations Ltd* (*supra*), and *Bramall and Ogden* v. *Sheffield City Council* (*supra*). The *contra proferentem* rule will not necessarily apply to contracts using standard forms such as the ICE or JCT forms of contract: *Tersons Ltd* v. *Stevenage Development Corporation* (1963) 5 BLR 54. The rule may be applied to particular amendments to a standard form imposed by the employer.

Extensions of time have perhaps been at the forefront of many disputes, most of which could have been avoided by care and attention to the matters which have been considered by the courts over many years. Later chapters will deal with some of these matters in greater detail.

1.5 Claims for additional payment: damages

Whenever there is delay, disruption or a change in circumstances or in the scope of the work, there is bound to be an effect on expenditure or income, either for the contractor or for the employer, or both. Subcontractors may also be affected. In some cases the risk is borne by the contractor (or subcontractor) and in others it may be borne by the employer. Where there is a breach of contract, or where there is a contractual provision to claim loss or damage, one party may have a claim against the other.

Claims relating to ground conditions are a regular feature in many building and civil engineering contracts. Numerous disputes have arisen as to the responsibility for information provided by the employer and upon whom the risk lies for unforeseen ground conditions. In *Boyd & Forrest* v. *Glasgow S W Railway Company* [1914] SC 472, the tendering contractors had only two weeks in which to tender for the work. The employer provided access to some information obtained by way of site investigations. The contractors claimed compensation for the losses caused by ground conditions which were not in accordance with the soil investigation information provided by the employer. It was held that the contractors were entitled to rely on the information provided by the employer and that the employer could not be protected against his own misrepresentation.

If employers were able to place the risk entirely on the contractor, the likelihood would be that tender prices would be much higher than if the risk was on the employer. The ICE and FIDIC forms of contract, being forms generally applicable to civil engineering contracts where a considerable amount of work is carried out in the ground, have provisions which recognise the problems associated with the uncertainty of ground conditions. Clauses 11 and 12 of these forms of contract have, in various editions over the years, provisions such as (quoting from the fifth edition of the ICE form of contract):

> '11 (1) The Contractor shall be deemed to have inspected and examined the Site and its surroundings and to have satisfied himself before submitting his tender as to the nature of the ground and sub-soil (so far it is reasonably practicable and having taken into account any information in connection therewith which may have been provided by or on behalf of the Employer) the form and nature of the Site, the extent and nature of the work...and in general to have obtained for himself all necessary information (subject as above-mentioned) as to the risks contingencies and all other circumstances influencing or affecting his tender.'
>
> '12 (1) If during the execution of the Works the Contractor shall encounter physical conditions (other than weather conditions or conditions due to weather

> conditions) or artificial obstructions which conditions or obstructions he considers could not reasonably have been foreseen by an experienced contractor and the Contractor is of the opinion that additional cost will be incurred which would not have been incurred if the physical conditions or artificial obstructions had not been encountered he shall if he intends to make any claim for additional payment give notice to the Engineer ....'

[The contract goes on to provide for an extension of time and additional payment.]

The above provisions appear to be a fair and reasonable attempt to ensure that contractors do not take the risk of *unforeseen* ground conditions and that employers are not exposed to unlimited claims. Notwithstanding these provisions, differences of opinion, ambiguity and deliberate tendering tactics have continued to provide an abundance of disputes and the results have often been against the interests of employers. Attempts have been made by the employer to escape responsibility for information on ground conditions provided by him.

In *Morrison-Knudsen International Co Inc and Another* v. *Commonwealth of Australia* (1980) 13 BLR 114, the employer disclaimed responsibility for the site investigation which he provided. It was held that the contractor was entitled to rely on the information provided and that the provisions in the contract were not an effective disclaimer. There may be a duty of care on the part of the employer in providing such information and the contractor may have a claim for misrepresentation: *Howard Marine & Dredging* v. *Ogden* (1978) 9 BLR 34.

Building contracts, by their nature, tend to be less vulnerable to claims involving ground conditions, but as can be seen from *Bryant & Sons Ltd* v. *Birmingham Saturday Hospital Fund* (*supra*), claims do arise from time to time.

The forms of contract in civil engineering recognised the concept of claims at an early stage and express provisions for additional payment in certain circumstances were a feature in these forms. The ICE conditions of contract use the term 'claim' whereas the RIBA and JCT forms of contract generally do not. Early RIBA forms of contract did not expressly provide for any additional payment over and above the contract rates except where it was appropriate under the variation clause. In the late 1920s and early 1930s the RIBA Model Form of Contract in general use contained no express provisions for 'delay and disruption claims' unless they could be dealt with as variations. Nevertheless it appears that architects and quantity surveyors of the time were of the opinion that there was power to make payment to the contractor without a variation being ordered. Horace W. Langdon Esq., F.S.I., wrote in *House and Cottage Construction* (*supra*):

> 'EXTRAORDINARY CIRCUMSTANCES
> At times during the progress of work, certain happenings may take place which involve the contractor in a much greater expense than he had anticipated, such as, for instance, not being given a clear site, as may have been first promised. Under such circumstances, it is obvious that the cost per unit of the particular work affected must be greater than would have been the case had he had a clear run. Such a matter cannot be dealt with by the quantity surveyor, whose business it is to ascertain actual measurements of work executed and to value same as previously described. Extraordinary happenings of the kind mentioned would be dealt with by the architect. If the contractor disagrees with the architect's ruling, he may have recourse to the clause appertaining to arbitration.'

The RIBA form of contract referred to by Langdon did not contain provision for the extra payment which appears to be contemplated, nor did it provide for an extension of time for the breach of contract which was used as the example to explain 'extraordinary circumstances'. Misunderstanding of forms of contract and the application of the law persists today and is one of the reasons for disputes and actions for negligence.

The 1939 RIBA form of contract did not contain any provisions intended to deal with failure to give possession of the site or other acts of prevention by the employer, but it did contain new express provisions for additional payment in clause 1:

> 'If compliance with Architect's Instructions involves the Contractor in loss or expense beyond that provided for in or reasonably contemplated by this contract, then, unless such instructions were issued by reason of some breach of this contract by the Contractor, the amount of such loss or expense shall be ascertained by the Architect and shall be added to the Contract Sum.'

Provisions of the type quoted above are to be found in later editions of the RIBA and JCT forms of contract. Bearing in mind the wide rules for valuing variations where there are changes in circumstances, this type of provision appears to be intended to deal with the consequential effects of architects' instructions on other work (which work may not in fact have been varied by an instruction). This type of claim which involves delay and/or disruption to the regular progress of the works is troublesome for a variety of reasons that will be dealt with in later chapters.

One important ingredient of delay claims is often interest or finance charges. As a general rule this head of claim did not succeed unless it could be dealt with as special damages. The most important cases which deal with this matter came before the courts fairly recently and are discussed in later chapters. However, as modern disputes sometimes take years to settle, or to be decided (unless decided by adjudication), interest on the claim itself was often the largest single element of it, although the recent trend of low interest rates has tempered this element of claim. Where interest is awarded in favour of the contractor, a nominal amount over and above the bank rate is usually the measure of damages. The benefit to the employer however is often the return earned by 'turning the money over several times per annum' which, even in a moderately profitable business, may be up to ten times the amount of interest awarded. This level of damages is not contemplated, but it is perhaps difficult to reconcile this fact with the 'absolute rule of law and morality which prevents a party taking advantage of his own wrong whatever the terms of the contract: *Alghussein Establishment* v. *Eton College* (*supra*).

An interesting feature of the 1939 edition of the RIBA form of contract was an optional clause (24(d)[A]) which provided for the retention fund to be deposited in a joint account in a bank named in the appendix to the contract. The interest which accrued was for the benefit of the employer, but as this was small compared with the return which could be gained by using the sum retained in a profitable business, the incentive for unscrupulous employers to seek to delay the release of the retention fund was reduced.

The JCT Standard Building Contract 2011 provides for the retention to be placed in a separate bank account at the request of the contractor and held by the employer on trust for the contractor (clause 4.18). This will provide a level of protection for contractors in the event of the employer's liquidation and it will prevent employers using retention funds as working capital. At the outset of every contract, contractors should ask employers for details of the trust fund and ensure that all retentions are held in the said fund.

A number of recent cases have shown that contractors are being more cautious and are insisting on retentions being placed in a trust fund. If employers resist, the courts may issue an injunction to compel them to place the retention fund in a separate account: *Wates Construction (London) Ltd* v. *Franthom Property Ltd* (1991) 53 BLR 23.

The JCT Standard Building Contract 2011 includes provision for a retention bond in lieu of the deduction of retention (clause 4.19).

1.6 Rolled-up claims

It is generally a requirement that the party making a claim should be able to illustrate that the damages claimed were caused by an event or circumstance which was a breach of contract or that it was a matter for which there was an express provision in the contract to make a payment therefor. It is not surprising that in complex building and civil engineering contracts, where many delays are occurring at the same time, it is difficult to allocate any particular element of damages to the appropriate event or circumstance which caused the damages claimed. In order to deal with this difficult problem, it was no doubt a common practice to formulate a general claim in which all of the damages which arose as a result of many interrelated causes were pursued as a 'rolled-up' claim.

This practice was challenged in *J. Crosby & Sons Ltd* v. *Portland Urban District Council* (1967) 5 BLR 121. In this case there had been some forty-six weeks' overall delay to completion due to various causes of delay of which thirty-one weeks had been held by the arbitrator as being attributable to causes of delay for which the contractor was entitled to compensation. The arbitrator proposed to award a lump sum to compensate for the delay of thirty-one weeks and the employer appealed claiming that the arbitrator should arrive at his award by determining the amounts due under each individual head of claim. The form of contract was the ICE fourth edition. It was held that, provided the arbitrator did not include an element of profit in the amount awarded, and that there was no duplication, then if the claim depends on 'an extremely complex interaction in the consequences of various denials, suspensions and variations, it may well be difficult or even impossible to make an accurate apportionment of the total extra cost between the several causative events', the arbitrator was entitled to make a lump sum award for the delay and disruption.

This type of claim appeared in the case of *London Borough of Merton* v. *Stanley Hugh Leach Ltd* (1985) 32 BLR 51, where the form of contract was the 1963 edition of JCT. The judge was persuaded to allow a rolled-up claim on the basis of the findings in the *Crosby* case.

In another case, *Wharf Properties Ltd and Another* v. *Eric Cumine Associates, and Others* (1988) 45 BLR 72, (1991) 52 BLR 1 PC, the employer (*Wharf*) pursued a rolled-up or global claim against his architect (*Cumine*) which relied on the same premise as both the *Crosby* and *Merton* cases. The Court of Appeal of Hong Kong did not accept the claim. On the face of it, there appears to be an anomaly which places doubt on the validity of this type of claim. However, in this case, there appears to have been a lack of evidence to link the damages claimed with the numerous alleged defaults of the architect. The *Wharf* case should not be regarded as the death knell for all claims of this kind. It should be noted that the judge in a subsequent case, *Mid-Glamorgan County Council* v. *J Devonald Williams & Partner* [1992] 29 ConLR

129, considered the previous cases involving rolled-up claims (including the *Wharf* case) and held that, provided the circumstances were appropriate, such a claim could succeed.

Global claims were again scrutinised in *Imperial Chemical Industries* v. *Bovis Construction Ltd and Others* (1993) 32 ConLR 90, where the plaintiff was ordered to serve a Scott Schedule containing:

- the alleged complaint;
- the defendant against whom the claim was made;
- which clause in the contract had been breached;
- the alleged failure consequences of such breach.

In *GMTC Tools & Equipment Ltd* v. *Yuasa Warwick Machinery Ltd* (1995) 73 BLR 102, the use of a Scott Schedule was raised again. The Judge had ordered that a Scott Schedule should be drawn up setting out the details and effects of each of the plaintiff's complaints.

The plaintiff had difficulty in preparing the Scott Schedule and failed to comply with the Unless Order. The matter was eventually dealt with on appeal where Lord Justice Leggatt ruled that a Judge is not entitled to prescribe the way in which the quantum of damage is pleaded and proved or to require a party to establish causation and loss by a particular method. His Lordship said:

> 'I have come to the clear conclusion that the Plaintiff should be permitted to formulate their claims for damages as they wish, and not be forced into a straitjacket of the Judge's or their opponent's choosing.'

In *British Airways Pension Trustees Ltd* v. *Sir Robert McAlpine and Son* (1995) 72 BLR 26, Judge Fox Andrews had ordered that the claim be struck out and the action dismissed on the grounds that the plaintiffs failed to properly particularise their claim. However, this decision was overruled by the Court of Appeal where Lord Justice Savill said:

> 'The basic purpose of pleadings is to enable the opposing party to know what case is being made in sufficient detail to enable that party properly to answer it. To my mind, it seems that, in recent years, there has been a tendency to forget this basic purpose and to seek particularisation even when it is not really required. This is not only costly in itself, but is calculated to lead to delay and to interlocutory battles in which the parties and the Courts pore over endless pages of pleadings to see whether or not some particular points have or have not been raised or answered, when in truth each party knows perfectly well what case is made by the other and is able properly to prepare to deal with it. Pleadings are not a game to be played at the expense of citizens nor an end in themselves, but a means to the end, and that end is to give each party a fair hearing.'

In *Amec Building Ltd* v. *Cadmus Investment Co Ltd* [1997] 51 ConLR 105 the judge appears to have taken the view that each case will be dealt with on its merits without laying down principles as to whether global claims will or will not be accepted.

In summary, in spite of numerous recent cases, it appears that little has changed since the principles laid down in *Mid-Glamorgan County Council* v. *J. Devonald Williams & Partner*. In practice, global claims should be a last resort, not just because it is difficult to particularise a number of claims but because particularisation is impracticable or impossible owing to complex entanglement with numerous overlapping and/or concurrent matters.

The Society of Construction Law has published a delay and disruption protocol which states that such claims are to be discouraged and are rarely accepted by the courts.

1.7 Notice

Most building and civil engineering contracts contain provisions which require the contractor to give notice of delay or of its intention to claim additional payment under the terms of the contract. It is usual for the contract to specify that notice should be given within a reasonable time, but other terms such as 'forthwith', or 'without delay' or within a specified period of the event or circumstance causing delay or giving rise to the claim may be used. The courts have had to consider the meanings of various terms and they have often been faced with the argument that the giving of notice was a *condition precedent* to the contractor's rights under the contract.

The ICE conditions of contract generally opt for a specified period within which notice should be given. Two cases involving the ICE conditions of contract are helpful in deciding if notice is a *condition precedent*.

In *Tersons Ltd* v. *Stevenage Development Corporation* (*supra*), the engineer issued a variation instruction for the first contract on 24 July 1951. The contractor carried out the varied work and gave notice of his intention to claim on 3 December 1951. In the second contract the engineer issued an instruction on 24 August 1951 and the contractor gave notice of his intention to make a claim on 6 February 1952. Work on the second contract commenced on 12 March 1952. The contractor did not submit his claims on a monthly basis.

The Court of Appeal was asked to decide whether the contractor's notices complied with the provisions of sub-clauses 52(2) and 52(4) of the second edition of the ICE conditions of contract. Sub-clause 52(2) required the contractor to give notice of his intention to claim a varied rate 'as soon after the date of the Engineer's order as is practicable, and in the case of additional work before the commencement of the work or as soon thereafter as is practicable'.

Sub-clause 52(4) provided for claims to be made monthly and 'no claim for payment for any such work will be considered which has not been included in such particulars. Provided always that the Engineer shall be entitled to authorise payment to be made for any work notwithstanding the Contractor's failure to comply with this condition if the Contractor has at the earliest practical opportunity notified the Engineer that he intends to make a claim for such work'. It was held that clause 52(2) only required a notice in general terms that a claim was being made and that clause 52(4) only related to payment in monthly certificates. The proviso in clause 52(4) which empowered the engineer to authorise payment, and the provisions of clauses 60, 61 and 62, which contemplated that the contractor's rights remained open until the final maintenance certificate had been issued were sufficient to show that the contractor had complied with the contractual provisions.

In *Crosby* v. *Portland UDC* (*supra*), the works were suspended by order of the engineer and the contractor did not give notice within the period specified in sub-clause 40(1) of the fourth edition of the ICE conditions of contract which contained the proviso 'Provided that the Contractor shall not be entitled to recover any extra cost unless he gives written notice of his intention to claim to the Engineer within twenty-eight days of the Engineer's order.' It was held that since the contractor had not given notice within the specified period the claim failed.

The distinction between the *Tersons* and the *Crosby* cases is best explained in *Bremer Handelsgesell-Schaft M. B. H.* v. *Vanden Avenne-Izegem P. V. B. A.* [1978] 2 Lloyds LR 109, in which Lord Salmon said:

> 'In the event of shipment proving impossible during the contract period, the second sentence of cl. 21 requires the sellers to advise the buyers without delay and the reasons for it. It has been argued by buyers that this is a *condition precedent* to the seller's rights under that clause. I do not accept this argument. Had it been intended as a *condition precedent*, I should have expected the clause to state the precise time within which the notice was to be served, and to have made plain by express language that unless notice was served within that time, the sellers would lose their rights under the clause.'

In the *Tersons* case neither of the ingredients stated by Lord Salmon were present, while in the *Crosby* case both ingredients were present (a precise time and clear language to bar a claim if notice was not served accordingly). If notice is to be a *condition precedent*, it is important to take account of these essential requirements when drafting the relevant provisions although the precise consequences of a failure to comply with a condition precedent do not have to be explicitly stated.

In the case of *Steria Ltd v Sigma Wireless Communications Ltd* [2007] 2008 BLR 79 the court held that provided the wording of the relevant clause was sufficiently clear, there was no requirement for the contract to set out the consequences of failing to comply with it.

Very little change has been made to subsequent editions of the ICE and FIDIC conditions of contract. Both ICE and FIDIC relaxed the *conditions precedent* with respect to suspension. However, the 1999 FIDIC contracts (Red, Yellow and Silver Books) now contain strict provisions to give notice within twenty-eight days for all claims (sub-clause 20.1). The giving of notice in accordance with this sub-clause (but not the requirements to provide particulars and accounts of claims) is a *condition precedent* to the contractor's rights to claim for delay or additional payment (see 4.9, *infra*).

The requirements to give notice in RIBA and pre-1980 JCT standard forms of contract were less stringent than the requirements in the ICE conditions. Notice of delay under the extension of time clause (clause 23 in the 1963 edition of JCT) is required to be given by the contractor 'forthwith'. The case of *London Borough of Merton* v. *Stanley Hugh Leach Ltd* (*supra*) dealt with a host of issues, one of which involved extensions of time if the contractor fails to give written notice upon it becoming reasonably apparent that the progress of the works is delayed. It was held that, if the architect was of the opinion that the progress of the works is likely to be delayed beyond the completion date by one of the specified causes of delay for which there was power to extend time for completion of the works, the architect owes a duty to both the employer and the contractor to estimate the delay and make an appropriate extension of time. The giving of notice of delay by the contractor was not a *condition precedent* to an extension of time. However, failure on the part of the contractor to give notice in accordance with the contract was a breach of contract and that breach may be taken into account when considering what extension should be made.

1.8 Interference by the employer

Most building and civil engineering contracts provide for the architect or engineer to be responsible for granting extensions of time and certifying payment of sums due under the contract. In carrying out these duties the architect or engineer is required to act fairly and impartially and the employer is not permitted to influence or obstruct them in the performance of their duties. Several early cases show that the courts have taken a consistent view in cases where the employer has sought to influence the person appointed by him to certify or value in accordance with the contractual provisions, even if there was no fraud on the employer's part: *Hudson's Building and Engineering Contracts, eleventh edition* at paragraph 6.142. In the case of *Morrison-Knudsen* v. *B.C. Hydro & Power* (1975) 85 DLR 3d 186, all of the contractor's requests for an extension of time were rejected and no extensions of time which were due to the contractor were granted. The contractor accelerated the progress of the work and the project was completed shortly after the contractual date for completion. It was subsequently discovered that the employer was instrumental in securing an agreement with a government representative that no extensions should be granted. The Court of Appeal of British Columbia held that the contractor was entitled to recover the acceleration costs which he had incurred as a result of the breach of contract. Further, the contractor would have been entitled to rescind the contract and sue for payment in *quantum meruit* if he had been aware of the breach.

In a Scottish case, the contractor claimed to be entitled to interest on a sum which the contractor claimed to be due but which had not been certified by the engineer. The contract was the ICE fifth edition which provided for interest to be paid in the event of failure to certify (clause 60(6)). The Judge held that the clause did not allow for interest if the engineer certified sums which were less than the sums which the engineer ultimately certified as being due. If the engineer had certified what in his opinion was due at the time, it could not be construed as a failure to certify.

However, it was discovered that the employer had instructed the engineer that under no circumstances should he certify more than a specified sum without the employer's permission. The engineer appeared to ignore the employer's instructions and prepared a draft letter to the contractor indicating that a sum exceeding the employer's ceiling was due. The employer sacked the engineer. The Judge held that the employer's interference was sufficient to deny effect to the engineer's certificates in which case there must have been a failure on the part of the engineer to certify within the meaning of clause 60(6) of the contract. In these circumstances the contractor was entitled to interest: *Nash Dredging Ltd* v. *Kestrell Marine Ltd* (1986) SLT 62. [This decision, on the general matter of interest payable in accordance with the provisions of clause 60(6) of the ICE conditions, should not be regarded as being applicable in England. See *Morgan Grenfell* v. *Sunderland Borough Council and Seven Seas Dredging Ltd, Secretary of State for Transport* v. *Birse–Farr Joint Venture* and other cases, (*infra*) Chapter 5.]

1.9 Claims against consultants

It has long been held that if a consultant acts negligently in the performance of his duties, and the employer suffers loss as a result, then the employer would have a claim for damages against the consultant. This was held to be the case in *Sutcliffe* v. *Thackrah and Others* (1974) 4 BLR 16. It appeared from the judgement in this case that the contractor may have a claim for damages against the consultant.

Several cases involving claims by contractors against consultants have been reported and the industry seemed to have a clear picture of the law in this regard when the contractor in *Michael Salliss & Co Ltd* v. *E. C. A. Calil and William F. Newman & Associates* [1989] 13 ConLR 68, successfully claimed damages arising out of the architect's failure to exercise properly the duty of care owed to the contractor. The law, as it appeared after the *Michael Salliss* case, was turned upside down in *Pacific Associates Inc and Another* v. *Baxter and Others* (1988) 44 BLR 33. In this case the Court of Appeal rejected the contractor's claim for damages arising from the engineer's negligence. The contractor had settled with the employer and sought to claim against the engineer on the grounds that

> 'By their continual failure to certify and by their final rejection of the claims the engineers acted negligently and alternatively were in breach of their duty to act fairly and impartially in administering the contract.'

As it now stands, contractors are unlikely to succeed in claims for damages against consultants if the claim is one which the contractor can make against the employer.

1.10 The future

The law relating to construction contracts has evolved rapidly in recent years and it looks set to continue at a similar pace in the future. Recent cases have put new interpretations on some aspects of the law but many grey areas still exist. The wide range of new or revised forms of contract will bring with them new problems that will need resolution. An increasing awareness of contract law and its application in modern contracts will be in evidence and new contractual provisions will be drafted to deal with the decisions of the courts. A considerable effort needs to be made in the direction of contracts administration, monitoring progress, claims formulation and presentation, and this is likely to be evidenced by the ever increasing number of seminars and training courses on the subject.

Resolution of disputes has become an increasingly costly exercise where the costs of arbitration are often no less than the costs of litigation. Procedures, extensive pleadings, tactics and joining of several parties have been the cause of escalating costs of managing an arbitration. Audication and mediation have found favour with all sides of the industry as means of resolving disputes as both are cheaper and less time consuming means of settling disputes than arbitration or litigation.

The end of the 1990s saw several changes in UK legislation. In particular, the Housing Grants, Construction and Regeneration Act 1996 (The Construction Act) incorporated (*inter alia*) mandatory provisions for resolution of disputes by adjudication (see 8.4, *infra*).

The Construction Act applies to all relevant construction contracts as defined by the Act.

Although very successful, the Construction Act had its problems and some of the flaws in the Act have been resolved by amendments which have been introduced by Part 8 of the Local Democracy, Economic Development and Construction Act 2009. In consequence of the amendments to the Construction Act, the Scheme for Construction Contracts, which implements many provisions of the Construction Act on a default basis, has had to be amended.

The Construction Act, as amended by the Local Democracy, Economic Development and Construction Act 2009 and the revised Scheme for Construction Contracts, are set out in Appendices D and E respectively.

The New Engineering Contract (NEC), now reissued as the Engineering and Construction Contract, provides for adjudication and standard UK forms of contract have followed suit. Internationally, all four of the 1999 Editions of the FIDIC International Contracts provide for dispute resolution by a Dispute Adjudication Board which may comprise a single member or three members.

While these moves towards resolution of disputes by adjudication are likely to improve cash-flow as a result of much earlier decisions, and also reduce the costs of settling disputes, it is likely that alternative methods will continue (Chapter 8, *infra*).

What may become evident is a potential battle between FIDIC contracts and NEC in the international arena. The NEC has been in use since the mid 1990s and has proved to be successful in the UK and as far afield as South Africa and Thailand where efficient management and fewer disputes are evident. The NEC encourages co-operation between all members of the construction team (taking on board many of the recommendations of the report *Constructing the Team*, published under the chairmanship of Sir Michael Latham (The Latham Report)). On the other hand, the new FIDIC conditions have continued to emphasise and tighten up the contractual machinery regarding notices and claims.

By way of example, the NEC requires the contractor or the employer's project manager (as the case may be) to give the other an early warning of any matters which may increase the price, delay completion or impair the performance of the works. For example, if the project manager is aware of any design delay on behalf of the employer, after giving the contractor an early warning of the problem, both parties can put their heads together to find the best possible solution which may involve rescheduling some of the work (very often at no extra cost). If the contractor is aware of a potential delay, such as late delivery of equipment, then following an early warning notice, both parties try to resolve the problem which may include the authorisation of alternative equipment. Properly used, these useful provisions may save time and money for both parties and avoid unnecessary delay and/or claims for additional payment. The employer also has a better chance of keeping the project on schedule.

The 1999 FIDIC Red, Yellow and Silver Books, intended for use on major international contracts (generally exceeding US$500000.00), only provide for an early warning to be given by the contractor to the employer. There appears to be no machinery for the employer to respond to an early warning by the contractor by way of a solution in the best interests of both parties. By way of contrast, in the Green Book, its contract for smaller works (generally less than US$500000.00), FIDIC goes part of the way to improve the matter by stating that both the employer and the contractor shall give an early warning. Unfortunately, the contract only provides for the contractor to '... take all reasonable steps to minimise these effects.' What are the employer's obligations?

It remains to be seen if co-operation (NEC) wins the day or if adversity (FIDIC) continues to stay in front in international contracting. No doubt the major funding agencies, such as The World Bank, will influence the outcome.

The Single European Market and the changes which have occurred in the 1980s and 1990s have lead to greater flexibility in contracting. Foreign firms often compete against British firms for work in the UK, and British firms are equally keen to compete in mainland Europe. There is still a long way to go. Harmonisation of products and standards is well advanced but differences in legal systems and forms of contract have not allowed any significant harmonisation in this area. Perhaps the NEC and FIDIC contracts will help to change the face of domestic contracting throughout Europe and that the days of having numerous different standard forms of contract in the UK will disappear. The NEC is already well established in the UK and overseas and there are no reasons why FIDIC contracts should not be used in the UK, France or Germany as a domestic contract. The NEC and FIDIC contracts go a long way to providing a solution to almost any type of contract under any contractual arrangements, thereby substantially satisfying the recommendations in the Latham Report (and in the Banwell Report of 1964) – that is, one form of contract for all types of building and civil engineering is desirable.

Choice of Contracts

2.1 The first steps

There are three main categories of client who require the construction of, or alterations, or extensions to, a building or civil engineering project. The first category consists of clients who embark upon a building or civil engineering venture only once or perhaps a few times. The second category consists of clients who regularly have the need to refurbish, alter or expand existing premises or develop new projects in the course of their business. The third category comprises a variety of speculative developers who construct projects for sale or lease.

Clients who embark upon any construction venture for the first time are often faced with a number of alternative routes but usually the first stop will be at the office of a qualified architect or engineer. For the majority of projects this approach may be sufficient. Most professional firms of architects and engineers are well versed in the use of standard forms of contract and, unless the client has unusual requirements, a standard form of contract will be available to suit most purposes. They are, however, not without their pitfalls and some architects and engineers fail to provide the necessary advice which may make the difference between ultimate client satisfaction and a potential claim for professional negligence.

Whether it is an architect, engineer, quantity surveyor, solicitor or a lawyer specialising in construction contracts, the best advice is usually given by someone who has had 'hands on' experience in administering or managing contracts and is well versed in contract law, including all of the recent developments in case law which affect the interpretation and application of standard forms of contract. An unamended standard form of contract may be more appropriate than a masterful piece of legal drafting which fails to take account of practical reality and commercial practice. In most cases a good contract will comprise the appropriate standard form suitably amended to rectify its deficiencies and incorporating reasonable client's requirements.

Clients who are familiar with the pitfalls of contracting often have their own amendments for use with a standard form or they may have a tailor-made form of contract to suit their own requirements. This is a step in the right direction but recent cases in the courts have shown that many amendments to tried and tested standard forms of contract, and some provisions in hybrid forms of contract, fail to contain the standard of clarity necessary to ensure that the draftsman's intentions are understood. The application of the '*contra-proferentem* rule' and other well established principles in English law may assist contractors when the terms of the contract are decided in the courts.

The criticism of contractual provisions introduced by major corporations and public clients suggests that some of them should approach the problems of contracting with equal caution to first time venturers. The vast sums of money which may be at stake merit special attention to the contract conditions and one of the first steps which ought to be taken by any client embarking on a major project should be to obtain expert professional advice from someone who is not a member of its own organisation. If this is done, the incidence of provisions which may appear to be in the client's interests, but which are likely to have the opposite result, may be reduced.

Some clients may be advised to proceed on the basis of an outline design brief and contractors may be invited to tender for the design and construction of the project. Independent advice is essential at all stages if this is to be adopted. If the client has confidence in a particular contractor, it may choose to go directly to the contractor to negotiate for the design and construction of the project. Only in exceptional circumstances should a client contract for work in this manner without the guidance of an independent professional throughout the contract.

2.2 Clients' objectives

The principal objectives of any client will be to have the project completed on time, within budget and to an appropriate standard of design, workmanship and materials. The priority or emphasis placed on these objectives will depend on a number of factors. Cost or time may determine the scope for design and specification for the work.

In view of the commercial pressures to minimise finance costs and to obtain revenue at the earliest possible date, priority may have to be given not only to a method of construction which is conducive to speed of erection, but to 'lead-in' times, phasing of design and construction, phased completion of the project, design by contractor and subcontractors, installation of client's equipment and many other factors depending on the complexity of the project. Major subcontractors or packages of work may have to be settled in advance of selection of the principal (or main) contractor. If a client has a generous budget, he may insist on the best quality and design while cost and time are secondary.

Whatever the client's objectives it is important to set out a master programme, showing the various anticipated design and construction phases, at an early stage. This may have a bearing on the type of contracting methods to be used and should not be overlooked. The most common causes of construction delay claims stem from insufficient time allowed for design and commencing on site before sufficient design and detailing has been completed.

2.3 Contracting methods

The most common method of contracting is where a contractor undertakes to complete the project for a lump sum according to the design prepared by an architect or engineer at the outset. This 'traditional' method of contracting envisages the design being complete subject only to explanatory details and limited provisional items. Any change to the original design will be dealt with by way of a variation. The size and complexity of the project may determine whether or not bills of quantities are to be used. In building contracts the bills of quantities are not generally subject to remeasurement (except for correction of errors in the quantities). In civil engineering it is generally accepted that the design may be dependent on factors outside the control of the employer (ground conditions) and the contract is subject to remeasurement.

This method of contracting, by its nature, contemplates substantial completion of the design by the designer at tender stage. That is not to say that every detail has been drawn. It envisages issuance of details which do not change the original design, but merely explain more fully what is shown on the contract drawings. In the normal course of events, provided the designer had considered the details necessary to make the overall design fit together, explanatory drawings should not constitute a variation to the original design.

It is often the case that some critical aspects of design cannot be properly represented on a drawing before the designer has drawn the details. This is fundamental drawing practice. Because of pressure to get tender documents together at the earliest possible stage, too many contracts get off to a bad start due to insufficient attention to detail before invitations to tender. In short, this type of contract envisages a design phase which is almost complete before the construction stage commences, and the only design to be done after commencement of construction is of an explanatory nature and variations to the original design for which there is machinery to adjust the contract sum and/or the contract period (see Figure 2.1).

Support for the view that a lump sum contract should be designed in all its essential elements at tender stage is found in *The Banwell Report* (*The Placing and Management of Contracts for Building and Civil Engineering Work*, HMSO, 1964). The JCT Standard Building Contract With Quantities 2011 used for this method of contracting clearly contemplates the design being substantially, if not wholly, complete at tender stage. The recitals of the JCT 2011 form expressly states that the employer 'has had drawings and bills of quantities prepared which show and describe the work to be done'. Clause 1.1 of the JCT Standard Building Contract With Quantities 2011 defines the drawings and the bills of quantities as Contract Documents, and clause 2.1 requires the contractor to 'carry out and complete the Works in a proper and workmanlike manner and in compliance with the Contract Documents'.

It has long been an accepted practice, and provided for in most forms of contract, that some work may not be fully designed at tender stage. This is usually dealt with by provisional sums or provisional quantities. In recent years the proportion of work covered by provisional items has increased beyond that for which this type of contract was intended. In some cases as much as forty per cent of the contract sum has been made up of provisional items, leaving the contractor unsure as to the scope of the work and the employer without a realistic budget for the project.

Other forms of abuse include the use of provisional sums under the guise of PC (Prime Cost) Sums. Very often the prime cost sum is no more than a provisional sum, whereas on the strict interpretation of the contract, a prime cost sum should be a reasonable estimate based on a design which was in existence at tender stage. This will be dealt with in more detail in later chapters.

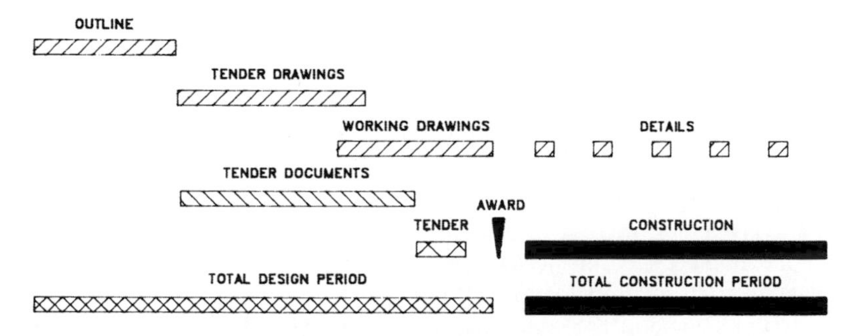

Figure 2.1 *Traditional contracting*

Some practitioners are bent on using a form of contract intended for use in the above circumstances, when it was known at the outset that the design stage would extend well into the construction phase. This practice may work if the designer co-ordinates the design into a master programme which is synchronised with the contractor's construction programme. However, there are many risks, such as under-estimation of 'lead-in' times for procurement, limitation on the flexibility in the contractor's programme (in the event that the contractor needs to change sequence for his own convenience) and an unacceptable incidence of variations caused by lack of foresight. All of these factors may lead to late completion and claims for compensation of one kind or another.

Another disadvantage of traditional contracting is that it does not usually permit the contractor to have an input at design stage. Many contractors are able to contribute to the design so that savings in cost and time can be made for the benefit of the employer. Sometimes contractors offer alternative designs, but very often this is so late in the day that it places more pressure on the design team to take account of the contractor's proposals in the overall design. Variants on the traditional forms of contract include an element of design by the contractor.

It is becoming increasingly popular for employers to move in the direction of design and build or turnkey contracts. A degree of competition may be introduced by a comprehensive design brief and a schedule of the client's requirements. It is important to ensure that firms bidding for work of this nature have a sound track record which can be verified and that a detailed inspection of previous projects is undertaken by the client's professional advisers. Care should be taken to investigate previous performance. Have the projects been completed on time and within budget? What are the maintenance costs? In addition to written testimonials from previous clients, it may be advisable to obtain permission to discuss the bidding contractors' performance and the quality of the buildings with clients and consultants for previous projects.

It is important to select a contractor in whom the client has complete faith and confidence. That is not to say that the client should go ahead without professional advice throughout the project. This may take the form of a project manager and possibly a quantity surveyor. An architect or engineer may also be engaged to advise on technical matters. A good project manager can make the difference between the success or failure of this method of contracting. It is essential that the person selected to carry out this role is given the freedom to act fairly and impartially. While the employer's interests must be given priority, it is very often counter-productive to adopt an adversarial position which creates distrust between all parties. Much more benefits can be obtained for the client if the project manager helps to preserve trust and confidence by showing authority, integrity and competence at all levels.

There are circumstances in which it is advantageous for the design stage of the project to overlap with a considerable period of the construction phase (see Figure 2.2). If this is carefully structured, it is possible to commence construction much earlier than in traditional methods of contracting. The total effect of this method of contracting may be to give rise to a higher overall expenditure on construction: however, if the client can get beneficial occupation earlier than it otherwise would have done by traditional contracting, there may be considerable savings or benefits such as earlier rental income and reduced finance charges.

There are several methods of contracting which are suitable where it is intended that the design stage and the construction stage overlap. Management contracting is one method which lends itself to this process. In its purest form it is based on the

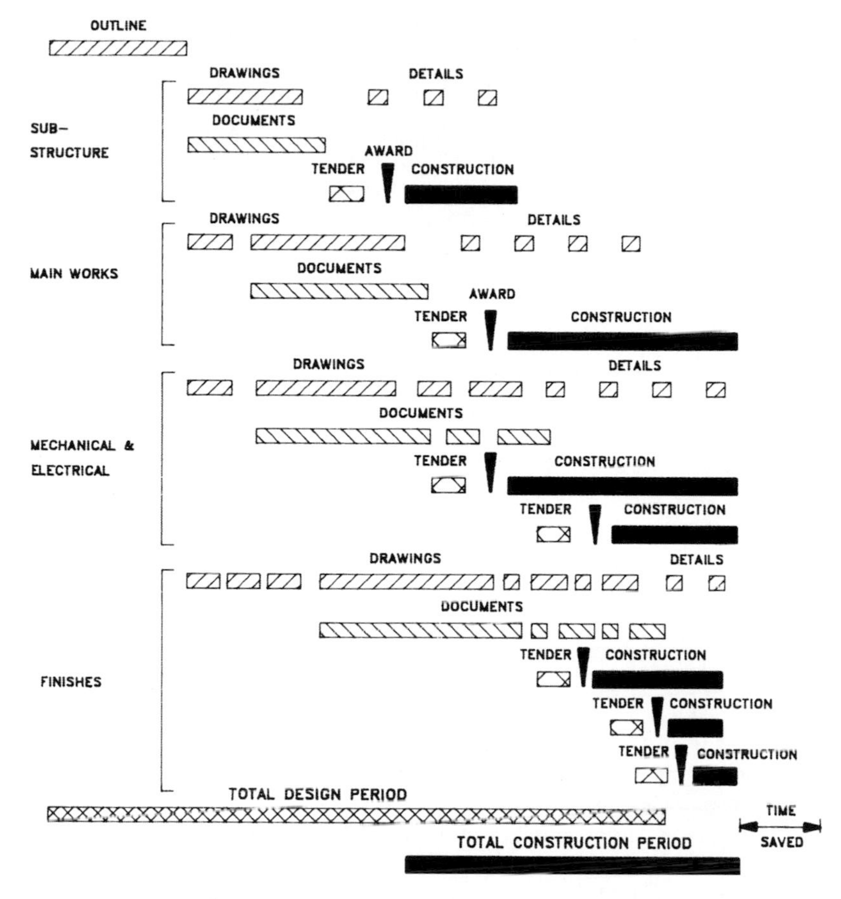

Figure 2.2 Phased design and construction

prime cost plus the fixed (or percentage) fee method of contracting which has been used for many years. The outline design of the project, together with a detailed brief, is prepared by the design team and bidding contractors are required to submit their proposals for the management and 'procurement of construction'. The criteria used as a basis for selection will include

- reimbursable costs of site management, supervision and general services (similar to 'Preliminaries' in traditional contracting);
- lump sum or percentage to be added to the prime cost of the project;
- management capability and resources;
- ability to contribute to the design of the project – 'buildability';
- programme and methods of construction;
- methods of ensuring quality control;
- systems for cost control;
- industrial relations;
- proposed packaging of work to be done by subcontractors;
- buying power and negotiation skills;
- previous track record.

The selected management contractor does not usually execute any work himself. His obligations are, in collaboration with the design team and the employer, to procure completion of the project on time and within budget, by subcontracting various parts of the work and by purchasing materials to be fixed by subcontractors. Balance will have to be made when considering the size and scope of work packages.

Large packages will not enable the employer to obtain the benefit of buying margins, but a lower management fee may be required. On the other hand, a large number of small work packages will usually reduce the prime cost, but the management fee and reimbursable costs may be higher to reflect the increased management, supervision and risk involved.

In this method of contracting, the management contractor enters into an agreement with the employer in the same way as the contractor in traditional contracting. The contracting structure is shown in Figure 2.3. It is often the case that the management contractor's liability for late completion is limited to any damages which it can recover from subcontractors. This can cause serious problems if the subcontractors are financially vulnerable. Subcontractors carrying out small work packages may be faced with damages for late completion which are out of proportion to the value of work undertaken by them.

In traditional contracting, the employer may recover all of the damages from the contractor without being concerned about which subcontractors were the culprits. In management contracting, the liabilities of several subcontractors responsible for overlapping delays can cause difficulties and may often lead to disputes and arbitration or litigation.

Some hybrid forms of management fee contracts place greater responsibility on the management contractor. It is possible to devise a scheme where the management contractor is also responsible for the execution of the work in the same way as the traditional contractor. The advantages are that the management contractor is involved in the design and selection of subcontractors, but once the subcontracts are awarded, the management contractor takes full responsibility as if the subcontractor was a normal domestic subcontractor in the traditional sense. The management contractor may also execute some of the work himself. The management fee is likely to be higher to reflect the greater risk in this form of contracting.

There are also many methods of project management or construction management which permit overlapping of design and construction. It is impossible to define these methods of contracting as there appears to be numerous variations on a theme. In very broad terms the project manager is responsible for co-ordinating and managing the design and construction of the project as part of the project team. The manager will enter into a contract with the client to manage the project, but he may not enter into subcontracts. Each work package is undertaken by direct contracts with the client and the work is carried out under the direction and supervision of the project manager (see Figure 2.4).

2.4 Standard forms of contract

Why use a standard form of contract? Firstly, it will have been prepared having regard to the nature of the work to be undertaken. Secondly, practitioners in the industry are more comfortable using a standard form of contract with which they are familiar and which is usually capable of interpretation by reference to readily available text books and case law. Thirdly, they are often drafted and agreed by recognised bodies

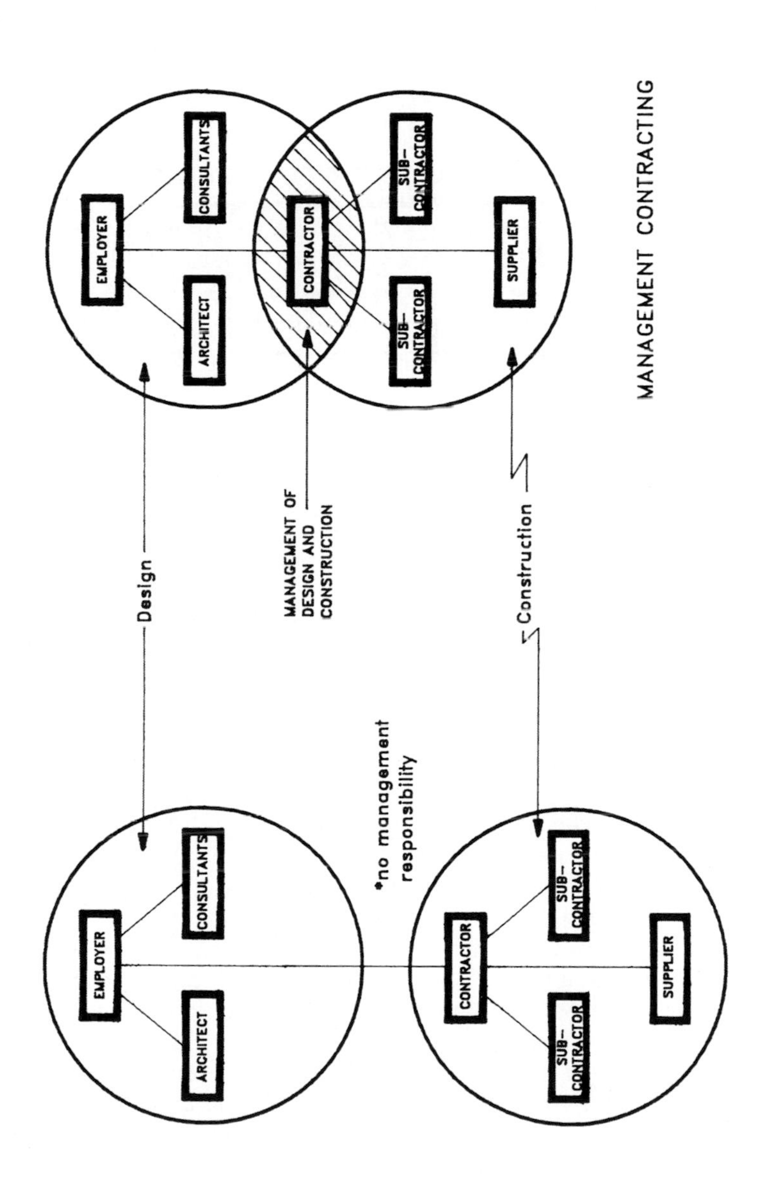

Figure 2.3 *Management of design and construction*

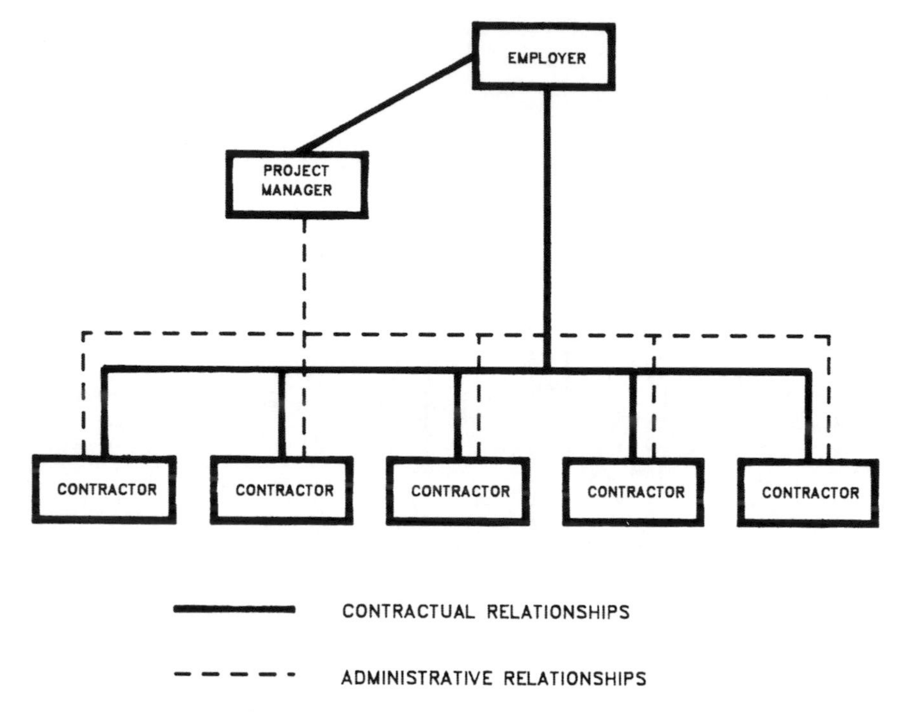

Figure 2.4 *Project management structure*

representing all sides of the industry which will be affected by them. This last point is to some extent a disadvantage in that a form of contract, 'by committee', is often a compromise containing some defective aspects of one form or another.

Standard forms of contract are available to suit contracts of almost any size and complexity and to suit most methods of contracting. Some practitioners select forms of contract with which they are familiar without having sufficient regard to their suitability or limitations. This practice is not to be recommended and should be regarded as 'short changing' the client. Any client embarking upon a construction project is entitled to expect sound advice from his professional advisers on all aspects of the contract, not least of which is the selection of the right form of contract for the purpose.

The methods of contracting discussed in this chapter will be a major consideration for many larger projects and for small or medium projects that require a considerable amount of preplanning. The type and size of contractors bidding for the job will also be important. For example, the use of a lengthy standard form, such as the JCT Standard Building Contract 2011, may not be appropriate when the tendering contractors are little more than 'one man' firms having no understanding of the complicated provisions in the contract. The use of this form of contract in such circumstances will increase the price and/or lead to all sorts of problems in administration of the contract. At the other end of the scale, the use of one of the simpler forms of contract may not be appropriate for a project with a high building services content.

In spite of the recommendations of both the Banwell Report of 1964 and the Latham Report of 1994 (that ideally, a single form of contract should be required for all types of building and civil engineering work), there are now well over one hundred different standard forms of contract for use in the UK alone. The NEC standard forms of contract stand alone in seeking to be suitable for all types of building and engineering contracts. In many cases so many amendments have been and continue to be made to standard form contracts that tender and contract documents often include not only the standard printed conditions, but numerous amendments which need to be referred to in order to fully understand the contract.

It is not possible to deal with all of the standard forms of contract in one chapter. However, some of the most common are considered very briefly.

2.5 The Joint Contracts tribunal standard forms of contract

The most commonly known standard forms of contract are those issued by The Joint Contracts Tribunal (JCT). The first standard form issued by the JCT was in 1963 which superseded the RIBA forms of contract. It was published in four main variants; the private and local authorities' versions, each with, or without, bills of quantities. The JCT forms were revised in 1980 and 1998. In 2005 they were revised again and their appearance is substantially different to their predecessors. The intention of the publisher is to publish a new edition each time the relevant standard form contract is revised rather than issue individual revisions to the existing standard form. Today there are a considerable number of standard forms for a variety of needs.

The standard forms of contract referred to below are the current editions of the most common forms of JCT contract and were introduced in 2011 in consequence of legislative amendments. There are other JCT standard forms available and they can be found in the guide referred to below.

The JCT publish a guide titled 'Deciding on the appropriate JCT contract' to assist to determine which form is suitable for particular circumstances. The guide is available on the World Wide Web but is only intended as a guide and professional advice should be sought as required.

The Minor Works Forms of Building Contract 2011

There are two versions of the Minor Works Building Contract, one with and one without, contractor's design. The 'with contractor's design' form of contract is not intended to be used where the contractor is to carry out the whole of the design of the works. It is only for use where the contractor carries out a portion of the design of the works.

It is a simple form of contract embodying the essential ingredients of a building contract. Suitable for a project where the work to be carried out is of a simple nature, where a contract administrator will administer the contract and where bills of quantities are not necessary. It is not suitable where nominated or named specialist subcontractors are contemplated. The recommended limits on its use are contained in the guide issued by the JCT. The guide does not form part of the contract. As the title implies, the form is intended to be used for minor works which can be adequately defined in drawings and specification.

The Intermediate Building Contract 2011

Like the Minor Works form, there are two versions of the Intermediate form, one with contractor's design and one without. Again, like the Minor Works form, the Intermediate 'with contractor's design' form of contract is not intended to be used where the contractor is to carry out the whole of the design of the works. It is only intended for use where the contractor carries out a portion of the design of the works.

This form of contract was drafted to fill the gap between the minor works form and the standard form of building contract and is intended for works of a simple nature involving basic recognised trades. It combines the simplicity of the minor works form of contract but many of the procedural provisions of the JCT Standard Building Contract 2011 are incorporated. The same form can be used for private and local authorities' use, and it contains alternative provisions so that it can be used with a specification, or schedules of work or bills of quantities. Limitations as to its intended use are printed on the cover of the form of contract and further guidance is given in the JCT guide.

The form contains supplementary conditions for use if it is intended to have partial possession or sectional completion. Without these supplementary conditions, difficulties may arise when applying the liquidated damages provisions. It also makes provision for 'named subcontractors' in the event that the employer wishes to specify subcontractors.

While this is a simplified form of contract when compared with the JCT Standard Building Contract 2011, it is contained in more than seventy pages, making it longer than the original JCT 63 standard form of contract. With very little amendment, it is an extremely flexible form of contract which finds favour outside of its intended limitations.

The Standard Building Contracts 2011

There are three primary forms of this standard contract, one with quantities, one without quantities and one with approximate quantities and each is available with or without contractors design.

This standard form is probably still the most widely used standard form for building works today but it is facing stiff competition from the NEC 3 Contract and the JCT Design and Build Contract. Notable amendments over its predecessor include the removal of nominated subcontractors, the ability to dispense with the necessity for collateral warranties by the introduction of third-party rights and the removal of arbitration as the default method of dispute resolution. The format of the contracts has changed considerably from earlier versions making it much easier to use.

The Prime Cost Building Contract 2011

This form of contract has replaced the Fixed Fee Form of Contract. This contract may be suitable where the design has not progressed sufficiently to accurately define the *Works*. However, the scope of the work to be done has to be defined and sufficient information to describe the *items of work to be done* is necessary. An estimate of the prime cost of the work to be done and a fixed fee form the basis of the estimated total cost to the employer. There is no provision to vary the scope of the work. The final cost to the employer is the actual prime cost ascertained from the contractor's accounts and invoices plus the fixed fee quoted by the contractor. There is provision for reimbursement of loss and expense caused by disturbance of the regular progress of the works.

The Management Building Contract 2011

The Management Form is intended for use on large scale projects requiring an early commencement on site prior to completion of the design. The employer is responsible for the design of the works and the management contractor manages the project for a fee and employs Works Contractors to carry out the construction works. The control of cost and time is dependent upon co-operation between all members of the design team and the management contractor. Clearly, any contract which envisages work commencing before the design is completed by a third party has potential for disputes to arise.

The form has lost favour in recent years, particularly to the benefit of the Design and Build Contract referred to below.

The Design and Build Contract 2011

This form of contract contemplates a reasonably detailed outline of the Employer's Requirements based upon which competitive tenders are invited, incorporating the bidding contractors' design solutions and price for designing and constructing the works. The same form of contract is often used as a basis for a negotiated contract.

While it is possible for the design to be complete prior to construction, the form of contract envisages design by the contractor during the contract period. Insufficient thought to design by the contractor prior to acceptance of the contractor's proposals by the employer often leads to disputes as to what constitutes a variation to the employer's proposals and what ought to have been contemplated by the contractor as part of the original design. Comprehensive and detailed proposals by the employer can reduce the scope for such disputes.

Unfortunately, the form is often heavily amended by employers. Amendments commonly include a requirement for contractors to adopt and be responsible for design work carried out on behalf of employers. It is an amendment that can cause problems with professional indemnity insurance as well as the simple problem of not knowing what criteria a designer was working to when employed by the employer.

The contract is not suitable where only a small amount of design work is to be carried out by a contractor.

2.6 Other forms of contract

Government forms of contract, such as GC/Works/1, are used extensively in the public sector. Amended versions exist for overseas projects. In the latest edition (1998) much of the administrative work falls on the project manager appointed by the authority (the employer). There are contractual provisions for acceleration. Variations and amendments to the standard publication enable alternative methods of contracting to be used, such as design and build.

Other standard forms of contract issued by professional bodies are available and are worth considering as alternatives to some of the better known standard forms of contract.

In the civil engineering field, the ICE forms of contract (traditional, design and build and minor works contracts) are well established in the UK. However the seventh edition is beginning to lose its almost universal recognition in the face of competition from the NEC 3 Contract.

The Model Form of General Conditions of Contract for use in connection with Home or Overseas Contracts for the Supply of Electrical, Electronic or Mechanical Plant – with Erection 1988 (MF/1) Revision 3 1995 is commonly used for major projects such as water or power plants.

FIDIC Contracts

The first contracts designed specifically for international contracts were probably initiated in the United States. These were largely defence project orientated and the most well known is probably the Corps of Engineers contracts. The Associated General Contractors of America and the Federation of Americana de la Industria de la Construction led the way for the US construction industry to move in the direction of the international conditions of contract known as FIDIC (Fédération Internationale des Ingénieurs-Conseils) which was based almost entirely on the pre-Fifth Edition ICE conditions of contract. The First Edition of FIDIC was published in 1956 and has gone through several revisions, the latest edition which followed the ICE format being the Fourth Edition (commonly known as the Red Book) published in 1987. This form of contract was intended for use where the design was done by the employer and construction was done by the contractor.

Because of a growing international demand for a variety of contracts to suit different methods of procurement, other standard international forms of contract issued by FIDIC up to 1999 were (for Electrical and Mechanical Works) the Yellow Book and (for Design–Build and Turnkey) the Orange Book. Apart from the changes giving emphasis to the nature of some of the specialist work in electrical and mechanical contracts, the main difference between these two forms is the degree of design responsibility placed on the contractor. Both the Yellow and Orange Books contemplate design by the contractor.

In 1999, FIDIC published a new family of contracts:

The Construction Contract (the 'Red Book')

Conditions of Contract for Construction for building or civil engineering works where the works are designed by the employer (or by his engineer) and where the contractor constructs the works in accordance with the design provided by the employer. However, the works may include some contractor-designed civil, mechanical, electrical and/or construction works.

The 1999 Red Book is intended to replace the 1987 fourth edition (also known as the Red Book).

The Plant and Design-Build Contract (the 'Yellow Book')

Conditions of Contract for Plant and Design–Build for electrical and/or mechanical plant, and for the design and execution of building or civil engineering works. Under this form of contract, the contractor designs and provides plant and/or other works, in accordance with the employer's requirements.

The 1999 Yellow Book replaces the previous Yellow Book.

The EPC/Turnkey Contract (the 'Silver Book')

Conditions of Contract for EPC Turnkey Projects for use in process or power plants, factories and the like, infrastructure or other types of development, where the employer requires a higher degree of certainty of final price and time, and where the contractor takes total responsibility for the engineering, design, procurement and execution of the project. Ideally there should be little involvement by the employer.

The 1999 Silver Book is intended to replace the 1995 Orange Book.

The Short Form (the 'Green Book')

Short Form of Contract for building or civil engineering works of relatively small capital value and/or of a repetitive nature or short duration. Under this form of contract, the contractor may construct the works in accordance with details provided by the employer or it may be used for contractor-designed civil, mechanical, electrical and/or construction works. FIDIC's guidelines for the use of the Green Book suggest that US$500000.00 (at 1999 prices) and twenty-six weeks should be regarded as reasonable limits on capital value and duration respectively, with the proviso that works of a repetitive nature may exceed these guidelines.

FIDIC also publish a contract for bank financed projects ('The MDB Contract') and a contract for dredging operations (the 'Dredgers Contract').

In spite of the criticism levied at the FIDIC contracts (*infra*), the new standard layout incorporating a great deal of common or 'core' conditions is welcome. Greater emphasis on definitions and a specific definition of '*force majeure*' is new. There are numerous minor changes to some definitions and clauses between the three contracts for major construction projects (Red, Yellow and Silver Books) but the principal changes appear in the following clauses:

Clause 3

In both the Red Book and the Yellow Book, these clauses are almost identical and deal with the powers and obligations of the engineer (the Red Book provides for the contractor to confirm verbal instructions of the engineer while the Yellow Book requires all instructions to be in writing). The engineer does not feature in the Silver Book where clause 3 deals with employer's administration.

Clause 5

In the Red Book this clause deals with nominated subcontractors. (In the Yellow and Silver Books there are very brief provisions for nominated subcontractors in sub-clause 4.5.) The same clause in the Yellow Book and Silver Book deals with design (by the employer). In the Yellow Book, the contractor may lose his rights to any claim in respect of incorrect information provided by the employer if he failed to properly scrutinise the employer's information in accordance with the contract and failed to give notice of the error within twenty-eight days. In the Silver Book, the contractor is deemed to have scrutinised the information provided by the employer before submitting the tender (before the base date) and shall be fully responsible for any error, inaccuracy or omission in the employer's information with the exception of

(a) information stated in the contract as being immutable or the employer's responsibility;
(b) definitions of the intended purpose of the works;
(c) criteria for testing and performance of the works; and
(d) information which cannot be verified by the contractor except as otherwise stated in the contract.

Clause 12

In the Red Book, this clause deals with measurement and valuation. In both the Yellow and Silver Books, clause 12 deals with tests after completion of the works.

The Red, Yellow and Silver books all have provisions for 'value engineering'. In the Red Book, the contractor and the employer share any saving that the contractor may be able to make or any benefit that the employer may receive as a result of

(a) accelerated completion;
(b) reduction in cost to the employer of executing, maintaining or operating the works;

(c) improved efficiency or value to the employer of completed works; or
(d) other benefits to the employer.

Under the Yellow and Silver Books, any such proposal (for value engineering) shall be treated as a variation. It is unlikely that value engineering will feature in the Yellow and Silver Books as most contractors ought to have 'value engineered' his design at the tender stage.

The Red, Yellow and Silver Books have much improved procedures for better management, monitoring and control of the project (see Chapter 4).

NEC 3 Engineering and Construction Contract 2005

The New Engineering Contract (NEC) (1991) has now been replaced by the Engineering and Construction Contract (EEC) but is confusingly still referred to as the NEC contract. The second edition was published in 1995 and the third edition was published in 2005. The Contract reflects a substantial move to recognise, and cater for, the various forms of contract which have been discussed herein. It is based on a core contract with flexible alternatives allowing the employer to choose the appropriate version to suit his needs.

The 2005 twenty three-document package consists of core contract provisions which are universal to all versions of the primary contract with various optional clauses which are intended to cover all types of construction and engineering contracts. The various versions are:

- Document A – Conventional Contract with Activity Schedule;
- Document B – Conventional Contract with Bills of Quantities;
- Document C – Target Contract with Activity Schedule;
- Document D – Target Contract with Bills of Quantities;
- Document E – Cost Reimbursable Contract;
- Document F – Management Contract.

The package also includes a short contract, two subcontracts, a professional services contract, an adjudicator's contract, a term service contract and a framework contract as well as guidance notes and flow charts.

The NEC contract includes philosophies not seen elsewhere.

Core clause 10.1 sets out the philosophy behind the contract:

> 'The *Employer*, the *Contractor*, the *Project Manager* and the *Supervisor* shall act as stated in this contract and in a spirit of mutual trust and co-operation.'

The NEC suite of contracts have become very popular and are now being used for works which were almost exclusively the domain of the JCT and ICE ranges of contracts.

In general terms, the project manager and the supervisor carry out the duties of 'the Engineer' in ICE and FIDIC contracts. The adjudicator settles disputes between the employer and the contractor.

There is provision for an 'early warning' to be given by the contractor or by the project manager (clause 16.1). The response to an early warning contemplated by the contract is refreshing and should be taken on board in any form of contract if the employer is really going to have the best possible chance of getting his project on time and within budget. Clause 16.3 states:

'At a risk reduction meeting, those who attend co-operate in

- making and considering proposals for how the effect of the registered risks can be avoided or reduced,
- seeking solutions that will bring advantage to all those who will be affected,
- deciding on the actions which will be taken and who, in accordance with this contract, will take them and
- deciding which risks have now been avoided or have passed and can be removed from the Risk Register.'

As stated earlier, the primary contracts (Options A to F) contain core clauses common to all. There are also main option clauses, secondary option clauses, cost components and contract data provisions in the contracts.

The main option clauses include provision for dispute resolution (options W1 and W2) by way of adjudication.

The secondary option clauses include provision for price adjustment (option X1), a parent company guarantee (option X4), sectional completion (option X5), bonus for early completion (option X6), delay damages (option X7), partnering (option X12) and low performance damages (option X17) to name a few. There is also provision for additional clauses required by the parties ('Z' clauses).

The parties are required to complete the Contract Data section of the contract without which the contract cannot be operated effectively.

The contract requires the parties to actively participate in its operation. It is not the sort of contract that can be signed and ignored until completion of the works (no contract should be but many are).

Build, Operate and Transfer Contracts (BOT)

These forms of contract (sometimes known as Build, Own, Operate and Transfer – BOOT) are becoming more common, particularly in countries where the government does not have sufficient public funds available to finance vital infrastructure, power or water projects and the like. While this method has seen most growth in developing countries such as India, Thailand, Malaysia, China and Vietnam, it is also popular in developed countries. In the UK, BOT or BOOT is the basis of the Government's Private Finance Initiative (PFI).

Projects which attract revenue by way of tolls or levies are candidates for this type of venture. A project is founded by the granting of a 'concession' for a period of years (say twenty to thirty years) to the promoter or concession company. The promoter will seek equity funding from interested investors and long-term finance from banks and financial institutions. Normally banks and financial institutions need to be satisfied on the debt:equity ratio and a minimum ratio may be set by the government. The promoter designs and constructs the project or it enters into a turnkey contract with a contractor for the design and construction of the works. Unlike a traditional contract, the concession company does not receive payment in stages or on completion, but relies on the income generated from tolls or levies throughout the life of the concession. The remuneration (and profits) are generated over the period of the concession by tolls or levies, out of which the capital and interest charges are repaid to the lenders, and dividends are paid to the investors. If there is delay to the construction of the project, then the promoter suffers a loss of revenue. Depending on the discount rate, one-year delay to completion of construction may require more than five years' extension to the concession period in order to recover the loss.

Any project which has the potential to earn revenue over a number of years which is more than sufficient to pay back loans and interest and produce a reasonable return for investors is suitable for a BOT scheme.

The contractual structure of a typical BOT scheme is shown in Figure 2.5 and the flow of expenditure and income for most models of BOT schemes is shown in Figure 2.6.

The comparison of costs incurred and the income does not, by itself, indicate whether or not the bid is profitable. The costs and the income must be brought back to a similar basis by discounted cash flow (DCF) techniques.

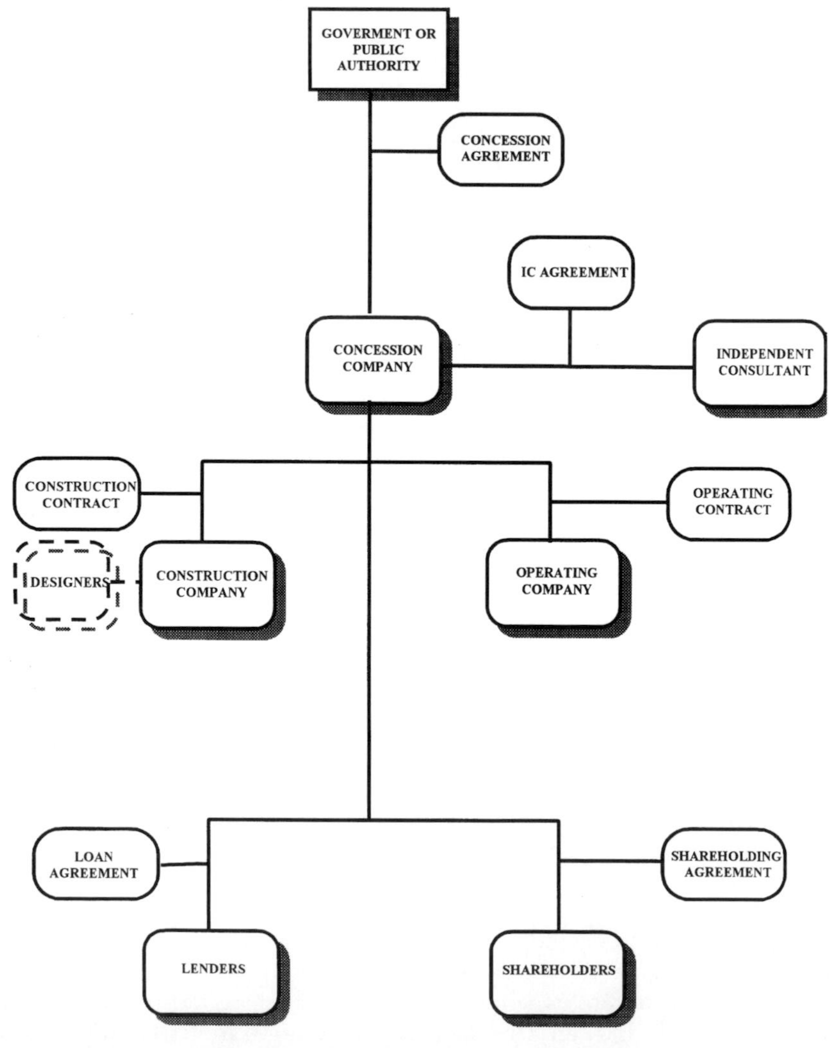

Figure 2.5 Contractual structure (BOT)

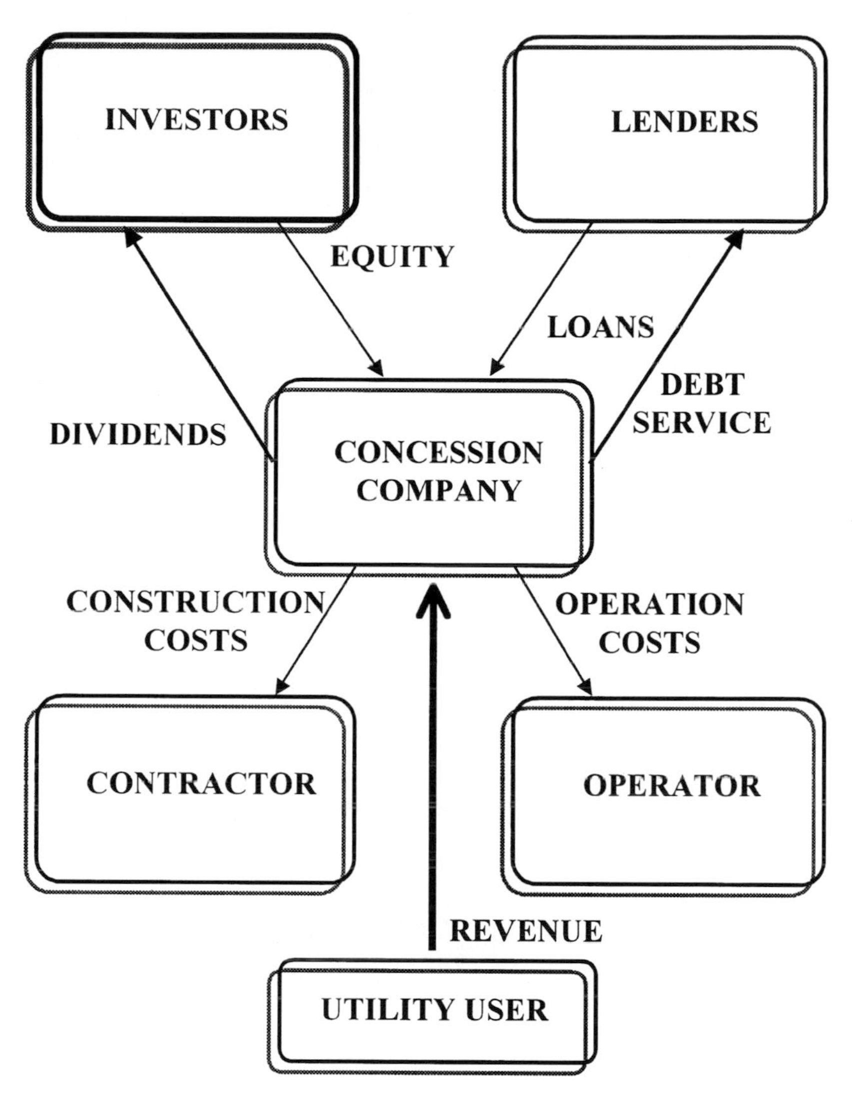

Figure 2.6 Costs and finance

Lenders to a project want to be sure that the project has a potentially satisfactory financial position. Lenders will measure the financial position of the concession company investors, for example, ROE (Return on Equity), and they expect to see a financially attractive scheme. Lenders fully realise that the project is more likely to succeed if the persons or bodies investing in the concession company have an excellent opportunity to earn a very good return.

In the early years of the operating period, all or most of the 'surplus' revenue will be used to repay loans – 'debt service and repayment of interest'. The ratio of debt to equity will diminish as years pass until, at a certain point, all the debt is repaid.

Diagrams showing how costs and revenue can be reconciled are given in Figures 2.7 and 2.8. Figure 2.7 shows expenditure and income and Figure 2.8 shows equity against dividends.

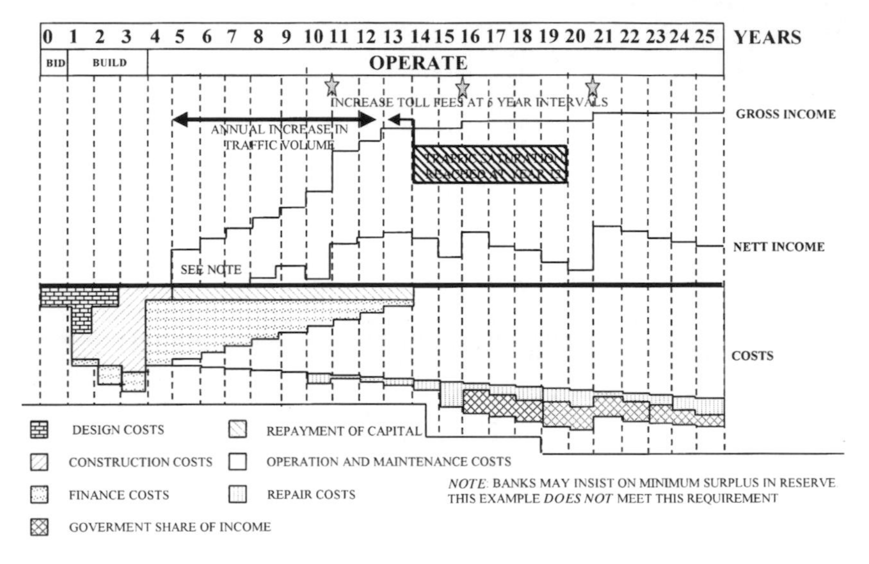

Figure 2.7 Expenditure and income (BOT)

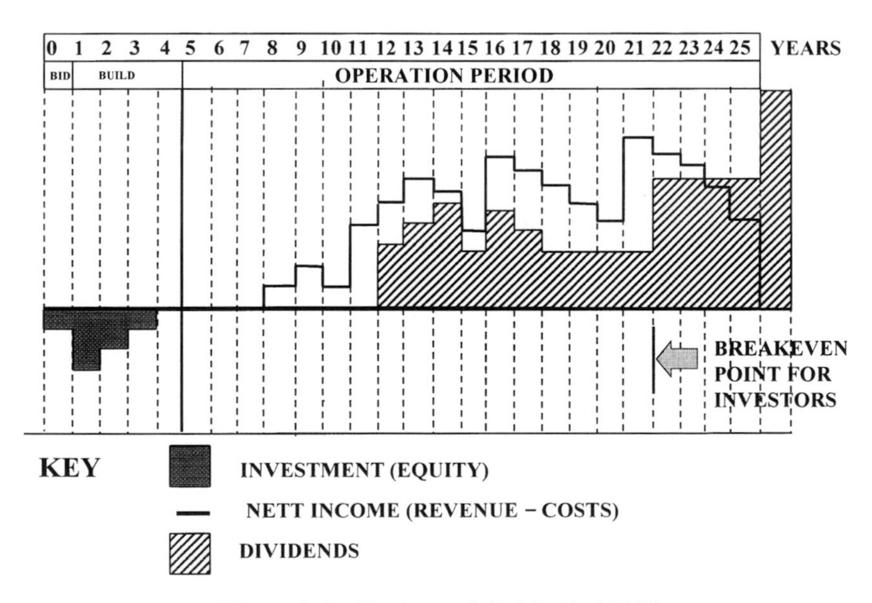

Figure 2.8 Equity and dividends (BOT)

The repayment of the loan and interest in Figures 2.7 and 2.8 assumes that the concession company must repay capital in equal instalments over nine years with interest on the reducing amount (commencing in the first year). Other options include a flat annual repayment, whereby the capital repaid in the first year is small and the interest is large. The amount of capital repaid each year increases and the interest decreases. In some cases, repayment may be deferred for three to twelve months (after commencement of operation). An important factor to be taken into account in some developing countries is the fact that much, and in some cases all, of the loans and equity will be provided in hard currency, but the revenue (out of which the loans have to be repaid and dividends paid) will be in local currency. The long-term effect of exchange rate fluctuations may be critical or even disastrous unless the concession agreement has a built-in remedy to compensate the concession company.

It should be noted that these 'financial models' in Figures 2.7 and 2.8 represent a poor investment on a number of grounds:

- ideally, there should be a reasonable surplus (income over expenditure) throughout the concession period (very often the lenders will insist on this);
- the breakeven point for investors should be within the first third of the total concession period;
- investors would normally expect dividends within a few years of commencement of operation;
- any delay to the project is likely to cause the project to be a failure because there is insufficient margin in the financial model (the breakeven point will disappear at the end of the twenty-five-year concession period if there is one year delay to the project with a ten per cent discount rate).

Where a BOT project involves the use of land or facilities owned by or controlled by government, it is necessary to pass specific legislation to cover the project. This may be done by enacting specific legislation governing the granting of a concession agreement and its terms for a particular project, or by enacting general legislation governing the terms of concession agreements and specific legislation for each particular project.

It will be seen from Figure 2.5 that there are a number of contracts between the various parties. There are no standard forms for BOT concession contracts (between government or public authority and the concession company). Likewise, there are no standard forms for operating contract, loan agreement or shareholder's agreement. The independent consultant agreement and the agreement between the contractor and the designer may be based on one of the standard forms, such as the FIDIC Consultants Agreement (1990).

While there are no standard forms of construction contract (between the concession company and the contractor), a number of standard forms of design and construct or turnkey contracts may be modified to suit the BOT model. FIDIC promotes its 1999 Silver Book as a form of contract suited to BOT (suitably amended). It is perhaps here that the debate over whether to use FIDIC or NEC (suitably amended) will be the hottest. On the one hand, there are good grounds to argue that a 'tough' contract such as FIDIC should be preferred. On the other hand, having regard to the fact that all parties suffer from increases in cost or delay in a BOT project, there is all the more reason for the parties to co-operate to ensure completion on time and within budget (hence the choice of NEC may be the better one). As long as the amendments to FIDIC take on board the constructive elements of NEC, it is probable that FIDIC will be equally, if not more, appropriate than NEC in these types of project.

One of the factors to be considered in any construction contract within a BOT model (FIDIC, NEC or any other) is how to deal with the contractor's conflict of interest (where, as is often the case, the contractor is a significant shareholder of the concession company). Such matters as loading construction costs, or errors in compiling the estimate of construction costs and variations which might have been avoided, need to be addressed by the use of deferred payment (but only if there is sufficient surplus in revenue). All shareholders and lenders should be aware that contractors will often look for short-term gains (profit in the construction contract) rather that long-term returns (dividends from the concession company).

However, the wise contractor will see that a sensible mixture of short-term gains (from construction) and long-term gains (from shareholdings in a number of concession companies) will be advantageous over several years, during which there may be cycles of 'boom and bust' in the construction industry.

2.7 Special conditions and contract documents

In many building contracts, the standard conditions of contract are intended to stand on their own to be used without amendment. Where partial possession or sectional completion of the works is intended, some forms of contract may need special attention to enable these provisions to be incorporated. Failure on the part of professional advisers to give sufficient thought to these matters is a common cause of dispute which is often resolved against the interests of the employer.

The general rule of law is that a specially written document which forms part of a contract will take precedence over a standard document. Many construction contracts have gone to considerable lengths to negate this rule.

The JCT Standard Building Contract 2011 provides that 'The Agreement and these Conditions are to be read as a whole but nothing contained in the Contract Bills or the CDP Documents [Contractor's Designed Portion], nor anything in any Framework Agreement, shall override or modify the Agreement or these Conditions' (clause 1.3).

It is self-evident, on the wording of the above-mentioned provisions, that intended amendments appearing in other contract documents, such as the contract bills (of quantities) may be of no or limited effect. It is also inappropriate to delete the relevant clause. The deletion may cause *everything* in the other contract documents to override or modify the standard conditions, which may not be the intention without the most careful drafting of the other contract documents. If other provisions are intended to take precedence over the standard document, such provisions ought to be incorporated by additional clauses in *The Conditions* [of Contract]. Alternatively, supplemental conditions of contract may be used with an appropriate amendment to the relevant clause such as clause 1.3 of the JCT Standard Building Contract 2011 (or the corresponding clause in other forms of contract) to give full effect to the supplemental conditions. Significantly, in its un-amended form, the NEC 3 Contract does not include an equivalent provision to the JCT form.

In *Barry D Trentham Ltd* v. *McNeil* (1996) SLT 202 it was held that the wording of clause 2.2.1 of JCT80 giving precedence to the conditions of contract (and appendix) over the contract bills, in the circumstances of this case, did not have the same effect as would have been the case under JCT63. The judge's reasons included the fact that the words 'or affect in any way whatsoever' which had appeared after 'modify' in JCT63 had been deleted from JCT80. This change, together with the same priority of the conditions and the appendix were sufficient to enable the employer to rely on the provisions for phased completion and liquidated damages for phase one which appeared in the contract bills. Reliance on this decision in all circumstances may not be sufficient to avoid problems when the intended amendments to the contract are set out in the contract bills, otherwise clause 2.2.1 (of JCT80) would be redundant.

For the avoidance of doubt, the contract documents should be clearly specified. In the JCT forms of contract, the contract documents are described in the contract (for example, see clause 1.1 of the JCT Standard Building Contract 2011). Sometimes other documents, such as exchanges of correspondence, are bound into the documentation with the intention of incorporating such documents into the contract. It is advisable to make the appropriate amendment in the conditions of contract giving full effect to other documents, setting out the order of priority in the case of ambiguity. If the latter is not done, it is likely that these other documents will take precedence (under the general rule). This may be acceptable if the entire contents of the other documents are to take precedence. However it is sometimes the case, after negotiation and clarification, that parts of the contents of such documents are not intended to apply. It is better practice to summarise any special provisions which may have been agreed in correspondence and incorporate such provisions in the contract. This will avoid the necessity to include correspondence in the documentation.

In civil engineering contracts, the contract documents are intended to be mutually explanatory of one another (clause 5 of the ICE (fifth, sixth and seventh editions)). The engineer is empowered to explain any ambiguities and make any necessary adjustment resulting therefrom. This is a potential cause of disputes, particularly where the drafting and editing of the contract documents (by the engineer who may be responsible for the ambiguities) are done without the necessary care.

In international contracts, the FIDIC conditions of contract provide for other documents to be incorporated by reference in the letter of acceptance or in the contract agreement. The order of priority of the documents forming the contract is specified (clause 5.2 of the fourth edition, and clause 1.5 of the 1999 Red, Yellow and Silver Books). This is a valuable feature which assists in dealing with ambiguities. Part II of the FIDIC conditions of contract contains the special conditions which take precedence over the standard conditions of contract. The use of this method encourages the standard of care necessary to draft clear and unambiguous contracts.

Other documents such as drawings, specifications and bills of quantities need careful attention to ensure that there are no ambiguities in, or between them. A common practice (to be discouraged) is the use of standard specifications or preambles which have not been edited to remove clauses which are not applicable to the work to be done. Every specification clause or preamble should be relevant to the work shown on the drawings. If it is decided to change the specification during the course of the project, then a new specification clause can be issued as part of a variation order. Some engineers and architects try to argue that contractors are required to carry out work which is not in the contract, at no extra cost, merely because it is mentioned in the specification.

Only the most careful editing of all of the documents forming the contract will minimise the exposure to claims arising out of ambiguities. Each contract should be treated as being unique and reliance on standard documents for all contracts should be discouraged in many instances.

Part II of the fourth edition of FIDIC contemplates a number of changes and additional clauses to suit particular circumstances. Unfortunately, it is common practice for employers or their professional advisers to modify the standard FIDIC conditions in such a way that the modifications go far beyond that reasonably contemplated. Some examples are:

- the deletion of contractor's rights to an extension of time for adverse physical conditions and delays by public authorities;

- contractor's rights to interest on late payment, suspension of work due to late payment (with extensions of time and additional costs) deleted;
- contractor's rights to determine his employment for non-payment changed from twenty-eight days to one-hundred days;
- almost all of the grounds for the contractor to terminate his employment due to the employer's default deleted;
- employer's additional rights to terminate the contractor's employment if the contractor fails to accelerate the progress of the works after being instructed to do so (even if the works had been delayed by matters for which the contractor would be entitled to an extension of time);
- deletion of all of the employer's risks and special risks: the contractor to be responsible for all of the risks described as employer's risks or special risks in clauses 20.4, 65.1 and 65.2 of FIDIC fourth edition.

The contractor is to be responsible for:

- existing ground conditions;
- existing underground services (whether or not they are shown on the drawings supplied by the employer);
- data provided by the employer;
- any design provided by the employer;
- general damages to apply in addition to liquidated damages (fortunately, this provision could not be enforced under the laws of the country in which this particular contract was to be carried out).

Some of the above revisions may be suitable for a turnkey contract (and some are in fact incorporated in the 1999 FIDIC Silver Book). However, they are not appropriate for a traditional 'Red Book' type of contract where the design is done by the employer and the contractor constructs the work in accordance with the employer's design.

Other examples of modifications to the fourth edition of FIDIC which illustrate a degree of incompetence on the part of the employer's advisers are:

- Contractor's rights to an extension of time due to the employer's failure to give possession of site deleted (see *Rapid Building Group Ltd* v. *Ealing Family Housing Association* in 1.4).
- Deletion of the standard extension of time clause (44.1) and its replacement with the text of the extension of time clause (23) from JCT63. Owing to the cross-referencing of another clause dealing with delays qualifying for extensions of time (in the standard FIDIC conditions) to the standard clause 44.1 of FIDIC (which refers to any cause of delay referred to in these conditions) and the fact that the replacement clause (23 of JCT63) does not include 'any other cause of delay referred to in these conditions', there may be some doubt as to how the revised provisions will be construed.

It remains to be seen if the 1999 FIDIC contracts will be subject to the same sort of abuse. Lessons may be learned from the fact that contractors sometimes conspire to boycott the contract by refusing to tender if the abuse justifies it.

3 Tender and Acceptance

3.1 Competition caution

In section 3.1 of the second edition of this book mention was made of the then common and historic practice of 'cover bidding', whereby a contractor, following consultation with other bidders, would submit an exaggerated bid with the intention of not being the successful bidder. The intention of an unsuccessful bid was usually to ensure that the contractor was not awarded a contract it was unable to fulfil and to avoid the client taking the contractor off future tender lists in consequence of the contractor declining to submit a tender. Irrespective of motive, the practice is now illegal throughout the United Kingdom and the European Union by virtue of section 2 of the Competition Act 1998 and Article 81 of the European Treaty respectively.

Often referred to as the 'Cartel Offence' infringement of section 2 of the Competition Act 1998 or Article 81 opens the offending contractor to a possible fine of ten per cent of annual turnover.

The matter does not end at a fine for the offending contractor. Under section 188 of the Enterprise Act 2002 the Cartel Offence is also a criminal offence and the officers of the contractor that committed the offence are liable to imprisonment of up to 5 years by virtue of section 190 of the Act.

In one investigation alone in the United Kingdom in 2008, one hundred and three construction firms were investigated for activities involving the Cartel Offence. The results of the investigation were published in September 2009 and fines totalling £129.5 million were imposed on the one hundred and three contractors.

Contractors are advised to make themselves fully acquainted with section 2 of the Competition Act 1998 and Article 81of the European Treaty. Tenderers are also advised to be aware of section 18 of the Competition Act 1998 and Article 82 of the European Treaty which prohibit the abuse of a dominant position.

Contractors should also be aware of public procurement law applicable to all European Union members. European Union public procurement is subject to EU Directives and the tendering processes required by the relevant law must be strictly adhered to.

The European Union Public Procurement Directives cover work in the public sector, which is to be carried out on behalf of Contracting Authorities (government departments, local authorities, nationalised industries and private sector bodies receiving more than fifty per cent of their funding from government and all bodies governed by public law), and the value of such work exceeds specified thresholds (which are subject to review).

The various Directives have been consolidated in the public procurement directive for works, supplies and services 2004/18/EC and implemented in England, Wales and Northern Ireland by the Public Contracts Regulations 2006 (SI 2006 No 5). In Scotland the relevant regulation is The Public Contracts (Scotland) Regulations 2006 (SSI 2006 No 1). The Regulations were amended to some extent by The Public Contracts (Amendment) Regulations 2009 (SI 2009 No. 2992).

Separate provisions apply to tenders for utilities, which are enacted in England, Wales and Northern Ireland by the Utilities Contracts Regulations 2006 (SI 2006 No 6) and in Scotland by The Utilities Contracts (Scotland) Regulations 2006 (SSI 2006 No 2).

The thresholds to which the various regulations apply are subject to change and were last amended on 1 January 2010.

The various regulations make provision for open, restricted and negotiated tendering procedures as well as for framework agreements, dynamic purchasing systems and electronic auctions. They also make provision for appeals by unsuccessful tenderers in the event that the regulations have not been followed.

3.2 Selection of tendering contractors: pre-qualification

Many mistakes and potential claims can be avoided if sufficient thought and planning is put into the pre-tender stage of a contract. A common mistake is to invite too many contractors, at the last possible minute, to submit a tender for a project. There have been cases of over twenty contractors being invited to bid for a project. In a recession, all or most of the invitees will oblige. This process may provide the lowest possible tender figure. However, it does not guarantee the lowest final account and very often completion of the project on time (if the contractor survives the course) may be in doubt because of the failure to resource the project properly.

Substantial benefits can be gained by early selection of contractors who are willing to submit a *bona fide* tender and who are capable of carrying out the work. This can be done by carefully selecting potential contractors, giving them reasonable notice of the proposed tender and inviting them to indicate their willingness to submit a tender for the project. The invitation should contain sufficient information to enable the invitees to consider their ability to submit a tender and execute the work, such as

- date for issuance of complete tender documents;
- date for receipt of tenders;
- date of award of contract;
- date for commencement of the work;
- contract period;
- form of contract (with or without bills of quantities);
- liquidated and ascertained damages;
- brief description of the project.

It should be made clear that any firms wishing to decline from submitting a tender would not prejudice their chances of being invited to tender for future work. Firms who accept the invitation should be given the opportunity to attend a preliminary meeting and view the drawings which are available.

If the above procedures are followed, the employer will be reasonably confident that he will receive serious bids from contractors. In the event of insufficient positive replies, the employer can widen his nett to make enquiries of other firms. In addition, each contractor will be able to prepare for the necessary staff to be available and it can begin to make enquiries of potential subcontractors and suppliers.

In the case of large complex projects it may be desirable to invite contractors to pre-qualify to tender for the work. The procedures described above will be equally applicable to this process. However, in addition to providing the information mentioned hereinbefore, the employer will wish to find out more about the potential tenderers' capability. Pre-qualification enquiries should cover

- previous track record on similar projects;
- proposed management structure and staff responsible for the project;
- financial standing of the firm;
- resources which can be made available for the project;
- details of any *joint venture* if tenders are to be submitted in the name of more than one firm;
- outline proposals for method of construction and programme.

In some circumstances it may be appropriate to include all of the matters described for management contracting in Chapter 2.

Pre-qualification enquiries should inform tenderers of the criteria to be used for selection. After receipt of pre-qualification documents from the invitees, a shortlist should be prepared according to the applicants' responses, measured against the relevant criteria. This should be followed by interviews of the shortlisted firms and the final tender list should be drawn up as soon as possible so that all firms can be notified without delay.

The criteria for selection of contractors may include evidence of capability and a proven track record for a given number of years, details of key staff, plant, equipment, labour and technical resources. References and financial information may be required. Failing to comply with certain laws, such as legal requirements to pay taxes and social security contributions, may be grounds for disqualification.

3.3 Time allowed for tendering

It is unreasonable to allow only a few weeks to tender for a construction project of any reasonable size. Nevertheless, this is often the case. It is understandable that employers wish to start construction as soon as possible and it is this pressure which leads to insufficient time being allowed to enable tenderers to prepare a tender properly. Insufficient time often leads to numerous potential errors. A survey carried out in the United States in the 1970s indicated the following incidence of bid mistakes (*Anatomy of a Construction Project* by Kris Nielsen, *International Construction*, November 1980):

- extension errors – 19 per cent (errors in multiplication to calculate quantities or price);
- lack of knowledge of work required – 16 per cent (insufficient attention to all of the work involved);
- lack of knowledge of contract administration requirements – 15 per cent (failure to identify risk or insufficient allowance for cost of administration);
- under-estimating escalation – 12 per cent;
- transposition errors – 10 per cent (transposing incorrect figures from one sheet or document to another);
- poor pre-bid planning – 9 per cent;
- poor resource planning – 9 per cent;
- incorrect measurement of quantities – 8 per cent;
- others – 2 per cent.

Given more time to tender for the work, the incidence and magnitude of errors ought to be reduced. A distinction must be drawn between mistakes in pricing by the contractor and mistakes on the face of the documents, such as incorrectly extending a rate for an item of work. It must be in the interests of both the employer and the contractor to avoid errors in the tender. A low bid due to one or more mistakes often causes the successful contractor to try every means to reduce costs and/or to pursue unmeritorious claims based on varying degrees of fiction.

However, it is not necessarily correct to assume that tenders will be higher if more time is allowed and errors are avoided. If competent contractors are given sufficient time to tender, they will be able to incorporate savings brought about by detailed studies into methods of construction, programming and procurement of plant and materials. Given that tenderers are in competition, some, if not all, of these savings will be passed on to the employer.

Many problems and mistakes can be avoided without delaying the date for receipt of tenders. Tenderers can be given more time if some of the tender documents are issued in advance of the entire set of tender documents. For example, drawings and sections of bills of quantities or specifications can be issued to tenderers before the preparation of the final tender bills is complete. A considerable part of a contractor's pre-tender planning and pricing will be based on the drawings. A detailed method statement will be prepared almost exclusively from drawings.

Tenderers often have to measure quantities of work from the drawings to determine plant size and other resources. This is the case even where bills of quantities are provided by the employer. Prices for special items are often obtained on the basis of the drawings. In many cases, tenderers may be able to establish, with reasonable accuracy, the cost of carrying out the works, before the final set of tender documents is issued. All that may remain to be done, during the relatively brief period allowed to submit the tender, is to thoroughly check the tender documents, obtain confirmation (or adjustment) of prices from subcontractors and suppliers, adjust costs where necessary, adjudicate on the final tender sum and compile the rates in the tender to arrive at the proposed tender sum.

A suggested timetable for the above is shown in Figure 3.1.

The Public Contracts Regulations 2006 and the Utilities Contracts Regulations 2006 (regulation 22) provide timetables for the various tendering processes referred to in the regulations. Generally, the open tendering procedure must allow a minimum of 52 days from dispatch of tender notice to receipt of tenders and the restricted (or selected) procedure must allow a minimum of thirty-seven days from dispatch of tender notice to receipt of applications to tender. An accelerated tender procedure may be permitted in some cases of urgency, in which case the periods may be reduced. Where no suitable tenders have been received during the normal tendering procedures, or where additional work is required in connection with an existing contract, direct negotiation with one or more contractors may be permitted.

3.4 Exploitation of poor tender documents by contractors

An increasing number of firms engage staff to scrutinise all of the tender documents to find ambiguities and other deficiencies that may be exploited to produce a lower tender and a potential claim for additional payment during the course of the project. It may be argued that all tenderers have the same opportunity to exploit such deficiencies, and the employer will end up paying no more, at the end of the day, than it would if the tender documents had contained no deficiencies.

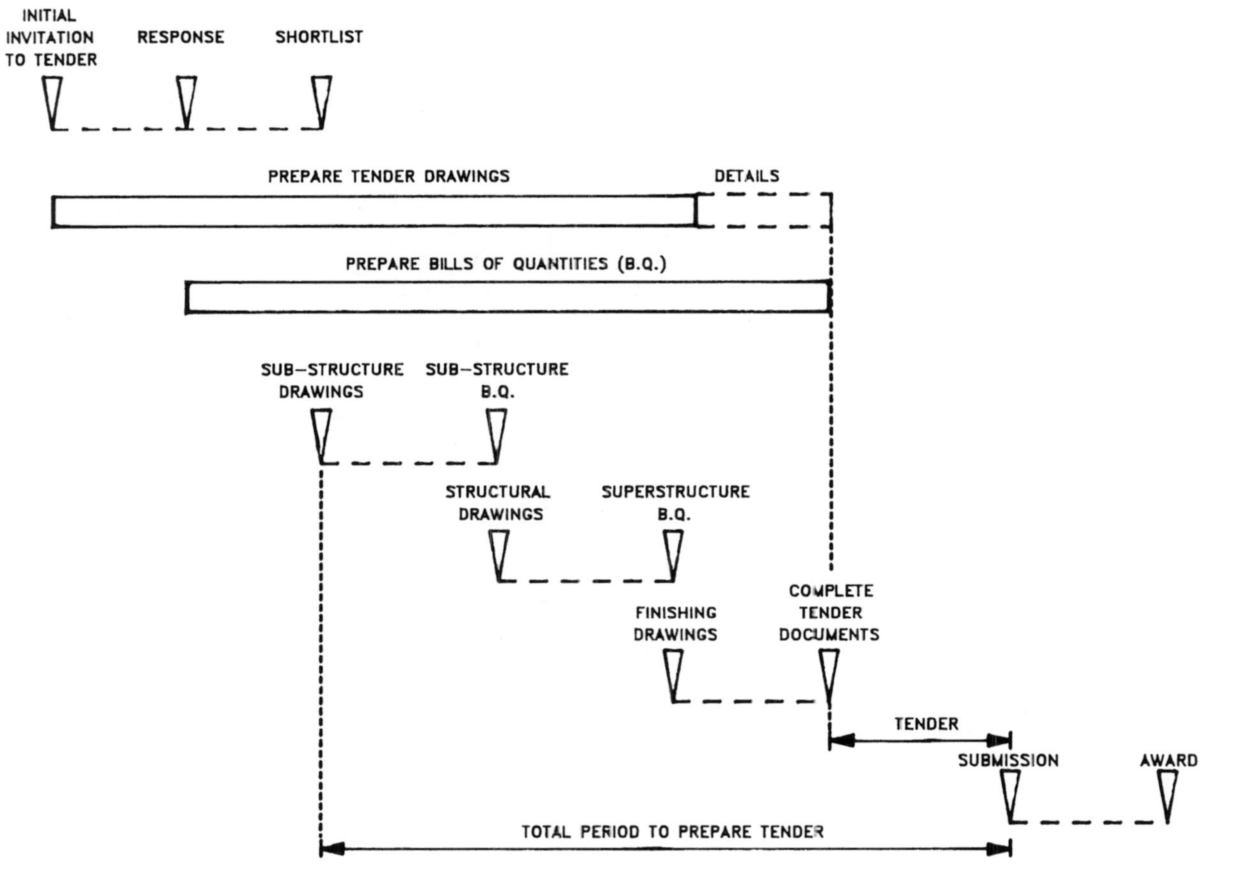

Figure 3.1 *Suggested timetable for phased issue of tender documents*

This is far from the case. The successful contractor will often recover more, by way of claims, than it would if all of the costs had been included in the tender sum at competitive rates. In addition, extensions of time for completion of the works may flow from these deficiencies, whereas no additional time would result if there had been no deficiencies. Claims which arise out of innocent misinterpretation of the contractual intentions, or exploitation, where there is an ambiguity or deficiency, are often the most difficult to resolve amicably, since they reflect on the competence of the employer's professional team.

Contractors can assist in avoiding problems that arise out of ambiguities by notifying the employer's professional team of any ambiguity discovered at pre-tender stage. These ambiguities should then be rectified and brought to the attention of all tenderers prior to submission of tenders. If this is done, all tendering contractors will be tendering on an equal basis and the risk of exploitation will be minimised.

The employer's professional team should take care when evaluating tenders so that any obvious pricing anomaly (between tenderers) is reviewed with the tenderers to establish the reason for it.

3.5 Preparing the estimate: the tender

The estimator's task is to accurately calculate the cost of carrying out the works and to apportion the cost to the various elements (or items in a bill of quantities) of the job. In order to do this he may have to rely on several other departments, or individuals, in the company. The cost of carrying out the works is very much determined by the method of construction and the programme for the project. The method of construction will determine the type of plant to be used and the productivity to be expected. The programme will determine the cost of time-related items such as external scaffolding, tower cranes and hoists. The amount of work to be subcontracted may determine the number of supervisory staff and the cost of attendance on each subcontractor. Compiling the estimate is a completely separate task from tendering. The estimator should not make decisions or allowances which are influenced by external market forces or post-contractual matters such as *front loading* the rates (increasing the rates for work executed early in order to improve cash flow). He may, however, advise management on such matters.

Once the estimate has been compiled and the cost of executing the work has been established, management will consider external factors such as the competition and the probable successful tender sum. The existing workload of the company and the requirement to obtain further work will also be considered, as well as the assessment of risk, staff resources, profit and possible savings in cost which can be made. After due consideration of all of these factors, the estimate will be converted into the tender for the works. The estimator will then make all of the necessary adjustments to the rates in accordance with the decisions of management. The form of tender will then be completed and submitted. In times of recession, the tender sum may be less than the estimate of cost for executing the works.

A typical estimating and tendering process is shown in Figure 3.2.

3.6 Qualified tenders

Subject to the regulations governing procurement of public contracts referred to earlier, (which preclude the acceptance of a qualified tender unless all tendering contractors are allowed to modify their tenders to incorporate the same terms and

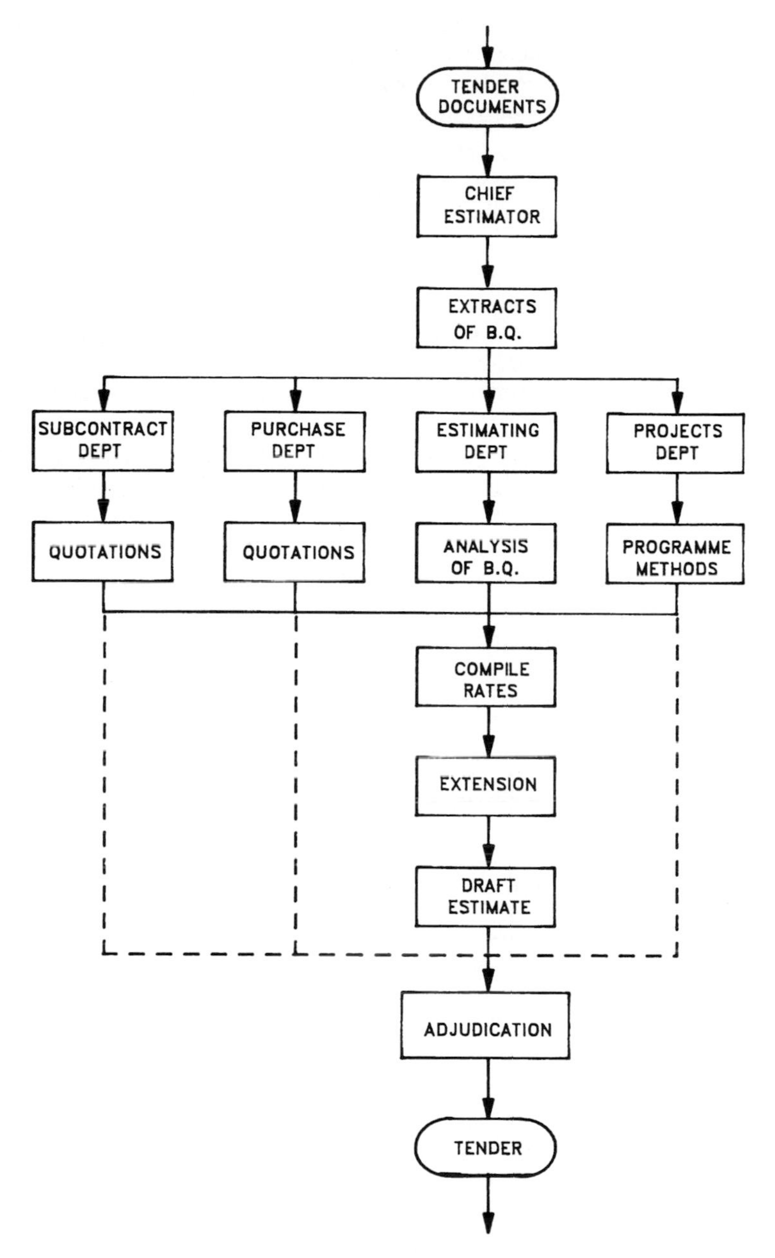

Figure 3.2 *Typical estimating and tendering process*

conditions and in some instances prohibit qualified tenders entirely), are there any reasonable grounds to qualify a tender?

Tendering contractors may suspect a risk if certain representations are made by the employer such as the availability of materials provided by the employer or as to the ground conditions. Careful examination of the proposed contract conditions or knowledge of the general law may render a qualification unnecessary, in which case none should be made as it detracts from what would otherwise be a complying tender. On the other hand, the proposed contract terms may be particularly onerous. The tendering contractor then has the option of pricing the onerous terms (which may not be possible without an element of gambling) or qualifying the tender in order to have the onerous terms modified or removed.

From a practical point of view, if the employer is properly advised, it may be sensible to invite a complying tender and an alternative tender incorporating certain changes which may be proposed by the tenderer. It could be a condition of tender that all proposed changes should be notified several days before the date for receipt of tenders, with the proviso that all tendering contractors will be informed of the proposed changes. If that is done, all tenderers will have the opportunity to submit an alternative bid incorporating those changes that they saw fit to adopt. If each adopted change was required to be priced individually as an omission from, or addition to, the complying bid, it would assist in evaluation of tenders and there would be no delay in making an award. If qualifications are permitted without prior notification on the date for receipt of tenders, there will almost certainly be delay caused by evaluation and possible re-tendering. By that time all of the tendering contractors will have a reasonable idea of the lowest tender, in which case there is room to make other adjustments in order to make the revised tender more competitive.

If a qualification is made to a tender, it is important to ensure that it is couched in terms which make it a condition and that it is incorporated in the contract. If extra costs are involved, the contract terms should clearly state how these extra costs are to be added to the contract price and in what circumstances. Qualifications contained in the tender, or in a letter attached to the tender, will only be effective if the tender (or letter) is a contract document: *Davis Contractors Limited* v. *Fareham U.D.C.* [1956] AC 696. Alternatively, the qualification should become a contract term by modifying the conditions of contract.

3.7 Tender programme

The preparation of a tender programme is essential. It is an important aid to the contractor when assessing cost and resources and to the employer when evaluating the tender. In many cases a simple bar chart will suffice. However, for complex projects, a detailed programme showing the logic and restraints is required. The programme should be realistic. All too often, the programme which is submitted is no more than a tool to form the basis of potential claims which may arise. The contractor is usually required to complete the project *on or before the date for completion*. Some contractors deliberately show early completion. If this is possible without a disproportionate increase in cost it is often in the interests of both parties to agree an earlier completion date. Problems can occur if the contractor's tender is accepted and completion is shown, on the programme, at an earlier date than the contractual date for completion (*Glenlion Construction Ltd* v. *The Guinness Trust* (1987) 39 BLR 89 and *Ovcon (Pty) Ltd* v. *Administrator Natal* 1991 (4) SA 71 – see Chapter 5).

The tender programme will not usually be a contract document, but it is often relied upon when formulating claims. For this reason it must be a document which is a genuine reflection of the contractor's intention and evidence to support this may be necessary. Estimated productivity, logic, proposed plant and methods are some of the matters which may have to be considered in detail to justify the contractor's programme.

Considerable areas of doubt may exist in any programme which relies upon prime cost and provisional sums for important elements of the project. The tendering contractor is required to allow for the completion of all of the work by the contractual completion date. It is good practice to indicate, on the programme, the sequence and duration of work to be done in respect of each and every prime cost and provisional sum. Ordering periods, relationship to other work and durations of the prime cost or provisional work, which have been assumed, should be clearly indicated. Wherever possible, the employer should inform all tendering contractors of proposed, or potential, nominated subcontractors and suppliers so that the programme requirements can be based on realistic information obtained from them. Any additional information regarding provisional items should be given to the tendering contractors so that the element of guesswork is reduced or minimised.

3.8 Evaluation criteria

Some public bodies are prohibited from accepting tenders on the basis of any other criteria than the lowest price (errors excepted). The lowest price does not guarantee the lowest final account, and a detailed analysis of tenders can sometimes indicate a possible exposure to a higher price than the tender sum.

Save where tenders are very close, the acceptance of the lowest tender may not be in the employer's best interests. A very low tender should not normally be accepted without first discussing every contentious matter with the tenderer. Errors should be dealt with in accordance with one of the codes of practice (which should be notified to tenderers prior to submission of tenders).

However, for some projects, price alone may not be the criterion which determines the best bid. The tender programme may indicate to what extent the tenderer has appreciated the complexity of the design. Proposed methods may indicate to what extent the tenderer has appreciated the details and co-ordination of services. It is essential that the employer sets out the criteria, giving each a standard, or yardstick, by which tenders are evaluated. Tendering contractors should be made aware of the evaluation criteria to be used so that the tender can be prepared accordingly.

Evaluation can be assisted if tenderers are required to submit additional information in support of the tender. This may include

- breakdown of major items into labour, plant, materials, overheads and profit;
- breakdown of costs related to time, volume, method and event;
- cash flow forecast.

Rates inserted in schedules, or bills of quantities, by the tenderers should be examined and compared to ensure that there are no obvious and significant departures from what is considered to be reasonable. Suspect rates may be due to ambiguous descriptions, mistake as to quality, failure to allow for materials or other causes. Inconsistencies in rates (between sections of bills of quantities) should be adjusted by agreement if it is appropriate.

Final selection should not take place before interview with the tenderer. Key staff proposed by the tenderer should attend the interview and all important matters should be discussed in detail to ensure that there are no problem areas that cannot be resolved.

The criteria for the award of contracts laid down in the EC Directives are lowest price or most economically advantageous tender. If the latter is to be adopted, the contracting authority is required to advertise the fact giving a list (and if possible, the order of priority) of the criteria to be used in evaluating tenders. Matters such as completion periods (which may be a competitive element), maintenance costs, costs in use and technical specifications may be used for evaluation purposes.

3.9 Rejection: acceptance: letters of intent

In the normal course of events (and subject to certain criteria laid down in the EC Directives), there will be no problem if a tender is rejected. However, in the event that a tenderer has been required to do a substantial amount of preparatory work which is outside the scope of that which is normally required, the tenderer may be entitled to payment. In the case of *William Lacey (Hounslow) Ltd* v. *Davis* [1987] 2 All ER 712, it was held that there was no distinction between work done which was intended to be paid for under a contract erroneously believed to exist and work done which was intended to be paid for out of proceeds of a contract which both parties erroneously believed was about to be made. Such work was not done gratuitously and a reasonable price must be paid for it. The same principle was applied in *Marsden Construction Co Ltd* v. *Kigass Ltd* (1989) 15 ConLR 116.

The EC Directives provide that tenders may not be rejected because they appear to be too low, without allowing the tenderer to give an explanation. In *Fratelli Costanzo SpA* v. *Comune di Milano (Municipality of Milan)* [1990] 3 CMLR 239, an unsuccessful tenderer commenced proceedings against the Municipality on the grounds that his tender had been rejected pursuant to the Municipality's formula which automatically rejected all tenders which were more than ten per cent lower than the average of all tenders. It was held that the tenderer had the right to seek enforcement of the Directive.

The Directives also forbid rejection on the grounds that the tender is based on equivalent alternative specifications which meet ISO standards. In *Commission of the European Communities* v. *Ireland* (1988) 44 BLR 1, an Irish company complained that its tender was rejected because the Spanish products offered by the tenderer did not comply with Irish standards specified in the tender documents. The Spanish products complied with ISO standards and it was held that the contracting authority (Dundalk Urban District Council) had failed to comply with Article 30 of the Treaty of Rome by excluding products of equivalent ISO standards. It should be noted that this particular contract was excluded under the threshold provisions of the Public Works Directive, but it was not exempt from the general provisions of the Treaty of Rome for non-discriminatory technical specifications.

Errors in tenders should not normally be cause for rejection. Where errors in the tender are discovered and dealt with in accordance with the relevant codes of practice, many potential problems can be avoided. In any event, if the employer discovers an error in the tender before acceptance, and the tender is accepted without adjustment, the contractor will not be bound by the error: *McMaster University* v. *Wilchar Construction Ltd* (1971) 22 DLR (3d) 9 – High Court of Ontario.

Tenderers are often asked to keep their tenders open for acceptance for a specified period. This does not prevent the tenderer from withdrawing his tender at any time. Tenderers may be bound by their tenders if there is consideration. The amount of consideration may only be nominal. Alternatively, a *Bid Bond* may be required by the employer. Once the employer has unconditionally accepted a tender within the time for acceptance of tenders (or within a reasonable time if there is no specified time) and provided that the tender has not been withdrawn, there is a binding contract.

Post-tender negotiations often take place, particularly in the private sector. Public tenders are less likely to be subject to negotiation. Current EC law does not cover post-tender negotiations. However, the Council of Ministers have issued a statement on this matter:

> 'The Council and the commission state that in open or restrictive procedures all negotiations with candidates or tenderers on fundamental aspects of contracts, variations in which are likely to distort competition, and in particular on prices, shall be ruled out; however, discussions with candidates or tenderers may be held but only for the purposes of clarifying or supplementing the content of their tenders or the requirements of the contracting authorities and providing this does not involve discrimination.'
>
> *Public Procurement Directives*, conference paper by Robert Falkner, 10 December 1990.

It is not unusual for acceptance to be conditional, usually by way of a letter of intent. Care should be taken by the employer when drafting a letter of intent. Equally, the contractor should carefully consider the terms of a letter of intent in order to understand fully to what extent he has been authorised to proceed and how payment for work done will be established. Matters to be addressed when drafting a letter of intent should include

- detailed instructions clearly describing the work which is to proceed, distinguishing between design, ordering, taking delivery and execution of work;
- full compliance with the tender documents so far as they apply to matters for which authority to proceed has been given;
- terms of payment to be made in respect of the matters for which authority to proceed has been given;
- provision for termination of contractor's rights to proceed pursuant to the letter of intent and the employer's liability for payment in the event of termination;
- provision for cancellation of orders placed pursuant to the letter of intent and the employer's option to pay cancellation charges or to take delivery of goods ordered;
- care of, and responsibility for, work and materials including insurance;
- goods and materials to be vested in the employer;
- provision to terminate the terms of the letter of intent in the event of award of the contract and provisions to credit payments made under the letter of intent against certificates issued under the contract;
- provision for settling disputes (usually retaining the same provisions as the proposed contract).

It is important that the letter of intent should make it clear that it is not an acceptance of the contractor's tender. It should, however, also make it clear that the employer has the option to accept the contractor's tender.

Even the most carefully prepared letter of intent may have its problems. In *British Steel Corporation* v. *Cleveland Bridge Engineering Co Ltd* (1981) 24 BLR 94, the courts had to consider whether, or not, a contract had been created by a letter of intent. It was considered that each case must depend on the particular circumstances. However, it was decided that if a party acted on a request in a letter of intent and was simply claiming payment, it did not matter if a contract was not created as payment could be based on *quantum meruit*.

In *C.J. Sims Ltd* v. *Shaftesbury Plc* (1991) QBD; 8-CLD-03–10, the court had to consider the payment terms of a letter of intent. The terms provided for reimbursement of reasonable costs, including loss of profit and contribution of overheads, 'all of which must be substantiated in full to the reasonable satisfaction of our quantity surveyor'.

At first glance it would appear that the above terms were reasonable commercial requirements for payment. The employer successfully argued that it was a *condition precedent* to any payment being made to the contractor that the costs should be *substantiated in full* and *to the satisfaction of the quantity surveyor*. The judge was not disposed to the view that the contractor should be paid something on account pending full substantiation (which, with respect, is what would normally be expected).

The terms of a letter of intent should never be exceeded. In *Mowlem Plc* v. *Stena Line Ports Limited* [2004] (unreported) Mowlem carried out work to a value greater than the maximum value of £10,000,000 set out in a letter of intent. The court held that Mowlem was not entitled to payment for work carried out in excess of the £10,000,000 stated in the letter of intent.

A potential disaster area exists when contracts proceed on the basis of protracted correspondence and exchanges of letters, all of which contain elements of change to previous documents and there is no clear definition of the terms agreed between the parties. In *Mathind Ltd* v. *E. Turner & Sons Ltd* [1986] 23 ConLR 16, the contract was intended to be JCT63. Exchanges of correspondence and an addendum bill of quantities dealt with phased handover. The works proceeded but the contract was never signed. Disputes arose over phased completion dates and liquidated damages. The court had to consider when and how the contract was made. In doing so it came to the conclusion that both parties had agreed to phased completion. As no contract had been signed the contractor could not rely on the words in clause 12(1) of JCT63 which prohibited modification to the standard printed form in the contract bills. (It should be noted that in *M.J. Gleeson (Contractors) Ltd* v. *London Borough of Hillingdon* (1970) 215 EG 165, provisions for phased completion were contained in the contract bills. The provisions were held to be ineffective on the grounds that the contract stipulated that nothing contained in the contract bills should override or modify in any way the contract conditions.) The effect of the provisions in the post JCT63 forms of contract regarding precedence of the contract conditions over the contract bills may be quite different (see *Barry D Trentham Ltd* v. *McNeil* in 2.7, *supra*).

It is not uncommon to agree to change the conditions, or specification or details, in the tender documents, prior to signing the contract. Failure to amend the contract documents to reflect the change may mean that the change, when made, is a variation to the contract despite the fact that the parties had agreed to the change prior to signing the contract. In *H. Fairweather & Co Ltd* v. *London Borough of Wandsworth* (1987) 39 BLR 106, the contract was signed after both parties had agreed that the specified *Clifton* bricks would not be used and that *Funton* bricks would be substituted therefor. There was delay in delivery of *Funton* bricks. The contractor claimed that the delay arose out of a variation and claimed an extension of time under clause 23(e) and loss and expense under clause 11(6) of JCT63. The architect granted an extension of time under clause 23(j)(ii) for unforeseen shortages of materials, and refused a claim for loss and expense. It was held that the substitution was a variation.

In view of the above, it is essential that all agreed changes to the tender documents should be reflected in the contract to be signed by the parties. Any agreed change which would otherwise constitute a variation should be reflected in revised contract bills. If any change affects the completion dates previously mentioned in the tender documents, the appropriate adjustment should be made in the contract documents prior to signature. If necessary, the tender (or contract) programme should be revised.

Finally, with the exception of essential key dates, it may be fatal to incorporate the contractor's programme as a contract document. Acceptance of a tender may be on the basis of the contractor's programme, but its use as a contract document can cause considerable problems. This aspect will be dealt with in Chapter 4.

4 Monitoring Delay and Disruption Claims: Prevention

4.1 Contracts administration

All forms of contract contain express or implied duties and obligations to be performed by the employer (or his agents) and the contractor. Contracts do not usually set out in detail how these duties and obligations should be performed. It is self-evident that the employer must give access to the site and provide information in sufficient time to enable the contractor to carry out the works by the due completion date. The contractor must give reasonable notice of delay or of any claim and the architect, or engineer, must decide and make extensions of time or certify additional payment.

Whatever the form of contract, it is important that all parties co-operate with each other in order to ensure that each is provided with sufficient information to enable them to carry out their respective duties and obligations. Too often, contractors believe that they have complied with their contractual obligations by giving notice of delay and very brief information (if at all) to support their contention that they are entitled to more time and/or money. It is not unusual for contractors to complain that no extension (or insufficient extension) of time has been granted by the architect or engineer. These complaints sometimes persist several years after the contract has been completed when the first pleadings are being prepared for litigation or arbitration. Even at this stage some contractors are unable to show what period of delay occurred and its effect on the progress of the works. Criticism of the architect, or engineer, for failing to make an extension which satisfies the contractor is hardly justified (provided of course that an honest attempt was made to assess the effects of the delay) if the contractor, himself, cannot illustrate the effects of the delay.

These problems can be avoided if all parties examine the contract terms to establish their express duties and obligations and the procedures that need to be adopted in order to ensure that these duties and obligations can be performed in accordance with the contract.

Whatever procedures are to be adopted, they should not become a costly and time-consuming burden so that resources are diverted from the main objectives of any building and engineering contract – to design and build the works.

4.2 Possession of site: commencement

Before award of the contract, the employer and the contractor should agree on the period of notice to commence, in order to allow for mobilisation and the taking of records and photographs showing the condition of access and of the site prior to possession by the contractor. Any restriction or limitation on the free use of the site should be recorded and the effects (if any) on programme or cost should be established as soon as possible. Contractual provisions which envisage possession of the site being given to the contractor within a short period (for example, seven days) should be avoided if possible. Consideration should be given to allowing the contractor to mobilise and set out even if there are outstanding approvals which are essential to commence construction of the permanent works. Early access to the site should be distinguished from the contractual date which is the commencement of the period for completion of the works.

4.3 Pre-commencement meeting

Prior to possession of the site (if practicable before award of the contract) the parties and their professional advisers should convene a meeting to discuss and record certain important matters. These should include

- the role and authority of each member of staff participating in the project;
- where the contract provides for delegating powers to other persons, these powers should be clearly established;
- status of the programme, key dates for information, periods for approval, long delivery periods and special problems;
- requirements for named, nominated and selected domestic subcontractors;
- works or materials to be provided by the employer;
- procedures for interim valuations and certificates;
- procedures for measurement, records, notices, particulars to be provided and response;
- procedures for monitoring the progress of the works, photographs, video, progress records and updating programme.

It is important that the representatives of both parties understand the need to recognise potential delays and to acknowledge that they may lead to claims from the contractor and subcontractors. Whatever procedures are adopted at this initial meeting, they should include measures to avoid or minimise delay by regular monitoring of design and detailing so that the construction of the works will not be affected by late issuance of essential information.

4.4 Regular progress meetings

Meetings should be kept to a minimum, but should be sufficient to satisfy the needs of the project. Each meeting, or series of meetings, should be designed to suit specific objectives, have the right persons present and take place at the right time or at sensible intervals.

Three categories of person should attend: those who can inform, those who can advise and those who can (and are authorised to) decide on the issues and delegate action.

The most important features of successful meetings are

- the correct agenda;
- accurate records of the meeting;
- decisions taken;
- identify responsibility for action;
- record of action taken (or outstanding) in respect of previous matters;
- accurate forecasts or projections;
- prompt distribution of minutes.

Where minutes of meeting are inaccurate, or where there are important omissions, it is essential that these are brought to the attention of the attendees and the necessary corrections made. Matters which require immediate attention should be dealt with in writing before the next meeting. Failure to follow these procedures causes major difficulties when trying to establish facts several years after the event. It is not unusual, when interviewing material witnesses in preparation for dispute resolution procedures, to be told that the minutes of meetings did not record what was agreed. Even if it is possible to verify such allegations, it is sometimes difficult to reconstruct the history of events.

Records of meetings can often mislead investigators searching to establish causes of delay several years after the event. A common practice adopted by contractors is to table a long list of alleged outstanding information at each meeting. Many items reappear week after week and month after month. It is often difficult to distinguish between information requested far in advance of being required and information which was essential but which was neglected by the architect or engineer. Each alleged outstanding item should be addressed during the meeting, or by written response before the next meeting, giving the status and anticipated date of issue, together with a note indicating the programme and progress of any work which may be affected by the outstanding information.

The agreed minutes including any amendments should be signed by authorised representatives as a true record of the meeting.

4.5 Instructions and drawing issues

Many instructions and drawing issues are of an explanatory nature to enable the contractor to construct the original works. Late issuance of information will lead to claims for delay and/or disruption. The designer must be able to understand the contractor's programme and make allowance for shop drawings (if applicable), obtaining quotations, ordering and delivery. The designer should not rely solely on the contractor's requests for information (sometimes the contract does not place an obligation on the contractor to make any such requests). It is essential to have regular meetings to determine when information is required in order to meet the programme or to prevent delay.

Few construction contracts proceed without changes of some kind. Revised drawings should clearly indicate the revisions so that the contractor can identify appropriate action without searching to find each revision. Such drawings should be accompanied by a variation order/instruction to facilitate cost monitoring and control, as well as indicating a possible review of the effects on the programme.

Some architects and engineers issue drawing under cover of instructions, letters, transmittal sheets and other forms, without distinguishing between explanatory details and changes to the original design. This practice does not facilitate control and often contributes to failure, by the contractor, to give notice of delay, or extra cost at the earliest possible time.

4.6 Site instructions: verbal instructions

There is an increasing tendency to design and detail the works as they proceed at site level. This indicates lack of knowledge of design and construction detailing. Projects which end in protracted disputes have often suffered from an unusually high proportion of design and detailing by way of verbal instructions and hand drawn sketches issued by the designer's site representative during a regular 'walkabout' on site. It is not unusual, when investigating causes of delay and disruption, to discover numerous references in minutes of meetings to the effect that the contractor was instructed to proceed in accordance with a sample, or method, agreed on site. Records of what was agreed are often difficult, or impossible, to find. Interviewing site staff months, or years, after the event sometimes assists in this exercise at considerable expense. A dimensioned sketch and/or photograph at the time of the agreement would avoid any misunderstanding about what was required and built.

Site instructions and verbal instructions should be used in an emergency only and not as a method of designing the works. Where verbal instructions are given, the architect, or engineer, should take the initiative in making sure that they are confirmed (whether or not there is provision in the contract for confirmation by the contractor which would give effect to such instructions).

Most JCT forms of contract, the ICE conditions of contract and the 1999 FIDIC Red Book all contain provisions for the contractor to confirm architects' or engineers' verbal instructions, and such instruction will be deemed to be architects' or engineers' instructions if not dissented from in writing within the period specified in the contract. In contrast (possibly a drafting error), there are no provisions for confirming verbal instructions in the 1999 FIDIC Yellow and Silver Books. All instructions must be in writing. Under these two FIDIC contracts, it is unclear as to what the contractor's obligations are if he receives an engineer's verbal instructions which are not promptly confirmed. The NEC 3 Contract provides for communications in a form that can be read, copied and recorded, which includes instructions and early warnings (clause 13.1). It does not make any provision for verbal instructions.

4.7 Form of instructions

Most contracts do not require an instruction, or variation order, to be in a particular form. A written site instruction, provided that it is issued by a person with the contractual authority to give instructions, is, for all the purposes of the contract, an instruction authorising the contractor to proceed. It is effective without the need for a standard form of instruction to confirm its contractual effect. Likewise, a drawing issued by an authorised person is an instruction in its own right, regardless of the form of the accompanying covering instrument (or if there is no accompanying covering instrument).

Without proper agreed procedures and consistency for the issuance of instructions, whether they are explanatory or variations, there is an increased probability that monitoring and control of cost and delay will be ineffective. Very often, the full effects of all of the instructions issued during the course of the project do not come to light until the final account is on the table and the contractor is reconstructing (with hindsight) the history of events in order to resist a claim for liquidated damages levied for late completion.

4.8 Programme and progress

With the exception of some of the more recent engineering forms of contract, and the latest editions of GC/Works/1, most standard forms of contract do not place sufficient emphasis on a construction programme. It is sometimes not even mentioned or required. Having regard to the sums of money spent on some modern projects and what might turn on events which affect the contractor's programme and progress, it is essential that a realistic programme showing how the contractor intends to construct the works should be available at the outset (see Chapter 3).

There may be problems if the contractor's programme becomes a contract document as failure to follow it in every detail may be a breach of contract. The contractor's obligations are normally to complete the works (or sections of the works) by given dates. Departures from the programme will be of no significance so far as the employer's remedies for performance are concerned. If there are good reasons for introducing key dates (for example, to facilitate installation of plant and equipment by the employer or specialists), these can be incorporated as contractual requirements, with appropriate remedies in the event of the contractor's failure to meet these key dates.

Another problem (when programmes become contractual documents) arises in the event of it being impossible to carry out the work in accordance with the programme. In the case of *Yorkshire Water Authority* v. *Sir Alfred McAlpine and Son (Northern) Ltd* (1985) 32 BLR 114, the contractor's programme and method statement became contract documents. The method statement, which was the contractor's own chosen method of working, provided for an outlet to a culvert to be constructed by proceeding upstream. The contract obliged McAlpine to execute the works 'in all respects in accordance with the contract documents'. It was found that this method was impossible and McAlpine successfully argued that it was entitled to a variation order to enable it to carry out the work. (It should be noted that the contract was based on the ICE conditions which provided, in clause 13(1), for the contractor to be relieved of its obligations to carry out work which is physically impossible.)

The 1987 fourth edition of FIDIC contains similar provisions regarding impossibility in clause 13. However, the 1999 family of FIDIC contracts do not provide for any similar relief in the event where the works are physically impossible to execute.

ICE and FIDIC contracts are well known for their 'clause 14 programme'. These provisions require the submission and acceptance of a first programme at the outset and regular updates in the event where actual progress departs significantly from the first or subsequent programmes. The 1999 FIDIC family of contracts continues this practice with much improved provisions in clause 4.21 (progress reports) and clause 8.3 (programme). The provisions of these two clauses, if put into practice, are likely to minimise delays and disputes. The NEC 3 Contract places great emphasis on the contractor's programme which must be in an accepted form and revised upon the occurrence of certain events (clauses 31 and 32). The programme is an important tool in the assessment of Compensation Events.

Having commenced work on the basis of a realistic programme, any significant departures from it should be monitored. Once delay has occurred which affects any important activities, it is essential that the effects of the delay are monitored, and that the programme is immediately updated to show the effects of the delay. If actual progress is monitored against a programme which is no longer valid, it is difficult, or even impossible, to establish the effects of a particular delaying matter on the overall programme and completion date. All progress, and delays, should be monitored against a programme which represents the contractor's proposed 'programme of the day', that is, a programme which has been revised to take account of all previous delays. As delays occur, these affect critical and non-critical activities. If regular updating is not done, the critical path may change, making the assessment of the effects of further delays a matter of guesswork. An example of how a critical path may change is given in Figure 4.1. In practice, this is no simple matter, and on contracts which have numerous, and often, continuing delays, it can only be achieved by additional staff and the use of various software and computers. It can be a costly exercise, and periodic updating may be a compromise which achieves reasonable results at an acceptable cost.

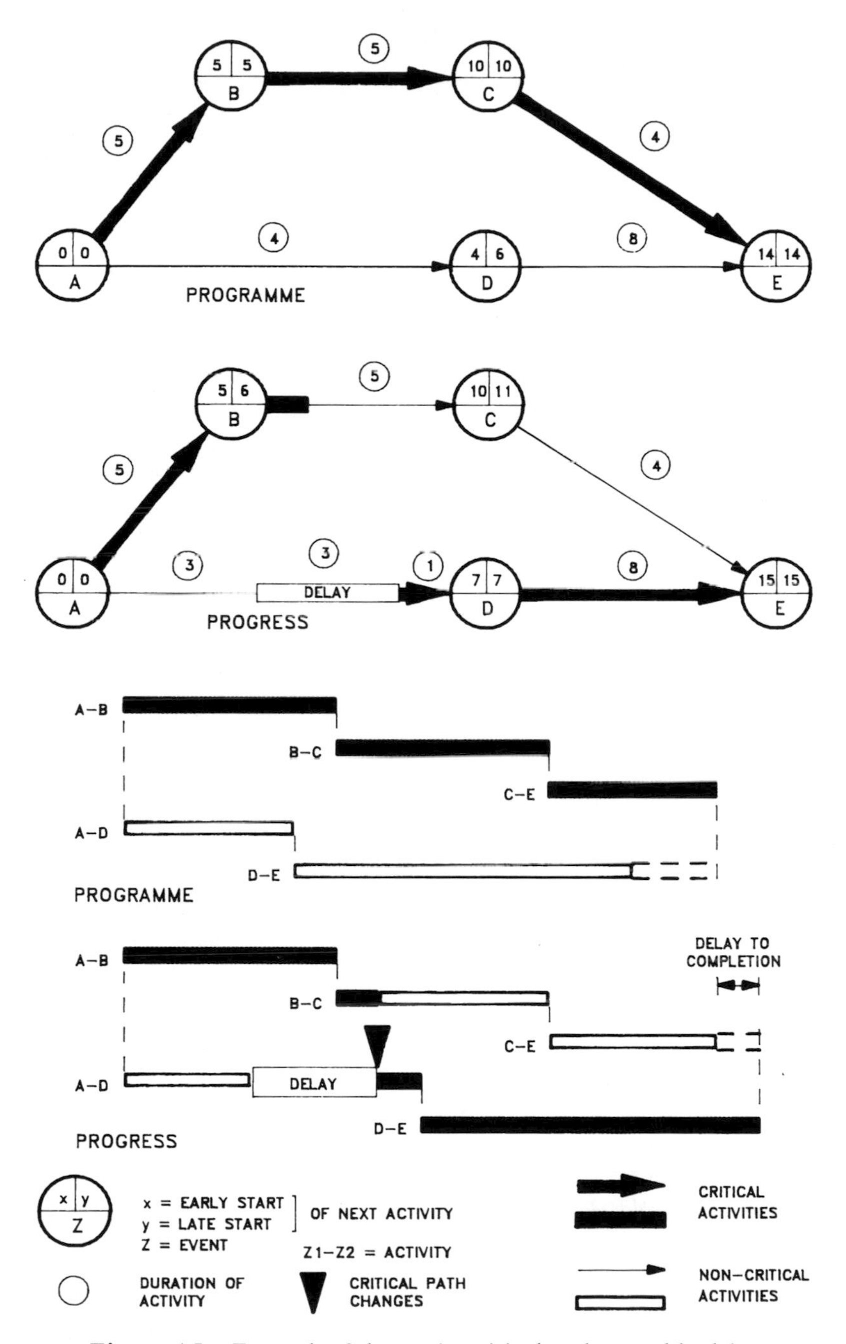

Figure 4.1 *Example of change in critical path caused by delay*

4.9 Notice: records and particulars

Many delay claims by contractors fail owing to lack of notice and/or failure to justify any (or sufficient) extension of time, or additional payment, because of a lack of records. No truer comment has been made than that of Max W. Abrahamson in his book *Engineering Law and the I.C.E Contracts, fourth edition* at page 443:

> 'A party to a dispute, particularly if there is arbitration, will learn three lessons (often too late): the importance of records, the importance of records and the importance of records.'

Whether, or not, there are contractual requirements to give notice of delay, or extra payment, contractors must, if they are to maximise the relief, or compensation, within the contractual remedies, give written notice of the delay or circumstance giving rise to the claim. Where the contractual provisions are stringent (and particularly where they are *conditions precedent*), contractors should ensure that each, and every, member of staff be made aware of these requirements and that each knows what role to play within contractual procedures designed to manage all delay and disruption claims. Where the contractor's staff have a good working relationship with the employer's staff, all notices should be clearly set out, identifying the contractual provisions under which the notice is being given, together with sufficient information to enable the recipient to be aware of the actual, or likely, effects of the matters in respect of which the notice is being given. In the unfortunate (and sadly, too frequent) cases where notice of any kind, no matter how well justified, produces a hostile reaction and continuous allegations aimed at 'muddying the waters', there may be some justification in couching the terms of any notice so that it is almost disguised. If this approach must be adopted, the significance of the notice must be capable of being understood in the light of other documents and the surrounding circumstances.

Having given notice, the contractor should keep contemporary records in order to illustrate the effects of the events, or circumstances, for which notice has been given. The recipient (the architect, or engineer) should also keep contemporary records. It is good practice to agree what records should be kept, to jointly monitor events and to agree facts during the progress of the works. Many contracts now contain express provisions for keeping records. Failure to agree facts is often caused by attempting, at the same time, to establish liability and entitlement. If both parties address their minds solely to agreeing *facts as facts*, leaving liability and entitlement for another day, agreement may be more readily achieved.

The most common records which ought to be kept are

- Master/Detailed Programme and all updates with reasons for each update (preferably showing delays to each activity);
- adverse weather conditions, including high winds and abnormal temperatures;
- Progress Schedule indicating actual progress compared with each revision of the programme;
- Schedule of Resources to comply with the original and each revision of the programme;
- records of actual resources used based on progress;
- cash flow forecast based on the original and each revision of the programme;
- records of actual cash flow;

- schedule of anticipated plant output;
- records of actual plant output on key activities;
- records of plant standing and/or uneconomically employed (with reasons);
- schedule of anticipated productivity for various activities;
- records of actual productivity on key activities;
- schedule of anticipated overtime (and the costs thereof) in order to comply with the original and each revision of the programme;
- records of actual overtime worked and the costs thereof;
- progress photographs and (where appropriate) photographs of work to be covered up;
- where appropriate, video records showing sequence and method of working;
- drawing register with dates of each revision and notes of amendments;
- site diaries and diaries of key staff;
- minutes of meetings and notes kept at meetings;
- cost and value of work executed each month (for the project);
- cost and value of work executed each month for all projects (company turnover);
- allowance for overheads and profit in the tender sum;
- cost of head office overheads each month (quarterly or yearly if not possible on a monthly basis);
- profit (or loss) made by the company for each accounting period.

Many contractors do not have the management information systems or procedures to keep all of these records. However, many of them are capable of being kept on site with the minimum of extra effort. It is important to specify what records should be kept by different members of staff. For example, the contents of the diary and records kept by the project manager will be different from those kept by a section foreman. Company policy should lay down procedures and guidelines so that there is the minimum of duplication (save where it is essential for verification) and that there are no gaps in the information to be collected.

The effect of failure to give notices and particulars varies from contract to contract. JCT, ICE and FIDIC contracts up to and including the 1987 fourth edition of FIDIC, for example, did not expressly provide for notices and/or particulars to be a *condition precedent* to the contractor's rights to a claim for extensions of time or additional costs. However, the 1999 FIDIC Red, Yellow and Silver Books have changed all that. These new contracts require written notice of all claims (for time and money) within twenty-eight days (clause 20.1). This provision is a *condition precedent* and the contractor will therefore lose his rights to such claims if he fails to give notice in accordance with this clause (see 1.7, *supra*). The 1999 FIDIC Green Book makes provision of an early warning a prerequisite to an extension of time or additional costs (clause 10.3), with the proviso that some relief may be given having regard to any reasonable steps that the engineer may have taken to reduce the effects if an early warning had been given.

As stated earlier, the NEC 3 Contract requires the Contractor and the Project Manager to give early warning notices of matters which could increase the total price, delay completion or a key date, or impair the performance of the works in use (clause 16). Further, clause 61.3 requires the contractor to give notice of events which may be the basis of a claim (defined as a Compensation Event) within 8 weeks of the event happening. In the event that the contractor fails to issue a notice within 8 weeks, the contractor loses his right to claim, unless the Project Manager should have notified the contractor of the event.

Clause 44(1) of the ICE 7th Edition requires the contractor to notify the engineer within twenty-eight days of a cause of delay or as soon thereafter as is reasonable. Clause 44(2)(b) allows the engineer to assess a cause of delay in the event that the contractor has failed to issue a notice.

Provisions, which effectively 'time-bar' claims if the contractual machinery is not followed, are extremely onerous. It may be easy to comply with a provision to give notice within twenty-eight days in some circumstances, but not in others. The demand on management resources to identify potential claim events in order to comply with the contract is likely to increase costs. Many notices will be for minor events which may not subsequently affect the works. Paperwork will increase unnecessarily and the resources required to deal with these notices and respond to them will also be increased. It should be noted that the courts have recently held that a contractual provision requiring notice of delay does not have to expressly state that failure to give notice will disentitle a party to an extension of time. See *Steria Ltd* v. *Sigma Wireless Communications Ltd* [2007] 2008 BLR 79 in which the court stated that provided the wording of the clause was sufficiently clear concerning the obligations of a party, there was no requirement to set out the consequences of failing to comply with that requirement.

In some civil law jurisdictions, contracts may not be permitted to oust a party's legal rights to a remedy or compensation by the incorporation of 'time-bar' provisions. In such jurisdictions, contractors may be able to claim even if there has been failure to give notice within a specified period. However, it is advisable to follow the contract whenever possible to avoid the potential high cost of finding out if a late notice is good enough. Where time-barring of claims is outlawed, it should not be seen as an excuse to leave all notifications to the last minute.

MF/1, a contract used extensively on major projects, contains very onerous provisions. While sub-clause 33.1 (extensions of time) contains requirements to give notice 'as soon as reasonably practicable', sub-clause 41.1(a) (notification of claims) requires notice to be given within twenty-eight days, failing which the claim will not be allowed (a *condition precedent*).

The requirements to keep particulars and submit accounts of claims in the ICE and FIDIC contracts (including the 1999 FIDIC Red, Yellow and Silver Books) are subject to the proviso that the contractor does not lose his rights to any claims if he fails to comply, however his entitlement may be severely prejudiced by such failure (clause 53(5) of ICE 7th Edition and clause 20.1 of the 1987 edition and 1999 edition of the FIDIC contracts).

The NEC 3 Contract allows the Project Manager to request the Contractor to provide quotations for compensation events, which should include details of delay and increased costs (clause 62). The project manager may also notify the contractor that he failed to issue an early warning notice (clause 61.5). Clause 63.5 provides that if the project manager notifies the contractor in accordance with clause 61.5 the event is considered as if the contractor had given an early warning notice.

On the employer's side of the fence, the architect, engineer, clerk of works and other staff should know what records they should each keep. If they are not kept jointly with the contractor, they should be agreed wherever possible. Keeping records for the purposes of defeating a claim in an arbitration may appear to be good practice, but it is more sensible to use them to settle contentious issues at the time so as to avoid costly disputes. In addition, if the contractor is aware that his grounds for a claim are doubtful (having regard to better records kept by the employer's professional team), it is more likely that the claim will be dropped and he will make an effort to get on with the job and possibly make up some lost time.

The employer's professional team should keep additional records to monitor delays by the contractor and delays for which no additional payment is payable.

Whatever records are kept, they are likely to be invaluable in the preparation of particulars in support of a claim. It should be remembered that particulars should, in addition to supporting the claim, be persuasive. It is all very well merely submitting all relevant records as particulars without some argument and illustration to set out the contractor's case and the entitlement sought, on the basis that it is the architect, or engineer, who is responsible for assessing the claim; but, once the architect, or engineer, has made their assessment, it is sometimes difficult to persuade them to change their minds. Their assessment may be insufficient because they did not appreciate the effects of some delays on the method, sequence or timing of an operation, or because they did not recognise the significance of some of the records submitted. Naturally, they may be reluctant to admit this fact, particularly if it will bring to light their inexperience, or emphasise that the delay was due to their own incompetence. Good particulars should, in addition to providing supporting records, illustrate the effects of the events, or circumstances giving rise to the claim. To this end, the contractor is well advised to provide details and diagrams indicating

- what ought to have occurred if there had been no delaying event, or circumstance;
- what actually occurred as a result of the delaying event, or circumstance;
- analysis of facts, calculations, explanations and arguments to show how the delaying event, or circumstance, was responsible for the change in the method and/or programme.

What a contractor should never do is fail to give sufficient notice in accordance with the provisions of the relevant contract simply because of an allegation that the contractor is being 'too contractual'. To do so is inviting trouble. In such circumstances the contractor should politely point out that it is simply complying with the provisions of a contract which the employer has entered into (and probably proposed).

4.10 Delays after the contract completion date

The best advice that can be given to any employer is not to cause any delay after the contractual completion date (extended, if applicable) has passed and when the contractor is in culpable delay. Very few contracts deal with delays by the employer after the completion date, and in many cases, once such a delay has occurred, the time for completion is no longer applicable and the contractor is allowed a reasonable time for completion of the works. Even where the contract does provide machinery for extending the date for completion in the event of such delays, there are few guidelines as to how the extension should be dealt with, and the effects on the employer's rights to liquidated damages. The Singapore Architects Standard Form of Contract contains very detailed provisions in clause 24 (see Figure 4.2). In this form of contract, it is intended that the employer may recover liquidated damages during a period of culpable delay by the contractor (even if a concurrent qualifying delay should occur

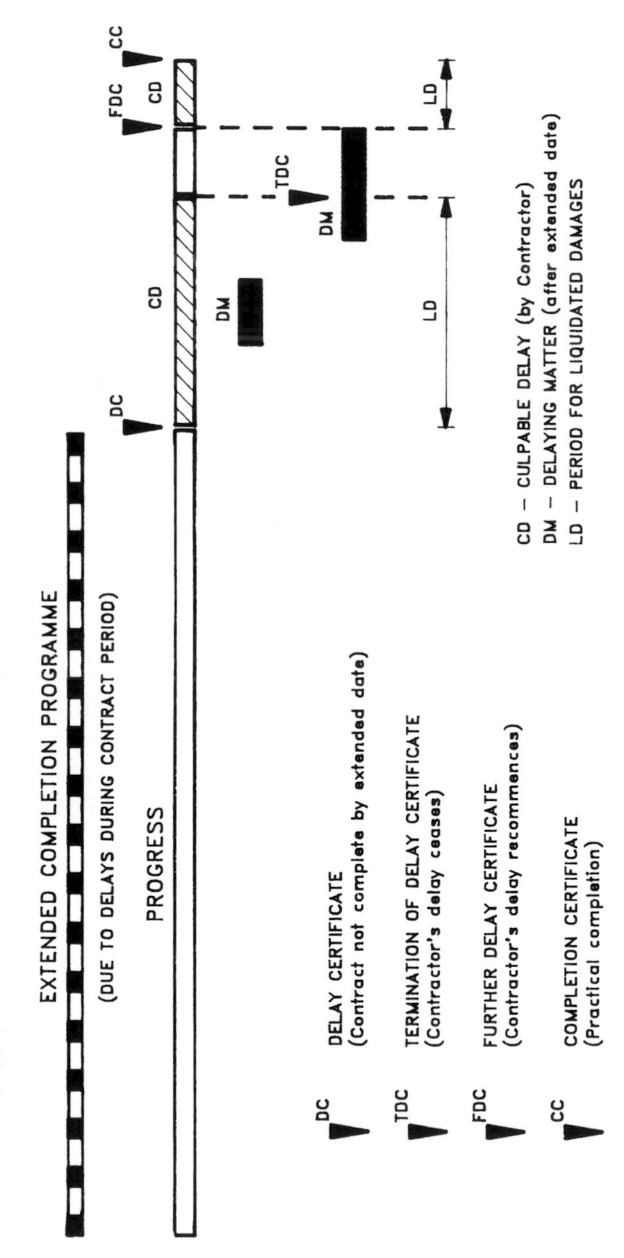

Figure 4.2 *Clause 24 – Delay in completion and liquidated damages; SIA form of contract*

during the period of culpable delay). Only if the contractor is not himself in delay is it intended that the employer's rights to recover liquidated damages be suspended during a further delay caused by a qualifying event or circumstance. However, with the greatest respect to the distinguished author of these provisions, they are unduly complicated, and they are likely to fail to protect the employer's rights to liquidated damages if the delay which occurs (after the completion date has passed) is one within the employer's control and which was caused by an event which would in any event have prevented the contractor from completing by the due date (provided of course that the employer was not relying on the contractor's progress in order to comply with a contractual, or statutory provision). Possible circumstances which give different results are given in Chapter 5.

If such delays cannot be prevented, careful monitoring and records are vital where there are several causes of delay after the completion date has passed.

4.11 Minimising exposure to claims: prevention

Stringent notice provisions and requirements to give particulars may be effective in avoiding claims by contractors who do not follow such provisions. However, this may increase the contract price and lead to conflict throughout the contract.

Whether, or not, there are sensible contractual provisions, and whether, or not, the contractor complies with them, the employer's professional advisors can minimise exposure to claims by ensuring that they do not cause delay by matters within their control (such as issuing late information). It is a mistake to assume that information can be delayed on the grounds that the contractor is in delay and is not ready for it. In many cases the contractor will be able to make out a case for an extension of time (or even time at large), particularly if the information is received at a time when it can be shown that it would have been impossible to complete the works by the due date having regard to all of the remaining activities (see Figure 4.3). Scheduling issuance of information in accordance with the contractor's progress is a recipe for disaster and to be avoided at all costs.

Where delay and/or disruption claims occur, careful attention to records and constant monitoring of the effects will enable the employer to minimise his exposure. Inflated, or exaggerated, claims can be refuted. Costs which are partly to be borne by the contractor can be identified and adjustments made (see Chapter 7 – concurrent delays). Even where delays on the part of the employer justify an extension of time, the contractor's claim for payment can be reduced, or disallowed, where it can be shown that the contractor was also in delay and the costs claimed would, in any event, have been incurred by the contractor.

Delays, and claims arising out of them, are almost inevitable in construction contracts. If this fact is acknowledged, and proper procedures are devised to deal with them, then claims would be more palatable to those having to pay for them. Usually, all parties are at fault to a varying degree, and adversity thrives on one or more parties attempting to place all of the blame on someone else. Contractual provisions do not, in themselves, avoid these problems. Education and training in contracts administration should be encouraged to improve the understanding of claims and how they arise.

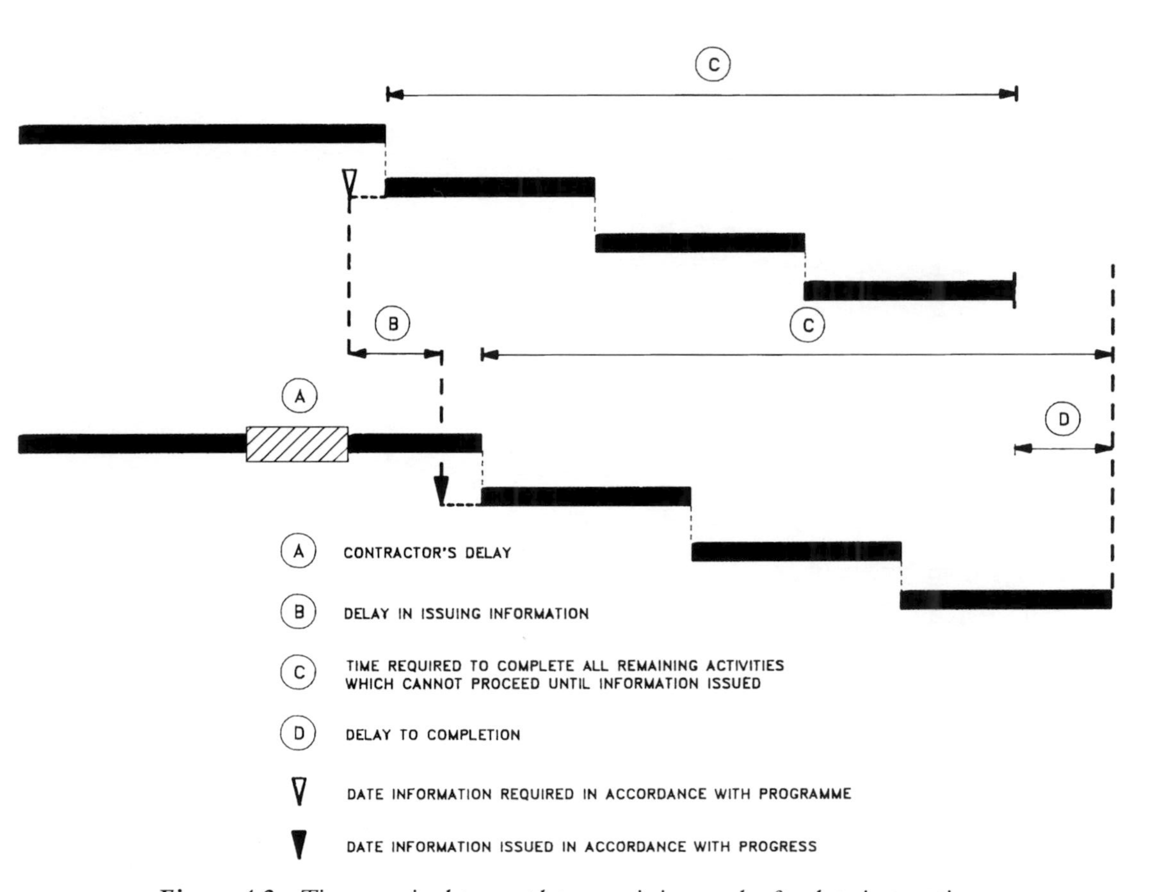

Figure 4.3 Time required to complete remaining work after late instruction

5 Formulation and Presentation of Claims

5.1 Extensions of time claims

All modern building and engineering contracts contain provisions for extensions of time in the event of delay. The nature of the work and the environment in which the work is carried out are such that it is almost inevitable that events and circumstances will cause completion of the work to be delayed beyond the original completion date. Notwithstanding, claims for extensions of time probably cause more disputes than any other contractual or technical issues. Major obstacles to prompt settlement of claims for extensions of time claims are

- the erroneous assumption that an extension of time is automatically linked to additional payment;
- late, insufficient or total lack of notice of delay on the part of the contractor;
- failure to recognise delay at the appropriate time and maintain contemporary records;
- failure to regularly update the programme so that the effects of delay can be monitored against a meaningful 'programme of the day';
- poor presentation of the claim to show how progress of the work has been delayed;
- insistence, on the part of the employer's professional advisers, that unreasonably detailed critical path programmes are essential in order to assess the effects of the delay;
- the probability that the cause of the delay will reflect on the performance (or lack of it) on the part of the employer's professional advisers;
- pressure, on the part of the employer, to complete on time, irrespective of delays which occur.

The first obstacle – delay means money – is understandable. Nevertheless, it should not be a consideration when dealing with extensions of time. It should be clearly understood that an extension of time merely enables the contractor to have more time to complete the works and the employer to preserve his rights to liquidated damages. An extension of time awarded for a cause of delay which appears to have a financial implication (delay within the control of the employer) does not necessarily lead to an entitlement to additional payment. If the contractor is, himself, also in delay, then the additional costs arising out of the extended period to execute the works may (in total or in part) have to be borne by the contractor (see concurrent delays, *infra*).

On the other hand, an extension of time awarded for neutral events (for example, adverse weather conditions) will not necessarily deprive the contractor of a claim for additional payment. The latter point was clearly illustrated in the case of *H. Fairweather & Co Ltd* v. *London Borough of Wandsworth* (*supra*). In this case the arbitrator had concluded that the architect had been correct in awarding eighty-one weeks' extension of time for the dominant cause of delay (strikes). The arbitrator had stated that the extension did not give rise to a claim for direct loss or expense. The contractor sought to establish that eighteen weeks' extension of time ought to have been granted for causes of delay which would give rise to a claim for loss or expense.

The contract was JCT63 in which some of the causes of delay (or disruption) in the loss and expense clause (24) are set out almost verbatim as some of the causes of delay in the extension of time clause (23). This is unfortunate and misleading and may be one of the reasons for some practitioners to assume a link between extensions of time and claims for additional payment. This misconception was cleared up by Judge Fox-Andrews QC in a hypothetical example which is summarised below:

> 'A tunnelling contract proceeds through the winter and is due to complete on 31 July. A variation instruction is issued in April which requires a further three months for completion of the works and for which an extension of time is granted up to 31 October. Two weeks before the revised completion date a strike occurs which continues until 31 March. The works cannot proceed and time passes through a second winter. On 1 April, the contractor recommences work, but due to the fact that it had not been able to protect its plant and equipment during the strike it takes two months to complete the remaining work. An extension of time for eight months for the strike (under clause 23(d) of JCT63) would not prevent the contractor from recovering loss and expense under clause 11(6)'. (See Figure 5.1.)

Nevertheless, in the circumstances of the case, the judge recognised the practical difficulties in the event of the extension of time not being made under the provision which linked the extension to the provisions of clauses 11(6) and/or 24(1) and he remitted the matter to the arbitrator for further consideration. The extension of time

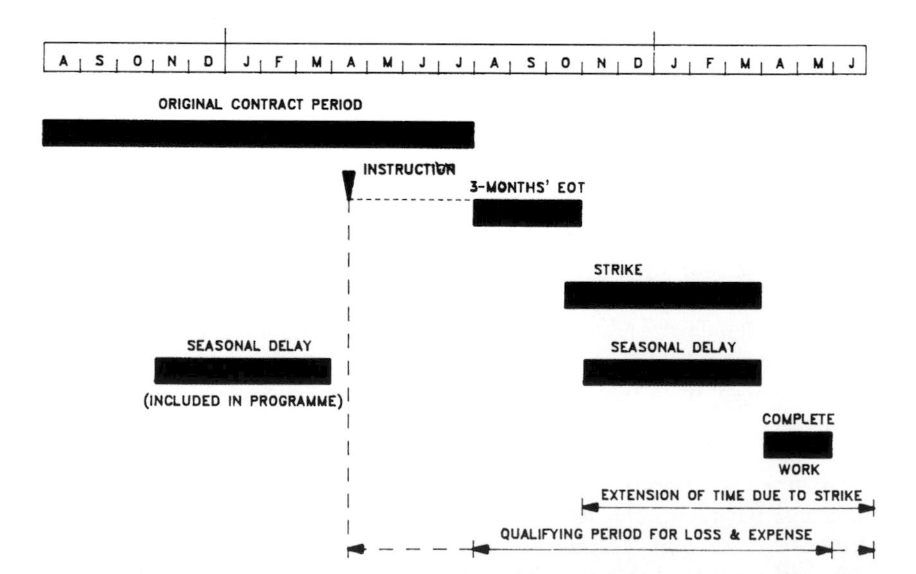

Figure 5.1 *H. Fairweather & Co Ltd v. London Borough of Wandsworth*

and loss and expense provisions of the JCT Standard Building Contract 2011 refer to 'Relevant Events' and 'Relevant Matters' which to some extent overlap (clauses 2.29 and 4.24 respectively) but practitioners should not be misled into assuming that an extension of time for the specified relevant events will bring with it an entitlement to additional payment.

The next three obstacles, notice, contemporary records and programme, are all practical matters which can only be addressed by ensuring that adequate contracts administration procedures are being followed from the date of commencement of the works. While the architect, or engineer, must do their best to estimate the length of any extension of time which may be due, irrespective of the lack of notice and particulars given by the contractor (*London Borough of Merton* v. *Stanley Hugh Leach Ltd*, *supra*, Chapter 1), contractors cannot complain if the extension made on the basis of inadequate information does not live up to their expectations.

5.2 Notice and presentation of extensions of time claims

Most contracts do not require the contractor to do more than give notice of delay, maintain records and provide particulars. However, as mentioned at paragraph 4.10 *(supra)*, a contractual provision does not need to expressly state that notice is a condition precedent to an entitlement to an extension of time, and failure to give notice in accordance with the relevant clause may disentitle a contractor to an extension of time (see *Steria Ltd* v. *Sigma Wireless Communications Ltd* [2007] (2008 BLR 79). See also *Multiplex Constructions (UK) Ltd* v. *Honeywell Systems* [2007] BLR 195 *(infra)*.

Notice provisions vary. Some examples are

- JCT Standard Building Contract 2011 – '...whenever it becomes reasonably apparent that the progress of the Works or any Section is being or is likely to be delayed the Contractor shall *forthwith* give notice...' (Clause 2.27.1).
- GC/Works/1 (1998) – Notice may be given at any time, but not '...after completion of the Works' (clause 36(4)). Clause 35 contemplates regular review of extensions of time, but there is no link to clause 36.
- ICE seventh edition – the Contractor shall '... within 28 days after the cause of any delay has arisen or as soon thereafter as is reasonable deliver to the Engineer full and detailed particulars...' (clause 44(1)).
- JCT Standard Building Contract 2011 goes on to require the contractor to give particulars of the expected effects of the delay and an estimate of the extent of any delay in completion of the works beyond the relevant completion date (clause 2.27.2).
- NEC 3 requires notice to be given by the contractor within 8 weeks of an event occurring, which he considers to be a Compensation Event (clause 61(3)). A Compensation Event can include financial compensation and delay.

None of the above provisions requires the contractor to show how it arrived at its estimate of the period of delay. Provided that the contractor has given details of all events, dates, what work is likely to be effected, or was affected, and the like (together with an estimate of the delay in the case of the JCT Standard Building Contract 2011), it appears that the contractual provisions have been satisfied and the onus is then on the architect, or engineer, to decide what extension is reasonable on the basis of the particulars provided and/or on the basis of further information obtained from other sources. Many contractors only provide information (often insufficient) and rely on the architect, or engineer, to make a reasonable extension of time. This tactic can be successful, but there is a risk that the extension made will be insufficient. Not all is lost, as the contractor can always present his case at a later date, hoping to persuade the opposition that more time is justified. The problems with this approach are

- it is usually more difficult to persuade someone to change their mind after they have made a written extension of time unless there is additional evidence which can be used to explain a change in the period of the extension;
- there will almost certainly be a period of protracted discussion during which the current (extended or otherwise) completion date and the progress of the works are inconsistent with a realistic programme and a subsequently revised extended completion date.

The 1999 FIDIC contract partially addresses the above problems.

Clause 4.21 of the 1999 FIDIC Red, Yellow and Silver Books requires the contractor to compare actual and planned progress and to show details of any event or circumstance which may jeopardise completion.

The two problems listed above must be avoided or their effects will be compounded, making it difficult to monitor future delays and to make realistic extensions of time having regard to all of the circumstances. The better approach, on the part of the contractor, is to present his claim for an extension of time, showing how he arrived at his estimate of delay and the effects on completion of the works. If the contractor has a detailed critical path programme using one of the well-tried software packages, or a tailor-made package, then this task can be simplified. Unfortunately, many contractors who use such packages become complacent, believing that the programme, and the software used, is the answer to all of their problems. Computer applications can only be truly effective if the delays are quickly identified and steps are taken immediately to monitor events and update the programme. In many instances, full-blown computer applications are not necessary. Carefully prepared linked bar chart programmes can be very effective provided that the original logic is right.

Example 1: A single cause of delay on the critical path

A linked bar chart showing how the contractor intended to complete the works in twenty-two weeks is shown in Figure 5.2.

A qualifying delay (D1) of two weeks occurred during weeks six and seven affecting progress of activity B–E (which is on the critical path – see Figure 5.3). In these circumstances it is a relatively simple matter to recognise that completion of the works was likely to be delayed by two weeks and an extension of time should be made for the full period of delay giving a revised completion period of twenty-four weeks.

The above example is straightforward as it deals with delay which is on the critical path and there are no concurrent delays. What is the situation in the event of delay which is not on the critical path? Some authorities exist which may be of some assistance (see Example 2).

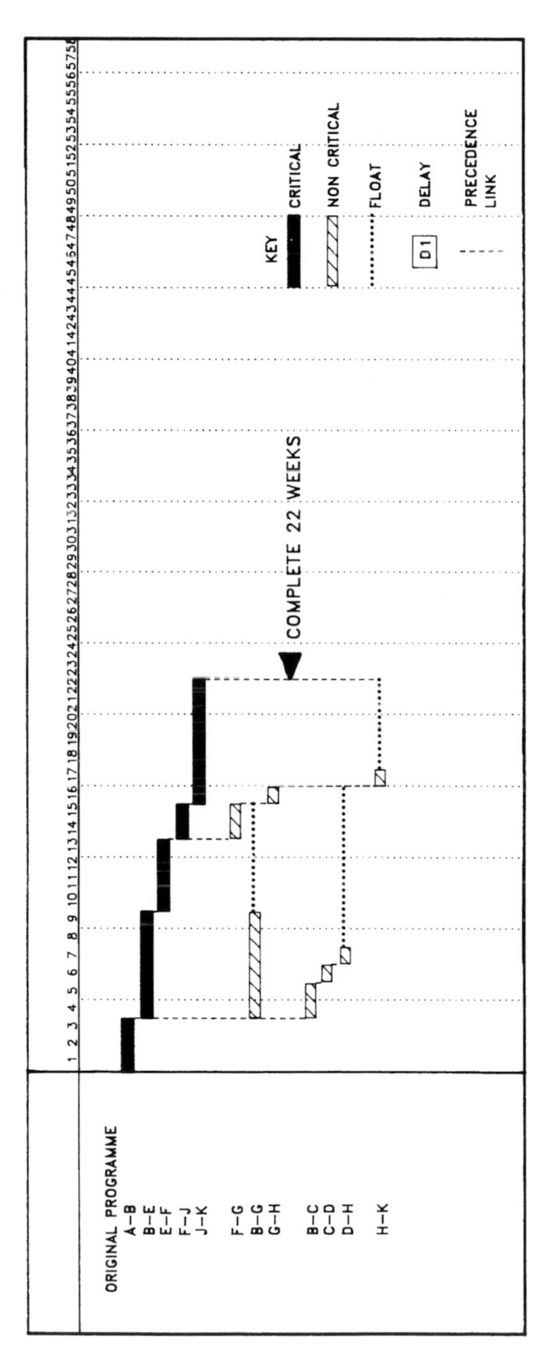

Figure 5.2 Precedence (linked) bar chart – original programme

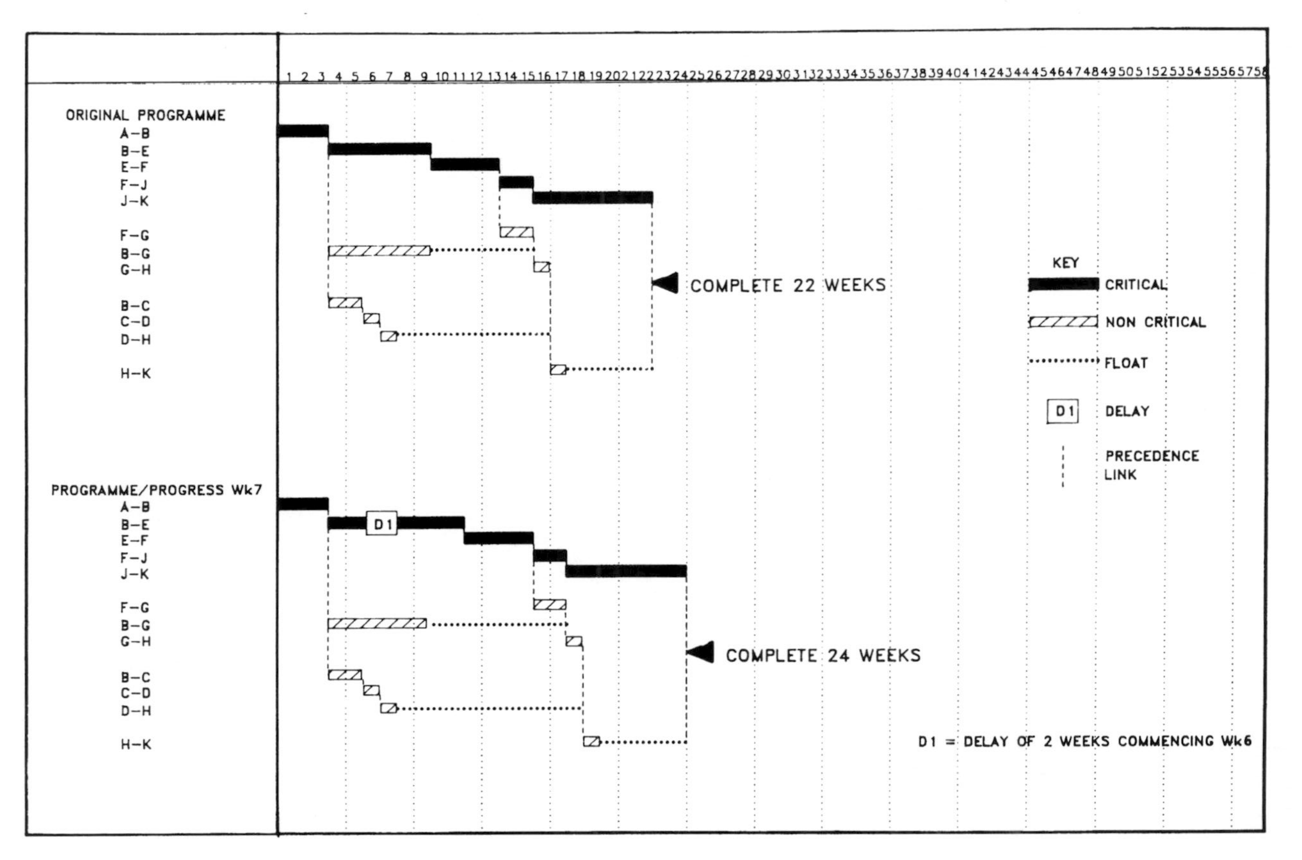

Figure 5.3 *Single cause of delay on the critical path*

Example 2: A single cause of delay – not on the critical path

Using the same linked bar chart in Figure 5.2, a qualifying delay (D2) of two weeks occurred during weeks six and seven which affected the progress of activity B–G (which is not on the critical path – see Figure 5.4). In these circumstances there is no effect on the completion date and no extension of time is necessary.

In *Glenlion Construction Ltd* v. *The Guinness Trust* (*supra*), the judge had to consider matters of extensions of time where the contractor had prepared a programme showing completion of the works before the contractual date for completion. Tenders were invited on the basis of a contract period of 104 weeks. *Glenlion* submitted an alternative tender for completion in 114 weeks which was accepted by *Guinness*. The completion date inserted in the contract was 114 weeks after the date for possession. The contract required *Glenlion* to produce a programme showing completion 'no later than the date for completion' and *Glenlion* complied by producing a programme which showed completion in 101 weeks. There were delays and disputes arose as to *Glenlion*'s entitlement to an extension of time. The crucial text of the judgement is (at page 104):

> 'Condition 23 [extensions of time] operates, if at all, in relation to the date for completion in the appendix. A fair and reasonable extension of time for completion of the works beyond the date for completion stated in the appendix *might* be an unfair and unreasonable extension from an earlier date.' [Emphasis added]

It must be concluded that if any delay occurs then it is not necessarily correct to make an extension of time equal to the period of delay. Some, or no, extension of time may be required. How much extension (if any)?

There is a widely held view that any float in the contractor's programme is for the benefit of the contractor and any delay on the part of the employer which reduces that float may have to be taken into consideration when considering the time required for completion.

This concept can be applied to *Glenlion* v. *Guinness* as shown in Figure 5.5. Bar A indicates the period for completion stated in the tender documents (104 weeks), bar B indicates the period for completion stated by *Glenlion* in the alternative tender (114 weeks, which was accepted by *Guinness*) and bar C indicates the period indicated in *Glenlion*'s programme (101 weeks). The programme shows completion thirteen weeks before the contractual date for completion.

Assume that a delay of five weeks occurs at the outset of the contract for which there is power to make an extension of time (that is, a qualifying delay or relevant event – bar D). This has the effect of reducing the contractor's float from thirteen weeks to eight weeks. No extension of time is necessary as completion is not likely to be delayed beyond the contractual date for completion.

A further qualifying delay of four weeks occurs during the contract period (bar E). Again, this only reduces the contractor's float from eight weeks to four weeks and no extension of time is necessary. Another qualifying delay of four weeks occurs towards the end of the contract which takes up the remaining float (bar F). Again, no extension of time is necessary.

Four weeks before completion, a further delay of four weeks occurs which does not qualify for an extension of time (for example, culpable delay on the part of the contractor). In these circumstances the contractor has need of an extension of

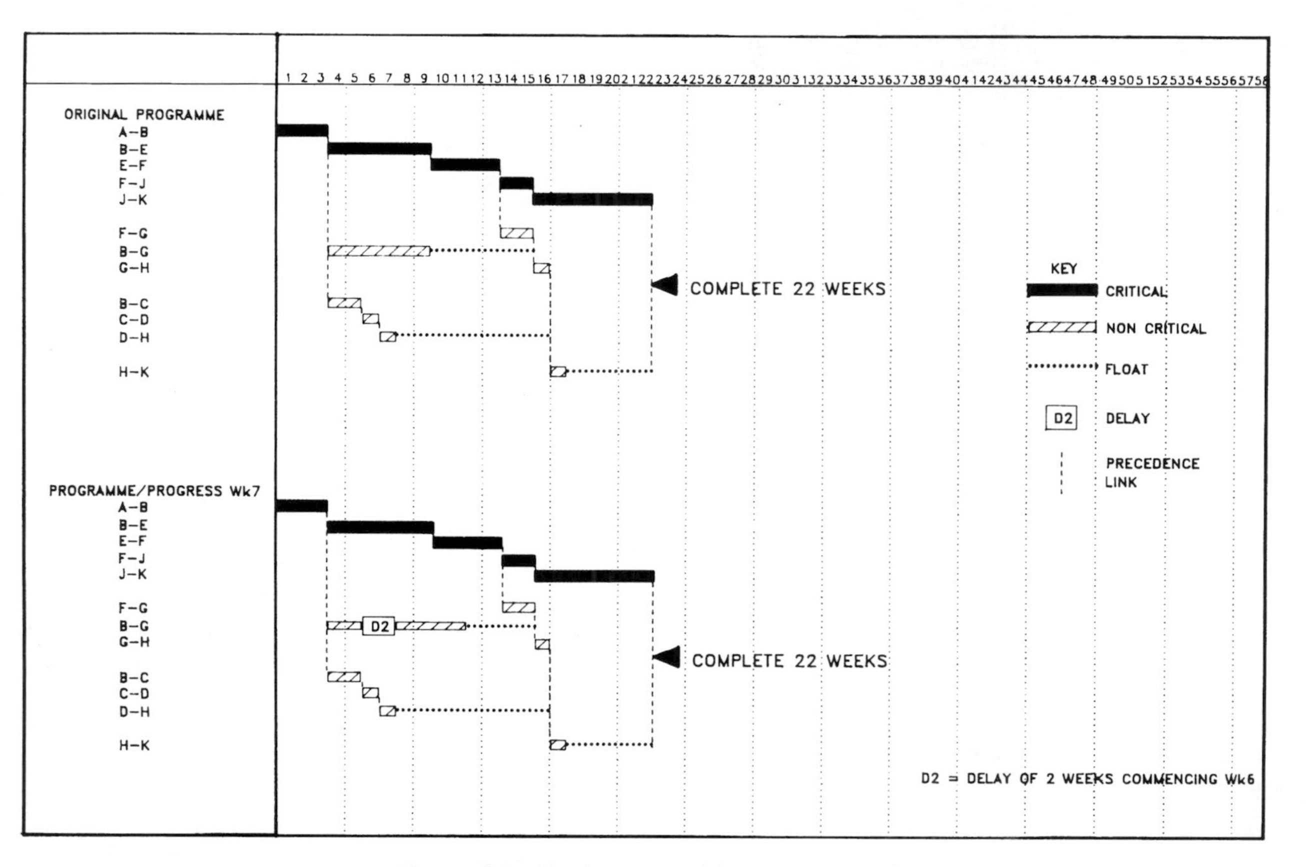

Figure 5.4 *Single cause of delay – non-critical*

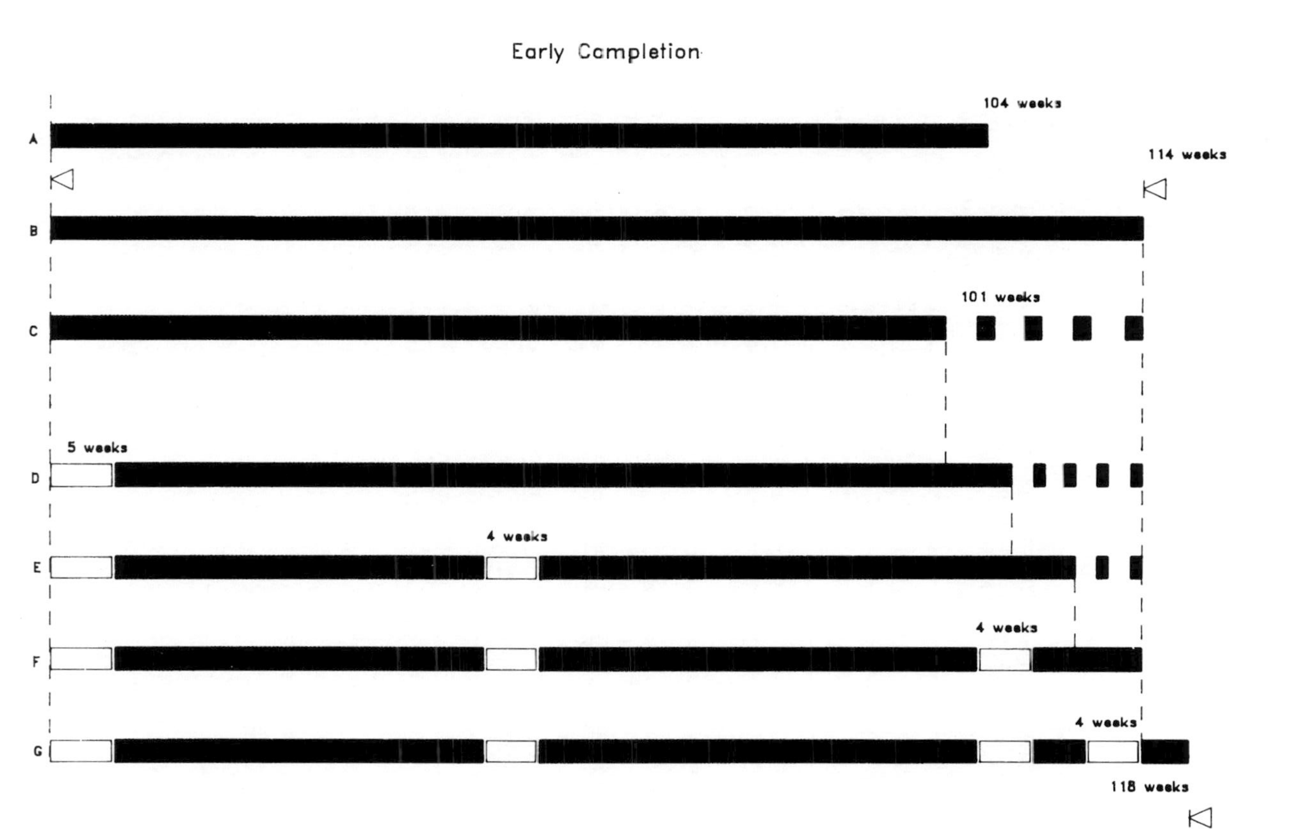

Figure 5.5 Glenlion Construction Ltd v. The Guinness Trust

time and it would therefore be reasonable to make an extension of time of four weeks. Difficulties may arise under the JCT Standard Building Contract 2011 because the extension of time clause (2.28.1) contemplates an extension of time being made if '...completion of the Works or any Section is likely to be delayed [by the relevant event] thereby beyond the relevant Completion Date...'. In the above example, completion of the works was delayed beyond the completion date by an event which did not qualify for an extension. However, the circumstances described in this example are probably covered by the provisions of clause 2.28.5.1 which empowers the architect to '...fix a Completion Date for the Works or for the Section later than that previously fixed if in his opinion that is fair and reasonable *having regard to any Relevant Events*' [Emphasis added]. Some may argue that clause 2.28.5.1 does not apply in these circumstances but the case of *Balfour Beatty* v. *Chestermount Properties* (1991) 62 BLR 1 suggests otherwise. Even if that view were to be correct, the employer would be unlikely to succeed in claiming liquidated damages for late completion when it has been partly responsible for the delay to the progress of the works. Regard may have to be paid to the nature of the contractor's culpable delay. Sheer dilatoriness on the part of the contractor may be viewed in a different light from matters such as a plant breakdown or failure to obtain materials in spite of taking all reasonable measures.

Those who resist making an extension of time in circumstances similar to the above example may be persuaded to change their view by considering the position if any (or all) of the delays in bars D, E and F had been due to the contractor's own delay and the delay in bar G had been due to a qualifying delay. In these circumstances, there is no room to argue that an extension of time is not required. This would appear to be the case even if the contractor's own delays had been due to dilatoriness, since the contractor would not be in breach of its obligation to complete until the completion date had passed.

The *Glenlion* case only dealt with delays and extensions of time when the contractor's programme showed early completion. The South African Case, *Ovcon (Pty) Ltd* v. *Administrator Natal* (*infra*), also dealt with delays when the contractor's programme showed early completion. However, the *Ovcon* case did not deal with extensions of time because the contractor's programme showed completion four months earlier than the contract completion date, and the contractor finished one month early as a result of a three months' delay by the employer (see Figure 5.6). The *Ovcon* case was concerned with the additional costs claimed by the contractor (which *Glenlion* did not have to consider) and is discussed in 5.9 (*infra*).

It should be noted that clause 63.3 of the NEC 3 contains the following provision:

> 'A delay to the Completion Date is assessed as the length of time that, due to the compensation event, planned Completion is *later than planned Completion as shown on the Accepted Programme*.' [Emphasis added]

It follows that if the accepted programme showed early completion, any qualifying delay which affected the planned completion date would merit an extension of time. Therefore, provided that the original float in the contract was not eroded by the contractor's own default, the period of float would be preserved.

Note – Clause 33(1) of GC/Works/1, (1998), requires the contractor's programme to be based upon 'a period for the execution of the Works to the Date or Dates for Completion'.

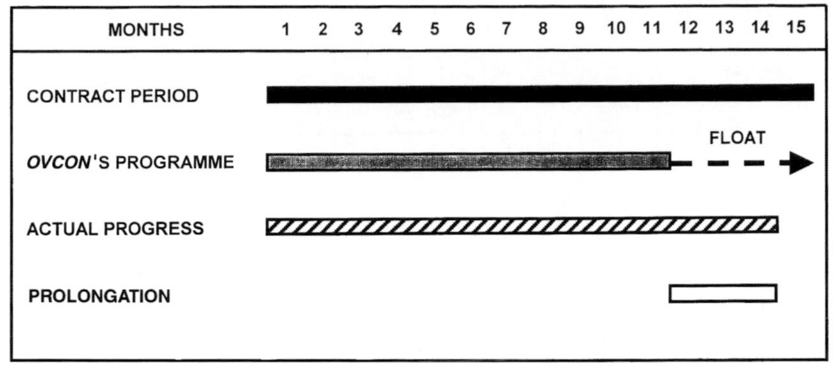

Figure 5.6 *Ovcon (Pty) Ltd v. Administrator of Natal 1991 (4) SA 71*

Example 3: Concurrent delays – critical and non-critical

Using the same linked bar chart in Figure 5.2, the delays referred to in examples 1 and 2 above occurred at the same time (see Figure 5.7). If both of the delays were qualifying delays, an extension of time of two weeks is necessary for the delay (D1) which affected activity B–E. If the delay to activity B–E is a qualifying delay, and the delay (D2) to activity B–G is due to the contractor's culpable delay, an extension of time of two weeks is necessary. This is the case even when it is clear that the concurrent delays are operating during identical periods. This would also be the case if the contractor's culpable delay (D2) to activity B–G was on a parallel critical path and therefore also delaying completion by two weeks.

If the delay (D1) to activity B–E was due to the contractor's culpable delay, and the delay (D2) to activity B–G was a qualifying delay, then no extension of time would be necessary.

It is very rare for concurrency to occur in the pure sense of delays commencing and ceasing at the same time. Usually, causes of delay will overlap.

In the case of *Henry Boot Construction (UK) Limited* v. *Malmaison Hotel (Manchester) Limited* (1999) 70 ConLR 32, Mr. Justice Dyson stated that 'it is agreed that if there are two concurrent causes of delay, one of which is a relevant event, and the other is not, then the contractor is entitled to an extension of time for the period of delay caused by the relevant event notwithstanding the concurrent effect of the other event'.

In *Royal Brompton Hospital NHS Trust* v. *Frederick Hammond & Others (No. 7)* [2001] 76 ConLR 148 HHJ Richard Seymour QC distinguished between delays which were truly concurrent and delays which overlapped. However, in *Balfour Beatty Construction Limited* v. *The Mayor and Burgesses of the London Borough of Lambeth* [2002] BLR 288 HHJ Humphrey Lloyd QC declined to follow the Royal Brompton decision.

In the Scottish case of *City Inn Ltd* v. *Shepherd Construction Ltd* [2010] Scot CSIH 68 the court reviewed the *Henry Boot* and *Royal Brompton* decisions (and others) and derived five propositions one of which is

> '... *where a situation exists in which two causes* [of delay] *are operative, one being a relevant event and the other some event for which the contractor is to be taken to be responsible, and neither of which could be described as the dominant cause, the claim*

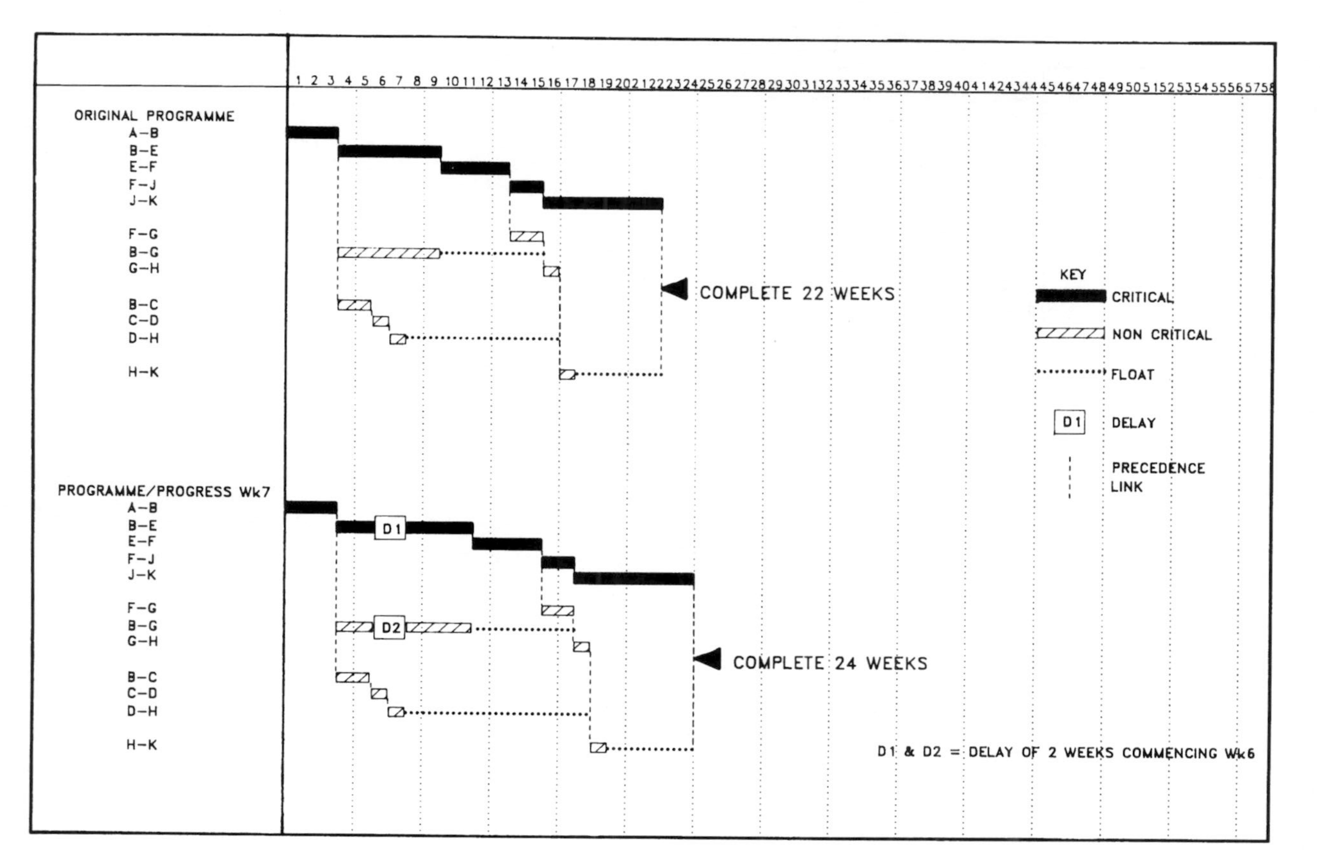

Figure 5.7 Concurrent delay – critical and non-critical

for extension of time will not necessarily fail. In such a situation, which could, as a matter of language, be described as one of concurrent causes, in a broad sense … it will be open to the decision-maker, whether the architect, or other tribunal, approaching the issue in a fair and reasonable way, to apportion the delay in the completion of the works occasioned thereby as between the relevant event and the other event. In that connection, it must be recognized that the background to the decision making, in particular, the possibility of a claim for liquidated damages, as opposed to one for extension of time, must be borne in mind and approached in a fair and reasonable manner.'

(Lord Osborne).

Example 4: Concurrent delays followed by subsequent delays

Using the same linked bar chart in Figure 5.2, the delays referred to in Examples 1–3 above were followed by further delays of seven weeks (D3) and five weeks (D4) to activities B–G and H–K respectively. If delays (D1) and (D2) were both qualifying delays (or if delay D2 was a non-qualifying delay), an extension of time of two weeks should already have been made (completion in twenty-four weeks). If delay (D3) was also a qualifying delay it would have the effect of delaying commencement of activities G–H and H–K, but no extension of time would be necessary because the float allowed for activity H–K is more than sufficient to absorb the delay (the float is reduced from five weeks to four weeks – see Figure 5.8).

However, for the reasons given previously, if delay (D4) occurred because of some event which did not qualify for an extension of time (for example, non-availability of materials, such as road surfacing, which could not be stored on site for use), an extension of time may be necessary *because the contractor had need of it* (see Figure 5.9). In these circumstances, qualifying delays (D2) and (D3) had reduced the contractor's float and non-qualifying delay (D4) had used up more than the remaining float, thereby causing completion to be delayed by one week (completion in 25 weeks). If delays (D2) and (D3) had not occurred, there would have been sufficient float remaining in activity H–K to absorb the delay (D4) and there would have been no delay to completion beyond the previously extended completion period of twenty-four weeks.

Numerous permutations may arise and each delay and its effects on the remaining float and the completion date need to be considered using the principles described above.

5.3 Delays after the contract completion date

Clause 61.3 of NEC 3 requires the Contractor to notify the Project Manager of a compensation event within 8 weeks of becoming aware of the event unless the Project Manager should have notified the Contractor of the Event. Clause 61.7 states that compensation events cannot be notified after the *defects date.*

The *Balfour Beatty* v. *Chestermount* decision (*infra)* suggests that the wording of clause 2.28.5 of the JCT Standard Building Contract 2011 does deal with delay occurring after the date for completion has passed. The extension of time provisions of JCT63 (clause 23) do not deal with delays which occur after the contract completion date (extended or otherwise) has passed and the contractor is in culpable delay. Indeed the clause is drafted in terms which appear to preclude making an extension of time for any delay which occurs after '… any extended time [date] *previously fixed….*'

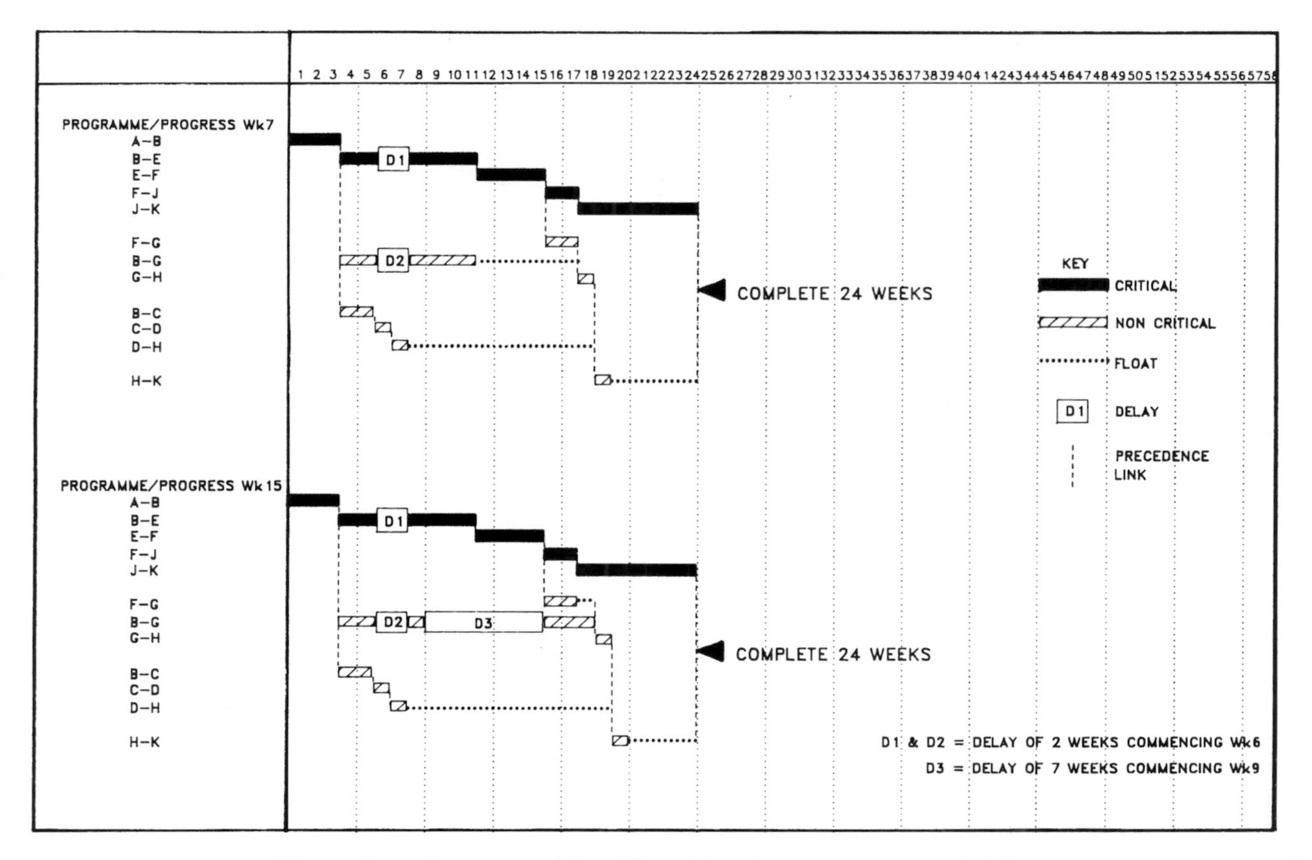

Figure 5.8 Concurrent delay followed by subsequent non-critical delay

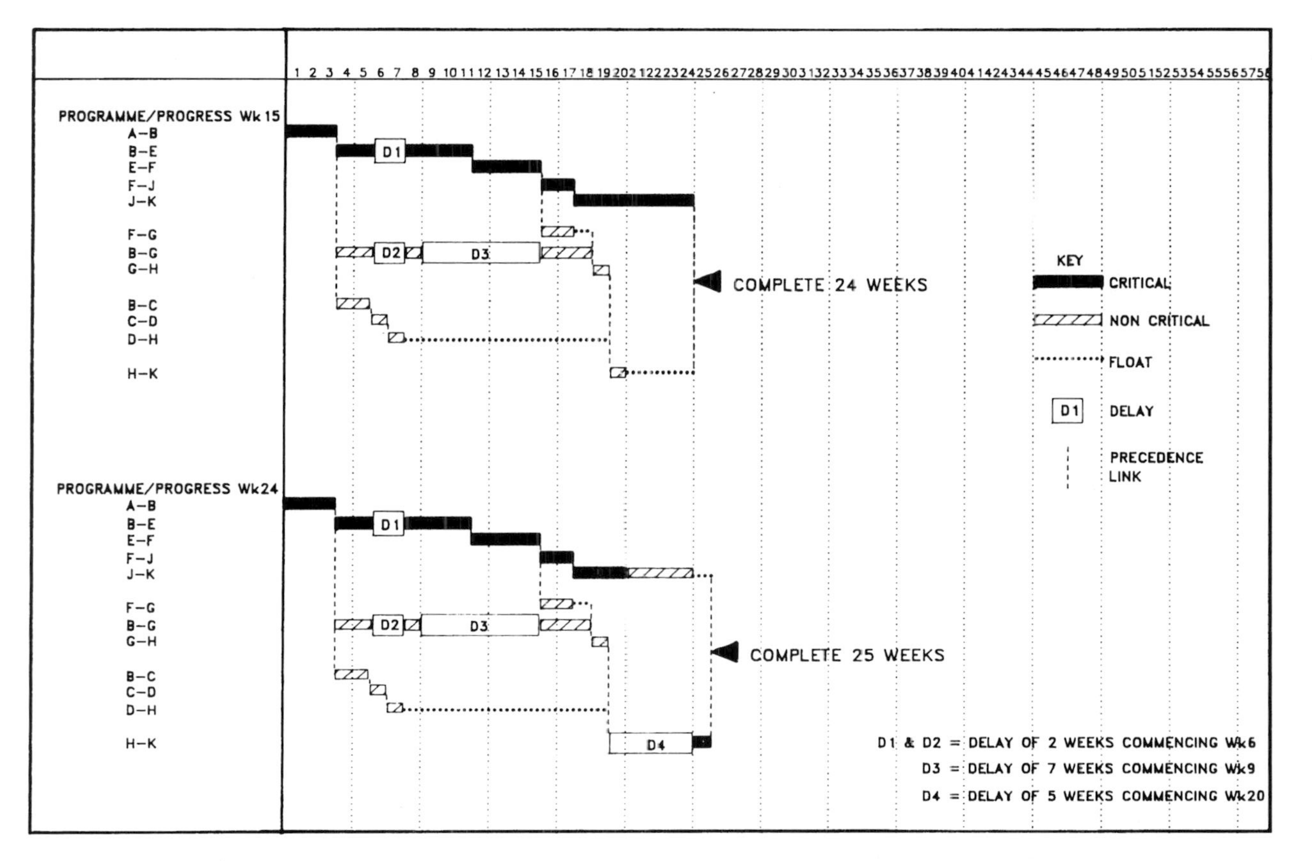

Figure 5.9 *Concurrent delays followed by subsequent non-critical and critical delays*

[Emphasis added]. That is to say, even if an extension of time ought to have been made for previous delays, if the extension has not been made by the (then) current extended completion date, and a new (otherwise qualifying) delay occurs, there is no power to extend time for completion. This situation does not appear to be capable of rectification by subsequently making an extension of time for the previous delay, thereby causing the new delay to occur before the subsequently revised extended completion date.

JCT63 has long been consigned to history in the United Kingdom. However, extensions of time provisions identical to JCT63 are still in everyday use in many parts of the world. Bahrain, Cyprus, Hong Kong and Jamaica are a few examples. Wherever these contracts are in use, it is therefore essential to make extensions of time for all known delays (whether, or not, notified by the contractor) before the existing completion date has passed. Failure to do so may cause time to be at large and invalidate the liquidated damages provisions.

Clause 24 of the form of contract issued by the Singapore Institute of Architects includes provisions to accommodate delays occurring after the date for completion. It is not considered to be necessary to deal with this clause at length in this chapter. However, a diagram showing how the clause is intended to operate is shown in Chapter 4 (see Figure 4.2, *supra*).

The Intermediate Building Contract 2011 expressly provides for extensions of time to be made for delays which occur after the completion date has passed, but there are no rules setting out how this should be done other than 'as soon as he [Architect/Contract Administrator] is able to …' (clause 2.19.2).

These problems are addressed in the following example (see Figure 5.10).

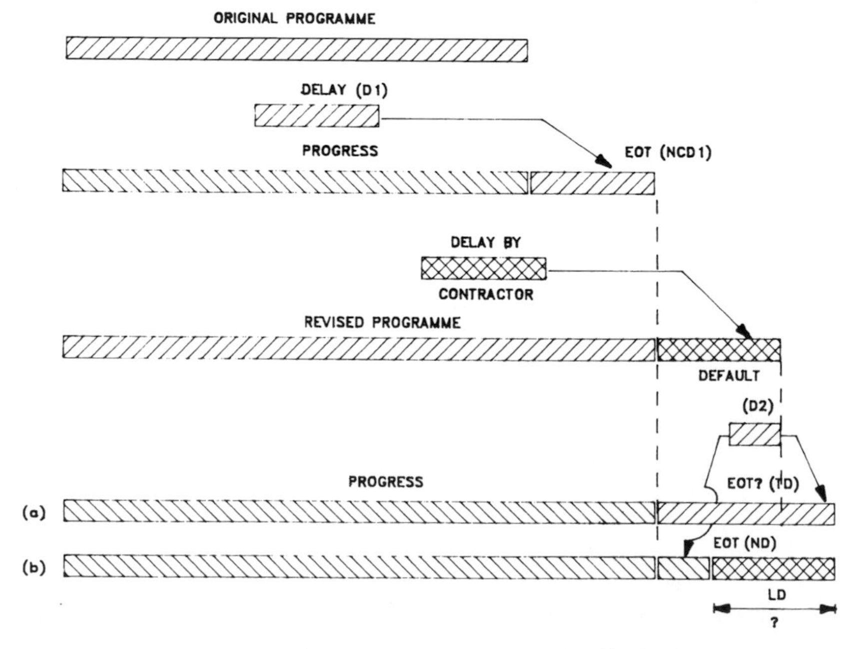

Figure 5.10 Delay by employer after completion date

In this example it can be seen that a delay (D1) which occurs before the contract completion date is capable of being dealt with by an appropriate extension of time. A new completion date (NCD1) can be fixed according to the circumstances.

When a new qualifying delay (D2) occurs after the completion date has passed and the contractor is in culpable delay, what period of delay should qualify for an extension of time? Should it be the total period of delay (TD) from NCD1 to the earliest completion date caused by the new qualifying delay, or should it be for the nett period of the new qualifying delay (ND)? Can liquidated damages be levied?

Consider two possible alternatives:

Alternative A

Eight weeks after the contract completion date, the contractor commences excavation for the final connections to the foul drainage. The work ought to have been carried out not later than two weeks before the completion date. With the exception of delay (D1), there have been no delays for any reason other than the contractor's failure to proceed in accordance with its programme. Unknown existing gas main and power cables are discovered which necessitate a variation to change the routing of the drainage and the construction of an additional inspection chamber. The additional work causes a delay of one week (D2) and completion of the works is delayed by one week.

In these circumstances, had the contractor not been in culpable delay, the necessity for a variation would have come to light before the completion date and an extension could have been made at the time. Therefore, if the contractor had been proceeding in accordance with his programme, one week extension of time (beyond the date already fixed as a result of delay D1 – NCD1) would have been reasonable (ND).

Alternative B

In the same circumstances as Alternative A, eight weeks after the completion date has passed, the contractor is instructed by the architect to cease work on the excavation for the foul drainage. The architect then instructs the contractor to vary the levels and diameter of the pipes and construct an additional inspection chamber and two additional branch connections for a future extension. The additional work causes a delay of one week (D2) and completion of the works is delayed by one week.

In these circumstances, the architect could, and ought to have, ordered the additional work in sufficient time to enable the work to be carried out before the completion date and without causing delay. The variations ordered by the architect were not dependent upon the contractor's progress and could not be attributable to the contractor's culpable delay. If the contract permitted an extension of time for delays which occurred after the completion date had passed, an extension of time for a period of ten weeks may be reasonable in the circumstances (TD).

In *Balfour Beatty Building Ltd* v. *Chestermount Properties Ltd* (1993) 62 BLR 1 it was held that on the wording of clause 25 of the 1980 Edition of JCT (The Joint Contracts Tribunal – JCT80) form of contract, an extension of time granted retrospectively, after the completion date, for delay caused by the employer was valid. This decision seems to have put an end to the uncertainty regarding delays which occur after the completion date has passed and the contractor is in culpable delay. Or has it?

This case is, and its implications are, of sufficient importance for consideration in detail. The facts of the case are summarised below (see also Figure 5.11).

The original agreed works to the core were intended to be completed by 17 April 1989. This date was extended on 11 October 1988 giving a revised completion date of 9 May 1989.

By February 1990, Balfour Beatty were several months late and already liable for liquidated damages. It was agreed that fitting-out works would be carried out by Balfour Beatty and a number of architect's instructions were issued between 12 February 1990 and 12 July 1990. It was agreed that these were variations to the original works.

Balfour Beatty completed the original works (core only) by 12 October 1990 and went on to complete the fitting-out works by 25 February 1991.

The architect granted two further extensions of time to account for the additional fitting-out works. The first, issued on 18 December 1990, extended the previously extended completion date by 126 days (to 12 September 1989) and the second, issued on 14 May 1991, made a further extension of 73 days (to 24 November 1989), that is, before the date of the first variation instruction.

Balfour Beatty's arguments were two-fold:

(1) That the effect of issuing the variation instructions for the fitting-out works rendered time at large, in which case Balfour Beatty were obliged to complete the works within a reasonable time and Chestermount would lose its rights to levy liquidated damages.
(2) Alternatively, if time was not at large, Balfour Beatty were entitled to an extension of time calculated by adding the period required for fitting-out to the date when the additional works were ordered. According to Balfour Beatty, this should have resulted in an extended completion date of 25 February 1991, that is 54 weeks after 12 February 1990 (when the work was actually complete).

Mr Justice Colman did not agree with Balfour Beatty. He held that clause 25.3.3 of JCT80 (the Form of Contract in this case) empowered the architect to grant an

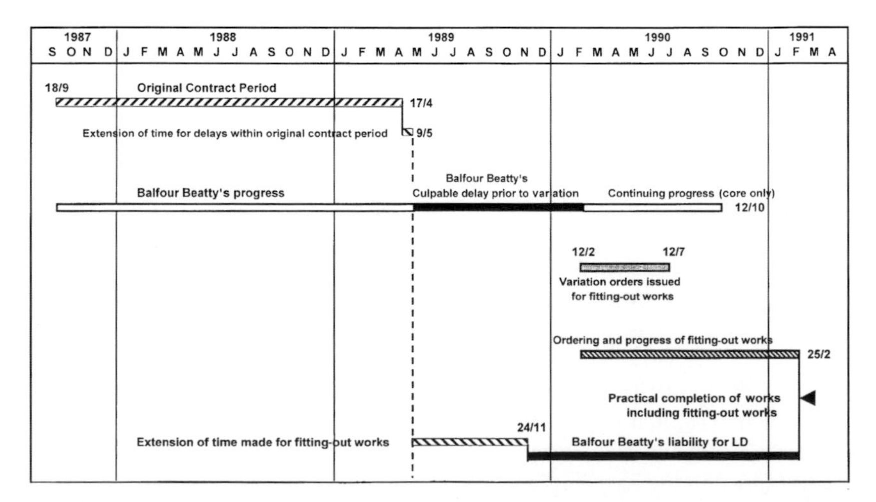

Figure 5.11 *Balfour Beatty Building Ltd v. Chestermount Properties Ltd (1993) 62 BLR 1*

extension of time after the completion date had passed and, when the contractor was in culpable delay, that it was right and proper to add a reasonable period to the previously extended completion date (of 9 May 1989) and that the final extended completion date of 24 November 1989 was reasonable even though that date was before the date when the additional works were ordered.

How does this judgement affect other forms of contract?

The Intermediate Building Contract 2011 provides for extensions of time to be granted for any qualifying causes (relevant events) occurring after the date for completion (clause 2.19.2). Many engineering contracts also contemplate extensions of time for delays after the completion date. Contracts such as ICE and FIDIC contemplate extensions of time which are fair and reasonable. The seventh edition of the ICE conditions also contain novel provisions regarding liquidated damages (*infra*).

The Balfour Beatty case left some issues unanswered.

While some modifications were made to the standard form of contract (it was said that these were of no consequence to the issue), there is no explanation as to why the completion date was originally 17 April 1989 and why the first extension of time was added to that date when, on 16 June 1988, the parties signed a contract which expressly included a date for completion of 16 June 1989. The extension of time clause in JCT80 contemplates extensions of time if completion is delayed beyond the completion date. It appears that the first extension of time made on 10 October 1988 (giving a new completion date of 9 May 1989) did not extend the completion date given in the appendix to the contract.

The second extension of time made on 18 December 1990 (giving a new completion date of 12 September 1989) appears to be outside the powers given to the architect in clause 25.3.1. However, this may be a sterile argument on the grounds that the architect could subsequently make a valid extension under clause 25.3.3.

Why did Balfour Beatty not ask for an acceptable extension of time before agreeing to carry out the extra works? (See *Fairclough Building Ltd* v. *Rhuddlan Borough Council* in 6.3, *infra*). Perhaps Balfour Beatty thought that instructions to carry out the extra works would get them off the hook for damages of any sort.

Why was it fair and reasonable in all the circumstances to grant an extension of time to 24 November 1989 (before the date of the instruction for the extra works) when it was clearly impossible for Balfour Beatty to complete the works until several months after the issue of the variation instruction?

With respect to the last of the above questions, close examination of the facts, as shown in Figure 5.11, indicates that it was highly likely that the commencement of fitting-out works was dependent upon Balfour Beatty completing a substantial part of the original core works. Chestermount may have been justified in wondering whether or not Balfour Beatty would ever finish. Had Balfour Beatty been proceeding according to programme and heading for completion by 9 May 1989 (the first extended completion date), Chestermount could have ordered the fitting-out works by (say) December 1988 and works could have been completed by 24 November 1989 (see Figure 5.12). If that was the case, the architect was fair and reasonable in making the extension to 24 November 1989. (This analysis and explanation is consistent with Alternative A above.)

However, what would have been the verdict if the cause of the delay (occurring after the completion date) had been independent of Balfour Beatty's progress as described in Alternative B above? Assume that on 12 February 1990, it was discovered

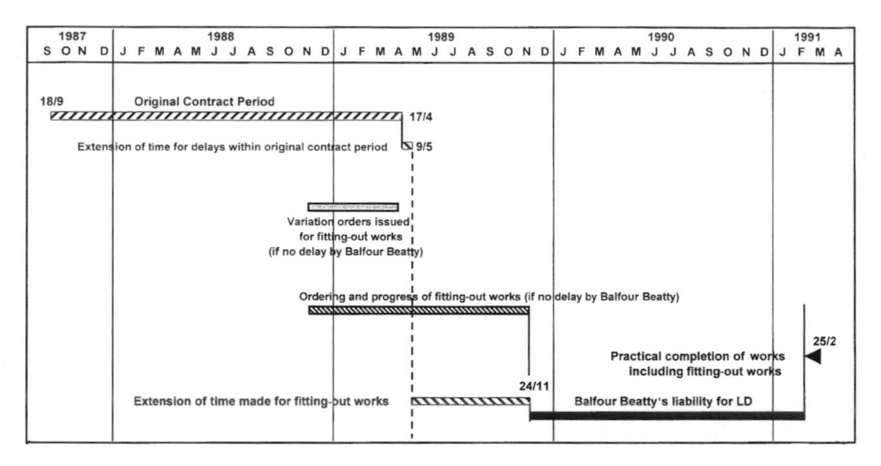

Figure 5.12 *Balfour Beatty Building Ltd v. Chestermount Properties Ltd (1993) 62 BLR 1: possible effect of variation to add fitting-out works if no delay by Balfour Beatty (explanation of extension of time to 24 November 1989)*

that the design of the structural core was defective on all floors. Discovery of the defect was not dependent upon Balfour Beatty's progress. Remedial works would have to be designed. Demolition of part of the completed works would be required and no further work on the remaining (incomplete) structure could proceed until the remedial works were complete. In these circumstances, perhaps it would be reasonable to make an extension of time on a gross basis, that is, until the date when the remedial works and remaining works could reasonably be expected to be complete. If a gross extension was not made, Chestermount would, in these circumstances, benefit from its own default by being able to levy liquidated damages during a period when the building could not, in any event, have been completed because of the design fault.

What would be the situation, if, for example, a strike had occurred after the completion date had passed? Would it have been reasonable to argue that if Balfour Beatty had completed on time, the works would not have been delayed by the strike (hence, no extension)? Imagine the difficulty if numerous delays such as strikes, adverse weather conditions and extra works were affecting the progress of the works after the completion date had passed.

It would seem, therefore, that each case may have to be looked at on its merits. The Balfour Beatty case dealt with a single delaying matter by the employer concurrent with Balfour Beatty's own delay. In practice, many delays may occur after the completion date. It may not be correct to rely on the decision in *Balfour Beatty* v. *Chestermount* to justify variations or late issuance of information after the completion date has passed. All that this case appears to do is set out what should be decided in the particular circumstances which arose in connection with this contract.

What are the alternatives?

It can be seen from Figure 5.13 that Balfour Beatty were ultimately given an extension of time to 24 November 1989 and were liable for liquidated damages after that date until completion on 25 February 1991.

Under the ICE conditions of contract, clause 47(6) provides for liquidated damages to be suspended in the event of such a delay occurring after the completion

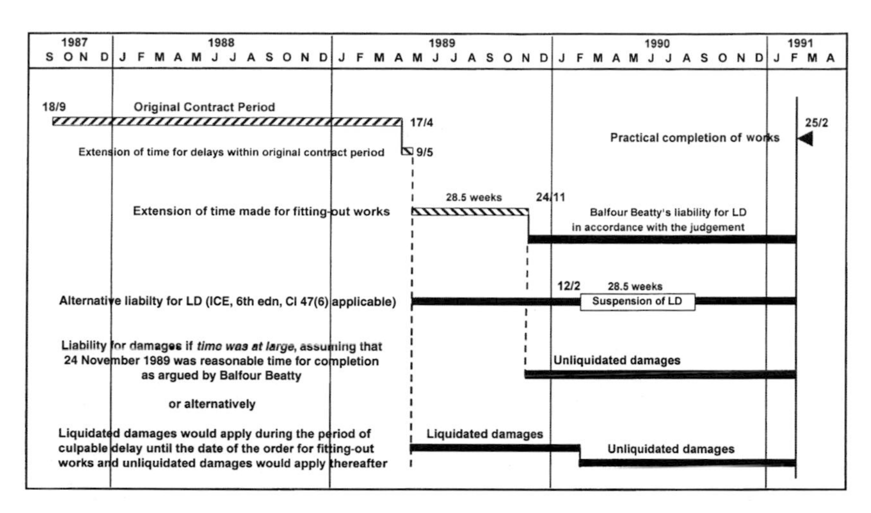

***Figure* 5.13** *Balfour Beatty Building Ltd v. Chestermount Properties Ltd (1993) 62 BLR 1: comparison of various liabilities for damages for late completion if extended date for completion (24 November 1989) is a reasonable date*

date when the contractor is already liable for liquidated damages. Therefore, had this form of contract applied, Balfour Beatty would have been liable for liquidated damages until 12 February 1990, after which its liability would be suspended for the period of delay (in this case twenty-eight and a half weeks). Thereafter, liquidated damages would continue until completion. The nett result would have been identical to that which arose using JCT80 (and would probably be the same under the JCT Standard Building Contract 2011).

Had Balfour Beatty been successful in arguing that time was at large and that Chestermount had lost its rights to liquidated damages, one of the following alternatives may have applied:

(1) Liquidated damages would be payable from 10 May 1989 to 12 February 1990 and general damages (damages which Chestermount could prove flowed from Balfour Beatty's default) would be payable thereafter, *or*
(2) General damages would be payable for late completion calculated from 10 May 1989.

If either of the alternatives (1) or (2) above applied, Balfour Beatty may have been no better off (or indeed worse off). There is at least the possibility that general damages could exceed liquidated damages: *Rapid Building Group Ltd* v. *Ealing Family Housing Association Ltd* (see 1.4, *supra*).

Often contractors argue that time is 'at large' meaning that there is no contractual time or date for completion and that they have a reasonable time to complete. It has often been seen as an easy option coupled with the argument that a reasonable time to complete was the time the contractor took to complete. In the case of *Multiplex Constructions (UK) Ltd* v. *Honeywell Control Systems Ltd* [2007] 2007 BLR 195, Honeywell successfully argued before an adjudicator that time was at large due to delay for which Multiplex were liable under the subcontract even though Honeywell had failed to comply with a condition precedent requiring them to issue a notice of delay. The matter was referred to the High Court and the court held that Honeywell's failure to comply with the notice provisions did not put time 'at large'. Despite the fact that Honeywell had grounds for an extension of time they were not entitled to an extension of time because they had failed to comply with the contractual machinery requiring them to give notice of delay.

Under the ICE contract, liquidated damages are suspended for the period necessary to complete the variation (clause 24). However, it appears that the suspension only comes into effect if the contractor is not also in default during the same period (see 4.10, *supra*). In the circumstances of this case, Balfour Beatty may have been liable for liquidated damages for the entire period from 10 May 1989 until completion on 25 February 1991, in spite of the fact that Chestermount ordered extra works requiring twenty-eight and a half weeks to execute.

Under the NEC, the matter appears to be simply dealt with in clause 63.3 (see 5.2, *supra*).

Under the NEC, there may be problems if the programme is not properly and regularly updated or if programmes are not accepted by the project manager.

5.4 Summary on presentation of extensions of time claims

In any claim for an extension of time, and whether or not there is a requirement to give details and particulars, it is good practice to include the following:

- a description of the cause of delay and the contractual provision which is being relied upon for the extension;
- the date when the delay commenced and the period of delay (giving details of intermittent effects, if appropriate);
- the date of notice of delay, specifying the reference of the relevant document;
- a summary of records and particulars relied upon (with copies included in an appendix);
- a narrative of the events and effects on progress;
- a diagrammatic illustration showing the status of the programme, progress and current completion date prior to the commencement of the delay;
- a diagrammatic illustration showing the effects of the delay on progress and the completion date (including subsequent delays which may have reduced the float in the programme);
- a statement requesting an extension of time for the delay to completion for the period shown on the submitted illustrations.

5.5 Recovery of loss and/or expense and/or damages

While failure to give notice of delay for extensions of time may not be fatal to a claim, failure to give notice in accordance with the contract with respect to additional payment may bar, or severely prejudice a claim.

There are good reasons for contracts to have provisions for the contractor to give notice. No employer will wish to have a substantial claim appearing 'out of the blue' at the end of a contract. In *J. and J.C. Abrahams* v. *Ancliffe* [1938] 2 NZLR 420, a contractor estimated the cost of building two residential units at $30000. Several months later the employer's architect issued a specification for the work and the contractor commenced work. It became evident that the specification provided for more expensive work than that which had been allowed for in the contractor's estimate. There were also problems in the foundations which increased the amount of work done and general building costs were escalating. The employer repeatedly asked the contractor for details of the expected costs but at no time did the contractor reply. When it came to settle the account, the employer argued that the contractor was in breach of a duty to give reliable information about the costs of building before the employer became committed to completing the units at an uneconomic cost. It was held that the contractor was under a duty of care to the employer in giving its original estimate and to inform the employer as soon as it was aware that costs were going to substantially exceed the estimate.

In most forms of contract, the onus is not entirely upon the contractor to keep the employer informed of increases in the contract price. In most instances, the employer relies to a great extent on his professional advisers. In varying degrees (according to the terms of the contract) there must be co-operation between the employer's professional advisers and the contractor so that any increase in the contract price can be ascertained at the earliest possible time: *London Borough of Merton* v. *Stanley Hugh Leach Ltd* (*supra*). Where there are no express terms, co-operation is usually implied. Most construction contracts have express provisions making it clear as to what form this co-operation should take.

In the UK, contractors may normally seek remedies under the common law in addition to, or alternatively to, rights under the contract: *London Borough of Merton* v. *Stanley Hugh Leach Ltd* (*supra*). However, under MF/1 (revision 3), no such alternative remedy is available since the contract excludes the contractor's rights under the general law (sub-clause 44.4). That is to say, the contractor's rights are limited to the rights set out in the contract: *Strachan & Henshaw Limited* v. *Stein Industrie (UK) Limited* (1998) 87 BLR 52. In some countries, it may not be possible to exclude rights under the general law.

Exclusion clauses

It should be noted that if there are no remedies for breach set out in the contract, or if a contractual remedy limits liability for breach of contract, a clause purporting to exclude liability may not be effective in the UK (an exclusion clause). In *George Mitchell (Chester Hall) Ltd* v. *Finney Lock Seeds Ltd* (1983) 1-CLD-05-18, it was held that a clause which limited the seller's liability to the costs of cabbage seed in the event of failure of the crop could not prevent the buyer from succeeding in a claim for full damages in the event of the crop being of no commercial value.

Similar provisions in construction contracts have arisen. In *Miller* v. *London County Council* (1934) 151 LT 425, the contract provided that there should be no allowance in respect of money, time or otherwise, other than such extensions of time as may be given. It was held, *obiter* (Du Parq J), that the clause did not include delay due to extras or interference by the employer or persons for whom the employer was responsible, that is the contractor may be entitled to compensation if the employer causes delay (whatever the clause says).

In some US jurisdictions 'no damages for delay' clauses are enforceable.

In *Hudson's Building and Engineering Contracts, eleventh edition*, the author writes at page 1101:

'Clauses of this kind would appear to be prime candidates for avoidance under the English Unfair Contract Terms Act 1976 [*sic*, 1977] or similar legislation elsewhere.'

Most civil law jurisdictions expressly prohibit contractual provisions which attempt to bar a remedy for breach of contract. For example, Section 373 of the Civil Code of Thailand states:

'An agreement made in advance exonerating a debtor from his own fraud or gross negligence is void.'

Under South African law, nothing prevents an employer contracting out of the consequences of his own breach. For example, extension of time clauses frequently provide for an extension of time but no monetary compensation. Where the extension of time arises from the employer's breach, such as failure to grant possession of site, the contractor would be entitled to the relevant time but nothing further.

5.6 Notice of intention to claim financial recompense

Most contractors do give notice of their intention to claim at some time during the contract. Some avoid any indication at all of their intention to claim financial recompense until after an extension of time has been made. The former may barely comply with the contract and may prejudice the contractors' entitlements to some extent or may have disastrous consequences if the principles established in *Steria* v. *Sigma Wireless* and *Multiplex* v. *Honeywell* (at paragraph 5.2, *supra*) are applied to financial claims. The latter will invariably be the beginning of an uphill struggle to obtain payment of substantially less (if anything at all) than might otherwise have been possible if the contractor had given prompt notice. Notice provisions in modern construction contracts vary considerably:

- JCT Standard Building Contract 2011 – Clause 4.23 merely requires the contractor to make an application '...as soon as it has become, or should reasonably have become, apparent to him that the regular progress has been or is likely to be affected [by the matters referred to]....' It may be difficult to decide whether or not an application is late in all the circumstances.
- NEC 3 – clause 61 requires notice of a Compensation Event to be provided by a contractor within 8 weeks of the occurrence of the event.
- GC/Works/1 (1998) – Clause 46(3) states that 'The Contract Sum *shall not be increased under paragraph (1) unless*, (a) the Contractor, immediately upon becoming aware that the regular progress of the Works or any part of them has been or is likely to be disrupted or prolonged, has given notice to the [Project Manager] specifying the circumstances causing or expected to cause that disruption or prolongation and stating that he is, or expects to be, entitled to an increase in the Contract Sum.'
- ICE seventh edition – Clause 53(2) requires the contractor to '... give notice in writing of his intention [to claim] to the Engineer as soon as may be reasonable and in any event within 28 days after the happening of the events giving rise to the claim.'
- 1999 FIDIC Red, Yellow and Silver Books – Clause 20.1 requires a notice within twenty-eight days. The giving of a notice within the stipulated period is a condition precedent to the contractor's rights to claim.

- MF/1 (revision 3) – Clause 41.1(a) requires a notice within thirty days, failing which the claim is time-barred.

5.7 Particulars and further information to support a claim

If proper notice has been given pursuant to the terms of the contract, both parties are aware of the claim and further steps can be taken to deal with it. Various provisions include

- JCT Standard Building Contract 2011 – *upon request*, the contractor *shall* submit appropriate information for the purposes of enabling the Architect/Contract Administrator to form an opinion as to whether or not the contractor *has incurred or is likely to incur* direct loss and/or expense (clause 4.23.2) and *upon request*, the contractor is required to provide details of the loss and/or expense (clause 4.23.3). No time limits are specified for the Architect's/Contract Administrator's or Quantity Surveyor's requests or for the Contractor's response, other than the Contractor is to provide the information '*upon request*'.
- NEC 3 – Under clause 62 the Project Manager may require the contractor to provide a quotation for a Compensation Event within 3 weeks of the request.
- GC/Works/1 (1998) – The Contract Sum *shall not be increased under paragraph (1) unless* '(b) the Contractor, as soon as reasonably practicable, *and in any case within 56 Days of incurring the expense*, provides to the QS full details of all expenses incurred and evidence that the expenses directly result from the occurrence of one of the events...' (clause 46(3)).
- ICE seventh edition – Requires the contractor to give a first interim account and details as soon as is reasonable in all the circumstances and thereafter further accounts at such intervals as the engineer may reasonably require (clause 53(4)).
- 1999 FIDIC Red, Yellow and Silver Books require particulars and accounts to be submitted within forty-two days but failure to comply will not bar the claim (clause 20.1).

It appears that, with the exception of GC/Works/1, there is no bar to a claim provided that notice and particulars are given within a reasonable time. MF/1 and the 1999 FIDIC Red, Yellow and Silver Books bar a claim if the initial notice is not given within the prescribed time, but failure to provide particulars and accounts on time will not be fatal to the claim. However, in the light of the *Steria* v. *Sigma Wireless* judgement it remains to be seen if the courts are taking a less lenient approach to the interpretation of notice provisions.

Notwithstanding the loose provisions which appear to prevail, contractors are advised to give prompt notice followed by detailed particulars backed up by adequate contemporary records.

The methods of illustrating delay and disruption in support of claims for additional payment are similar to those used for illustrating claims for extensions of time.

5.8 Prolongation claims

Qualifying delays on the critical path will usually support a claim for prolongation costs for the period of delay (if such delays are matters which give rise to additional payment). For the purposes of claims for additional payment, the term 'qualifying delay' means delay which brings with it the right to additional payment (some qualifying delays for extensions of time, such as adverse weather conditions, do not normally give rise to additional payment). Typical heads of claim arising out of prolongation of the contract period are:

Site overheads or preliminaries

It is surprising how many claims are submitted on the basis that the extra site overhead costs due to prolongation are those incurred after the original contract completion date and up to the extended (or actual) completion date. This is, of course, incorrect, but it may explain why some contractors wait until the end of the project to give notice and submit a claim. The following example illustrates how prolongation costs may be significantly understated using the above assumption.

The qualifying delay on the critical path (D1) shown in Example 1 (see Figure 5.3) has caused the completion date to be delayed by two weeks. The actual weekly costs of the contractor's general site establishment (time-related costs) are shown in Figure 5.14.

It will be seen that the weekly costs incurred during the two-week period of overrun (CD) are much lower than the weekly costs during the period of delay (CO). It is the cost incurred during the period of delay which should be the basis of the contractor's claim for prolongation costs. A claim based on the costs incurred during the period of overrun will normally be substantially less than the actual costs incurred during the period of the delay.

The costs incurred during the period of delay may not reflect the true *additional* costs of the delay. For example, the contractor may have recruited an electrical engineer to commence on site in the ninth week to supervise the electrical installation. There may be no other site at which the engineer can be usefully employed and it may not be possible to postpone his employment. The delay may have caused the commencement of the electrical installation to be delayed by two weeks, in which case the contractor is faced with paying the salary of the engineer for two weeks (weeks nine and ten) when there is no work being done which requires the engineer's supervision. This additional cost is a direct result of the qualifying delay and ought to be recoverable. However, the cost of the engineer is not included in the costs incurred in weeks six and seven (the period of delay). In order to overcome such problems, the contractor should show the periods when every time-related resource was on site (and their costs) and when they ought to have been on site (save for the delay) – see Figure 5.15.

In practice, some qualifying delays may occur in isolation (as in the previous example) and/or numerous qualifying delays may occur over a period in which each qualifying delay overlaps with other qualifying delays. The nett result of all of the qualifying delays may cause prolongation of the contract period. Provided that there are no major concurrent delays by the contractor (which would be a matter of evidence) it may be reasonable to base a claim for prolongation costs on the costs shown in Figure 5.16.

In the above example, the cost of the isolated delay (A) may be established using similar principles as those in the previous example. The costs arising out of the numerous continuing delays during the period (B) may be taken as four-tenths of the total costs incurred during period (B). Some adjustments may have to be made for special circumstances such as the case of the electrical engineer used in the previous

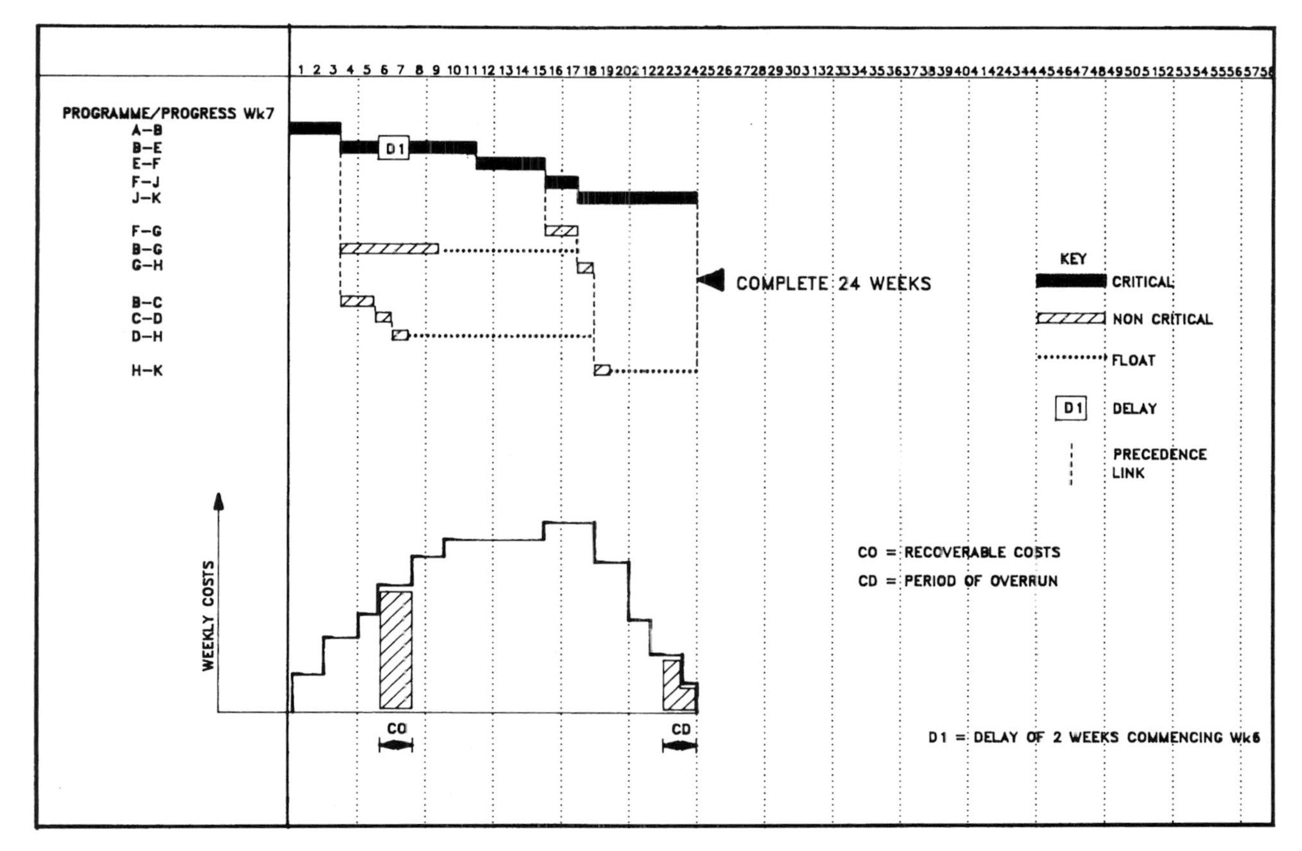

Figure 5.14 Recoverable site overheads

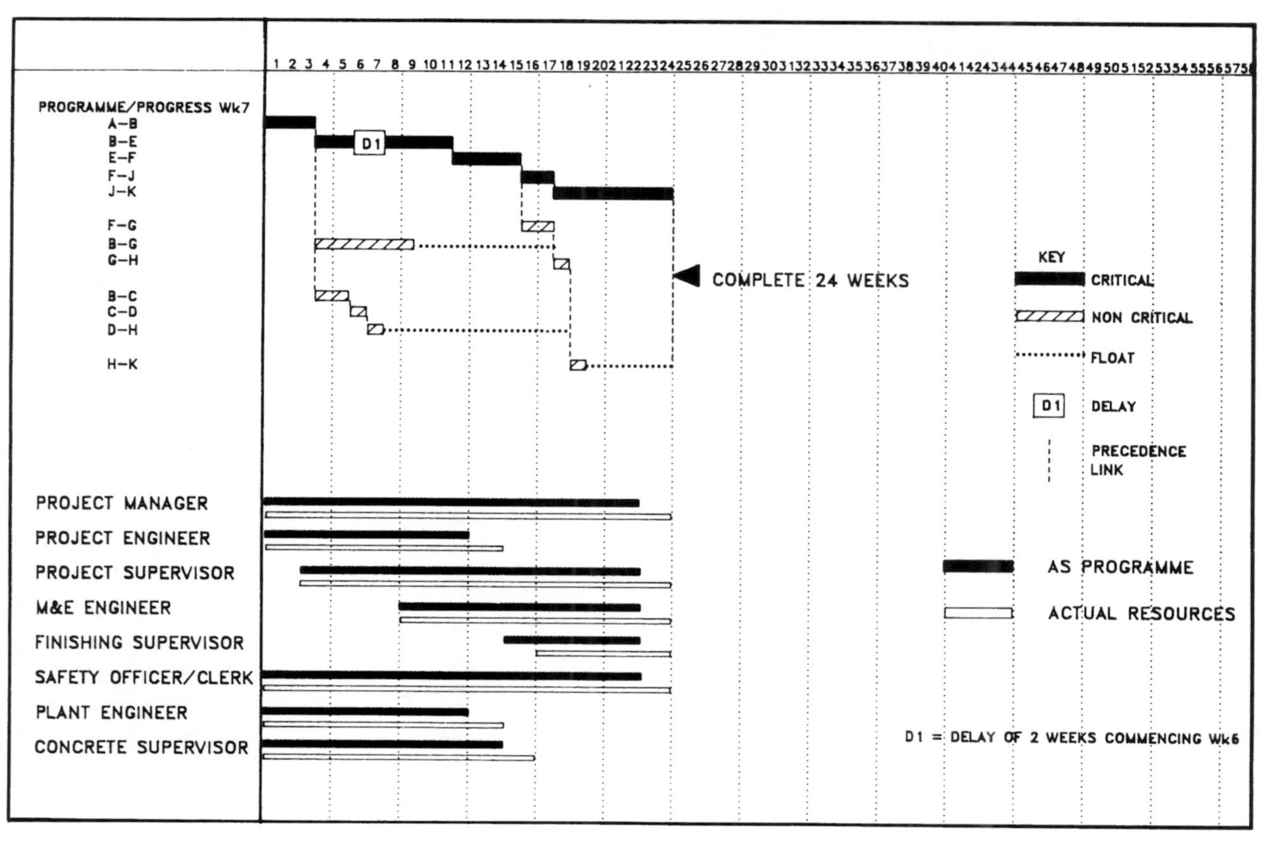

Figure 5.15 *Anticipated and actual resources*

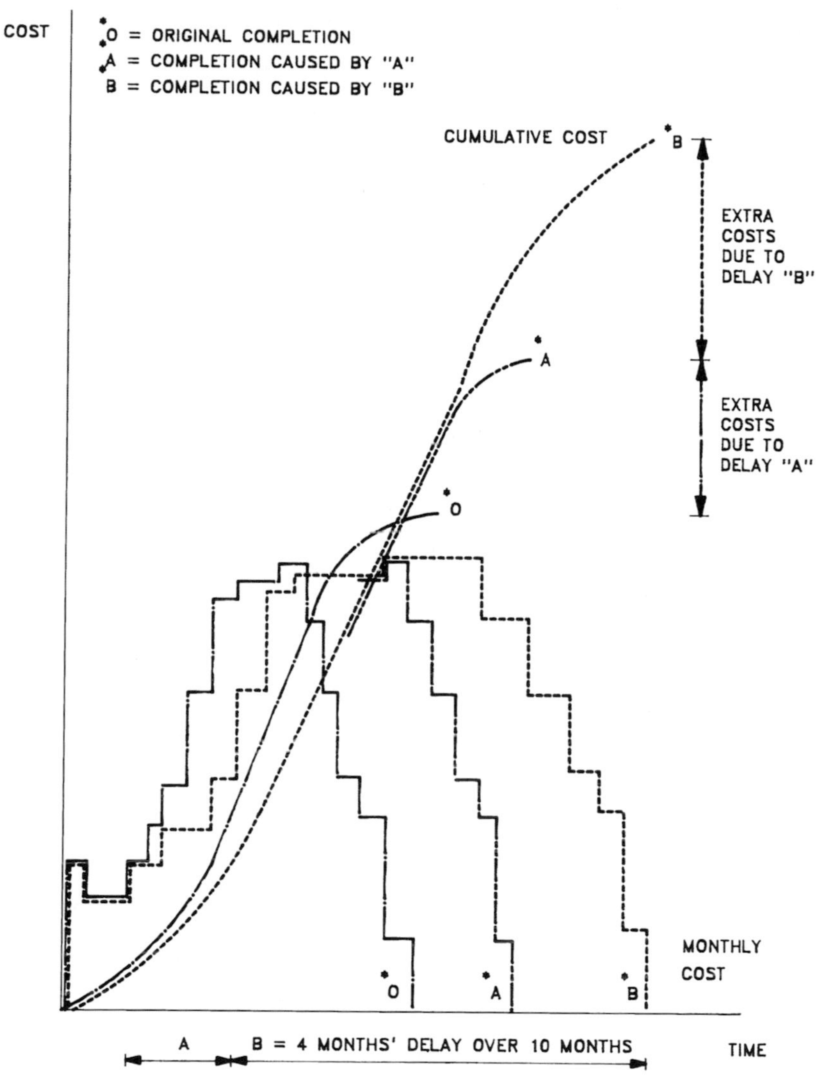

Figure 5.16 *Extended preliminaries*

example. Alternatively, comparison between the resources which were utilised on site and the resources which ought to have been utilised (save for the delay) may give a more accurate result.

In any event, it is not the comparison between the actual resources and those included in the contractor's tender which form the basis of the claim. If the contractor can show that it was reasonable and necessary to employ more weekly resources than those allowed in the tender he may be able to claim on the basis of the increased resources. However, if there was no good reason to employ additional resources, the contractor's claim may be limited to the costs of resources which were consistent with the contractor's tender assumptions. If the contractor's actual resources were less than the tender provisions, the employer would not expect to reimburse the contractor any more than the actual costs incurred.

Prolongation of individual activities

Some delays may not be on the critical path, in which case there will be no general prolongation costs. However, some time-related costs may be solely attributable to a particular activity. If delay (D2) in Example 2 (see Figure 5.4) is in respect of an activity which requires scaffolding for its total duration, then the cost of the scaffolding for the period of the qualifying delay of two weeks would be recoverable. Supervision and other plant and equipment utilised solely for the activity may also be recoverable. This is particularly valid where the activity is for work carried out by a subcontractor. The subcontractor will have a prolongation claim against the contractor and the contractor will seek reimbursement under the relevant provisions of the principal contract.

Valuation at cost or using contract rates for preliminaries

If the delay was caused solely by a variation, it could be argued that the valuation of the variation should take into account the time-related rates in the contract bills (see Variations, *infra*). Account would have to be taken of significant changes in actual costs when compared with the time-related rates in the contract bills. If the delay was caused by breaches of contract, such as late issuance of drawings and details, the remedy is by way of damages, thereby requiring the loss to be based on the contractor's actual costs irrespective of the contract rates. If the delay was caused by variations and breaches of contract, and the periods of delay for each cause cannot be disentangled, it is suggested that actual costs should be used as the basis of any claim.

Head office overheads in the event of prolongation

Various formulae may be used. However, some doubt was cast upon the use of a formula in *Tate & Lyle Food Distribution Ltd and Another* v. *Greater London Council* [1982] 1 WLR 149. It should be noted that in this case very little evidence (if any) was put forward to establish the extent of disruption and delay and there was no evidence presented to support the percentage claimed. The *Tate & Lyle* case did not establish a general rule of law that formulae should not be used and there are many judgements which refer to and rely on the use of formulae.

The logic behind the use of a formula is shown in Figure 5.17. Line a–a represents the contractor's anticipated or actual head office overheads (depending upon the formula used). Line b–b represents the contractor's anticipated turnover on all projects. Profile c–c represents the contractor's anticipated turnover on the present project. Profile d–d represents the contractor's actual turnover on the present (delayed) project. Profile e–e represents the contractor's actual turnover on all projects.

It will be seen that the delay has caused the actual turnover on the project (d–d) in the early months of the project to be considerably less than would have been the case if there had been no delay. Accordingly, the total actual turnover (e–e) has fallen below anticipated level (b–b). During the latter months of the project, the actual turnover on the present project (d–d) continues during the period of prolongation (making up for the shortfall in the earlier months). In theory, the actual turnover on all projects during the period of prolongation should increase (see x–x) because the

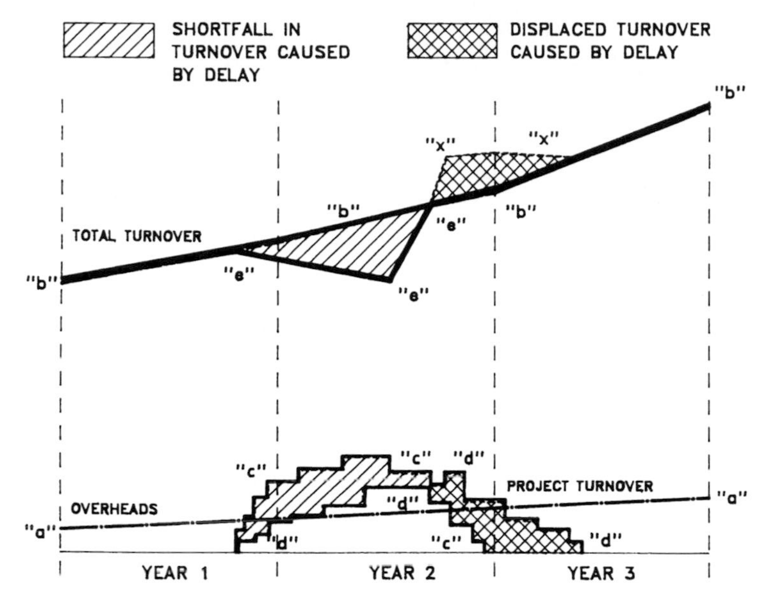

Figure 5.17 Overheads and turnover

turnover on the delayed project in the latter months was not included in the planned turnover for the same period. However, this increase can only be achieved if the resources on the present delayed project can be released to generate more work on a new project. Unless the contractor can take on more resources, it will have to forego new work which it could otherwise have taken. Therefore, as a result of the shortfall in turnover during the delay, the contractor is unable to recover sufficient overheads from the delayed project to make the requisite contribution to its total overheads.

In *St Modwen Development Ltd* v. *Bowmer and Kirkland* [1996] 38 BLISS 4 the arbitrator awarded head office overheads based upon a formula method of recovery. The employer appealed, not with respect to the formula itself, but on the basis that no evidence had been presented to prove that the contractor was unable to use his head office resources elsewhere during the period of prolongation to generate overheads and profit as a result of the delay.

The Court appears to have been influenced by *Hudson's Building and Engineering Contracts, tenth edition*:

> 'However, it is vital to appreciate that both these formulae (Hudson and Eichleay) were evolved during the 1960's at a time of high economic activity in construction. Both assume the existence of a favourable market where an adequate profit and fixed overhead percentage will be available to be earned during the delay period. Both also very importantly, assume an element of constraint – that is to say that the contractor's resources (principally of working capital and key personnel, it is suggested) will be limited or stretched, so that he will be unable to take on work elsewhere.'

The Court rejected the appeal on the grounds that both expert and evidence of fact had been heard on which the arbitrator was entitled to base his award.

In *Amec Building Ltd* v. *Cadmus Investment Co Ltd* [1997] 51 ConLR 105, Mr Recorder Kallipetis QC had this to say:

> '[I]t is for the plaintiff to demonstrate that he has suffered the loss he is seeking to recover … [and] … this proof must include the keeping of some form of record that the time was excessive and their attention was diverted in such a way that loss was incurred … [and he must] … place some evidence before the Court that there was other work available which, but for the delay, he would have secured … thus he is able to demonstrate that he would have recouped his overheads from those other contracts and, thus, is entitled to an extra payment in respect of any delay period awarded in the instant contract.'

It follows that in order to succeed with delay claims involving loss of overheads (and profit) using a formula, the contractor must be able to show that

(1) the anticipated turnover was adversely affected by the delayed project, *and*
(2) he was prevented from earning a contribution to overheads (and profit) as a result of the delay (see possible methods under 'profit', *infra*).

The various formulae used will enable the contractor to calculate the loss of contribution to its head office overheads as a result of the delay. As the contractor has been unable to release his resources to earn the contribution to overheads on another project, he must earn a similar contribution by making a claim on the delayed project.

It will not normally be necessary for the contractor to submit a graphical representation of its turnover and overheads in the above manner as the use of formulae are well known. Where there is resistance to the use of a formula, illustrations using actual data may be persuasive.

However, when a project goes seriously wrong, the use of a formula may produce a substantial underestimate of the costs of prolongation. A contractor may have to increase the time spent by its managerial and supervisory staff of its head office to cope with the particular problems of the project. Numerous variations and other delaying matters may place greater demands on managerial staff including purchasing, planning, costing, quantity surveying and administration staff. It may be necessary to place a director, in a full time role, to deal with the overall management of the project (where none would have been necessary if the project had gone according to plan).

Before leaving overheads, it is worthwhile considering the different circumstances between the *Tate & Lyle* case and those cases where a formula was accepted as a fair means of calculating overheads to be reimbursed.

In the *Tate & Lyle* case, the court was considering the cost of managerial time spent on work done to remedy an actionable wrong. It had nothing to do with a delayed project. In the cases which approved the use of a formula, the courts were concerned not only with the cost of managing a project which was delayed, but they were also considering the *loss of productivity* (loss of contribution) of the contractor's overhead resources. That is to say, because of the delay, the managerial time could not be used to earn the required contribution to overheads on the delayed project, nor could it be used to earn the required contribution from other existing projects (as this would mean recovering additional expense from other employers who were not in default) or additional projects (which could not be undertaken on account of key resources being retained on the delayed project).

It is thought that where a contractor can show evidence of delay, and the extent of it, and where there is evidence to support the contention that resources were prevented from earning a contribution to overheads and the percentage to be used, then one of the recognised formulae may be used. In the case of *Property and Land Contractors Limited* v. *Alfred McAlpine Homes North Limited* [1995] 47CLR74 HHJ Humphrey Lloyd QC stated (at page 81) that

> 'the *Emden* formula, in common with the *Hudson* formula ... and with its American counterpart the Eichlay formula, is dependent on various assumptions which are not always present and which, if not present, will not justify the use of a formula. For example the *Hudson* formula makes it clear that an element of constraint is required (see Hudson para 8.185) ie in relation to profit, that there was profit capable of being earned elsewhere and there was no change on the market thereby affecting profitability of the work. It must also be established that the contractor was unable to deploy resources elsewhere and had no possibility of recovering cost of the overheads from other sources, eg from an increased volume of work. Thus such formulae are likely only to be of value if the event causing delay is (or has the characteristics of) a breach of contract.'

The Hudson formula

This formula was put forward in *Hudson's Building and Engineering Contracts, tenth edition*, 1970 (page 599). It is set out at paragraph 8.182 of the eleventh edition but is only referred to in the twelfth edition (at para. 6.072). It uses the percentage in the contractor's tender for overheads (and profit, if applicable) as a basis for the contractor's loss of contribution to overheads (profit), as a result of delay, in the following formula:

$$\frac{\text{Head office overheads (profit)\%}}{100} \times \frac{\text{Contract sum}}{\text{Contract period}} \times \text{Period of delay}$$

Hudson's formula found favour with the judge in *Ellis-Don* v. *Parking Authority of Toronto* (1978) 28 BLR 98. In this case, the judge stated that neither counsel before him had been able to think of a better approach.

The *Hudson* formula was also referred to in the Scottish case of *Beechwood Development Company (Scotland) Limited* v. *Stuart Mitchell t/a Discovery Land Surveys* [2001].

Emden's formula

This formula can be found in *Emden's Building Contracts and Practice, eighth edition, Volume 2* (page N/46) by Bickford-Smith. The formula is identical to the *Hudson* formula, save that the head office overheads percentage (and profit) used in the formula is the actual percentage based on the contractor's accounts and is arrived at as follows:

$$\text{Head office overheads (profit)\%} = \frac{\text{Total overhead cost (Profit)}}{\text{Total turnover}} \times 100$$

Emden's formula was approved in the case of *Whittall Builders Company Ltd* v. *Chester-le-Street District Council* (1985) unreported. The judge clearly stated the principles behind *Emden's* formula as follows:

> 'What has to be calculated here is the contribution to off-site overheads and profit which the contractor might reasonably have expected to earn with these resources if not deprived of them. The percentage to be taken for overheads and

> profits for this purpose is not therefore the percentage allowed by the contractor in compiling the price for this particular contract, which may have been larger or smaller than his usual percentage, and may not have been realised. It is not that percentage (i.e. the tendered percentage) that one has to take for this purpose but the average percentage earned by the contractor on his turnover as shown by the contractor's accounts.'

In *J.F. Finnegan* v. *Sheffield City Council* (1989) 43 BLR 124, the judge endorsed *Emden's* formula as follows:

> 'I infinitely prefer the Hudson Formula which in my judgement is the right one to apply in this case, that is to say, overhead and profit percentage based upon fair annual average, multiplied by the contracts sum and the period of delay in weeks, divided by the contract period.'

Note – The judge referred to the *Hudson* formula, when in fact it ought to have been *Emden's* formula.

The *Emden* formula was also approved by HHJ Anthony Thornton QC in the unreported case of *CFW Architects (a firm)* v. *Cowlin Construction Limited* [2006].

Eichleay's formula

A similar formula to *Emden's* formula was developed by *Eichleay* in the United States in *The Appeal of Eichleay Corporation*, ASBCA 5183, 60–2 BCA (CCH) 2688 (1960) and this has found approval in the US courts: *Capital Electric Company* v. *United States* (*infra*). This formula uses the actual overheads (and profit) in a similar manner to *Emden*, but the total value of all certificates (the final contract price, including remeasurement and variations) is inserted in lieu of the contract sum.

Profit

The principles behind a claim for loss of profit arising out of a delayed contract are similar to those applicable to a claim for overheads. It should be noted that some contractual provisions only provide for recovery of additional cost or expense. Where that is the case, a claim for loss of profit is not permissible under the terms of the contract. However, unless there are clear terms to limit the contractor's remedy to those contained in the contract (that is, excluding a common law claim), the contractor may be able to make a claim for loss of profit under the general law. The JCT forms of contract permit reimbursement of loss of profit.

Having established that there is a contractual, or common law, right to recover profit lost as a result of delay, what level of profit is reasonable and what standard of evidence to support a claim for loss of profit is required?

It is an impossible task to show that, save for the delay, the contractor would have been successful when tendering for a particular project (which he declined, or submitted a deliberately high bid) and that, having been awarded the contract for the project, he would have made a profit on it. If that was the appropriate test, no claim for loss of profit would succeed.

However, it will be necessary for the contractor to show some evidence that he was given the opportunity to tender for other projects and that he could not reasonably take advantage of these opportunities because of the fact that his resources were retained on the delayed project. In formulating a claim for loss of profit, the contractor would be advised to keep a record of the following:

- all tenders submitted and awarded (so that a success ratio can be established);
- all projects for which the contractor was invited to tender, but which were declined or a deliberately high tender submitted (this may cover a period of several months before the present delayed project has overrun, since decisions to decline new work may have to be taken in advance as soon as the overrun is anticipated).

The former is relatively easy to illustrate. The latter may need some analysis to establish that any bids were deliberately high. This should be possible by a bid ratio technique (a system of recording the nett cost included in each tender as a percentage, or factor, of the successful tender). As regards the latter, it should be remembered that, as stated in Chapter 3, competition law throughout Europe prohibits tenderers from discussing their tenders for the purpose of submitting a deliberately high tender.

Example

Nett cost for constructing a project = *C*, say £100000
Successful tender sum = *T*, say £105000
Bid Ratio = T/C = £105000/£100000 = 1.05

Any tenders with a bid ratio above an established competitive bid ratio would qualify for deliberately high pricing. This technique may require statistical analysis and adjustment for 'rogue' bids and errors.

Other evidence, such as proximity of the submitted tender to the competitive range of other tenders, may suffice. Further, a general analysis of construction activity during the period of overrun may be acceptable. Limitations on the contractor's bonding facility may also be a factor.

If the contractor can demonstrate that, on the balance of probability, he would have been able to obtain other contracts during the period of overrun, that alone ought to be sufficient to establish the claim in principle. In a United States case, the employer, the United States Government, contended that the contractor was required to prove that he was capable of taking on the extra work which he alleged was lost as a result of the government's delay and that he could have made a profit on it. It was held that the contractor had produced unrebutted evidence that he could not have taken on any large construction jobs during the various delay periods owing to the uncertainty of delays and limitation on his bonding capacity. The mere showing of these facts is sufficient to transfer to the government the burden of proof that the contractor suffered no loss or should have suffered no loss: *Capital Electric Company* v. *United States* (Appeal No. 88/965, 7.2.84) 729 F.2d 743 (1984).

A very simple approach was adopted in *Whittall Builders Company Ltd* v. *Chester-le-Street District Council* (*supra*). The judge was satisfied that there was sufficient activity in the construction industry at the relevant time that it was reasonable to assume that *Whittall* would have been able to obtain other profitable work.

Hudson, Emden or Eichleay? Percentage to be used: period for calculating the relevant percentage

A great deal will depend on the nature of the delay. If the sole reason for a particular delay is extra, or additional work, contemplated by the variation clause in the contract, it may be appropriate to use *Hudson's* formula (see Variations, *infra*). If the reason for delay is breach of contract, or if periods of delay caused by variations cannot be disentangled from periods of delay caused by breaches of contract, it is suggested that the remedy is by way of damages, in which case *Emden's* formula may be more appropriate.

At tender stage, the contractor will be looking at historical data (based on several years' expenditure on overheads and the recorded turnover for the same periods). Some adjustment may be made for anticipated changes in turnover in the future overheads. In any event, the percentage for overheads in the contractor's tender should be a realistic estimate of the probable apportionment of overheads in the rates for the work in the contract. The level of profit in the tender may have no relationship whatsoever to historical data, but it will depend on the profit (or loss) which the contractor anticipates should be allowed, having regard to external market factors and operating turnover requirements. Where a positive profit has been allowed in the tender, and where there has been no substantial change in the market, the *Hudson* formula may be fair to both parties where delay is caused by variations.

Where a negative profit has been allowed in the tender, adjustment to the percentage may be considered, particularly if the delay is out of proportion to the value of additional work and/or there had been an improvement in the market (part *Hudson*, part *Emden*). Where the delay was not unreasonable, having regard to the value of variations, adjustment for overheads only (ignoring the negative profit percentage) may be the applicable solution. This would depend on the terms of the contract and the circumstances of the case.

Where a formula is used, there may be some difficulty in deciding upon the appropriate period to be taken for establishing the turnover and overheads and profit in the formula (see Figure 5.18).

Period a (prior to commencement with possible adjustment for anticipated changes) represents the period used for *Hudson's* formula.

Period b (the original contract period) represents the period used for *Eichleay's* formula (see *Construction Contracts: Principles and Policies in Tort and Contract* by I.N. Duncan Wallace at page 128). However, period c (the extended contract period) would appear to be equally appropriate.

Period d (prior to commencement of the qualifying delay) would appear to be the most appropriate for *Emden's* formula, since it is the most contemporary period before the percentage is distorted by the qualifying delay (which would normally reduce turnover and increase the percentage for overheads).

Period e (the period of the qualifying delay) would normally be too short for useful figures to be obtained and it would suffer from greater distortion than period d.

Period f (from commencement of the qualifying delay until completion) may be appropriate in certain circumstances but may be subject to distortion.

Period g (period of overrun) is most suitable for the loss of profit element (since this is the period in which the profit ought to have been earned on a new project). However, it is normally too short. Profit from the nearest year's accounts may be appropriate as a basis of assessment.

Contractors may seek to use the period which gives the most favourable result. In practice, the nearest accounting periods which include period d are likely to be the appropriate periods for calculating the percentage for overheads, while the nearest accounting periods which include period e are likely to be the appropriate periods for calculating loss of profit. However, since the use of a formula does not purport to produce an accurate result, it is suggested that period c should be appropriate

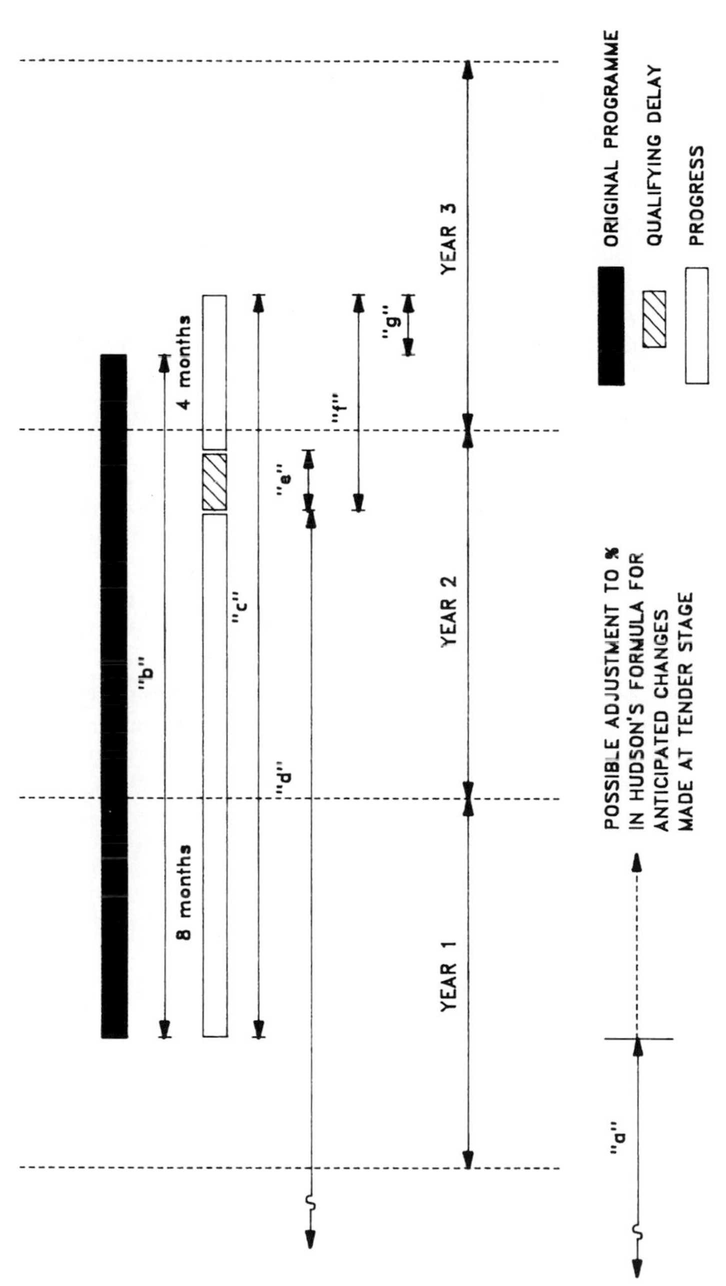

Figure 5.18 *Periods for calculating percentage overheads and profit*

(for overheads and profit) in most cases. If claims are to be settled prior to such information being available, the most recent accounting periods may have to suffice.

The accounting periods will not usually coincide with the actual period, in which case an adjustment may be made. For example, assuming that c has been agreed as the appropriate period, the percentage overheads and profit may be calculated as follows:

	Year 1	*Year 2*	*Year 3*	*Total*
Turnover	£1800000	£2000000	£2400000	
	× 8/12	× 12/12	× 4/12	
	£1200000	£2000000	£800000	£4000000
Overheads and profit	£240000	£300000	£300000	
	× 8/12	× 12/12	× 4/12	
	£160000	£300000	£100000	£560000
% overheads and profit	13.33%	15.00%	12.50%	14.00%

A more accurate assessment may be made by graphical means or by using monthly or quarterly figures.

One pitfall when using actual audited accounts is that they may not include any (or the correct) provision in them for the recovery to be realised by payment of the claim on the delayed contract (and possibly other contracts). Provisions in previous years' accounts may have been under or over-estimated and amounts received in the years used for calculation may distort the real figures. Adjustment may be possible if good management accounts are kept. However, unless there are unusual circumstances, it is suggested that these factors will be self-compensating in the long term.

It has been said that a formula produces a result which includes overheads and profit on the overheads and profit included in the contract sum. However, this is not the case if the overheads and profit are expressed as a percentage of the turnover income (and not annual cost), as can be seen from the following example:

Annual cost of all projects	= £60000
Overheads and profit	= £5000
Annual turnover	= £65000
Overheads and profit	= 8.333% of cost or 7.692% of turnover
Contract sum of delayed project	= £345000
Less overheads and profit	(7.692%) = £26537
Cost of delayed project	= £318463
Original contract period	= 300 days
Period of delay	= 70 days

Overheads and profit during period of delay (using contract sum and overheads and profit as percentage of turnover income in the formula)

$$= \frac{7.692}{100} \times \frac{£345000}{300\text{ days}} \times 70\text{ days} = £6192$$

Overheads and profit during period of delay (using contract cost and overheads and profit as percentage of annual cost in the formula)

$$= \frac{8.333}{100} \times \frac{£318463}{300\text{ days}} \times 70\text{ days} = £6192$$

This example illustrates that there is no mathematical problem when the percentage for overheads and profit included in the tender is the same as the average percentage for overheads and profit on all projects. Adjustment may be necessary if different percentages are evident (as will almost certainly be the case using *Emden's* formula). If this is so, it is a simple matter to convert the percentages so that they are expressed as a percentage of cost, in which case the formula becomes:

$$\frac{\text{Overheads \%}}{100} \times \frac{\text{Contract cost}}{\text{Contract period}} \times \text{Period of delay}$$

In most cases the traditional use of the formula will be sufficiently accurate. Only where there is a significant difference between average profit and the profit on the delayed project will any adjustment be necessary.

A formula may also produce a suspect result (over-recovery) if the delay being considered is at the end of a project, when most of the work has been done and few key resources are retained on site. The opposite (under-recovery) may occur when the delay takes place during the peak months and the maximum resources are on site. All of the resources should earn a contribution to the overheads and this can be catered for by sensible adjustments to the formula. For example, the following factor may be suitable in some circumstances:

$$F = \frac{\text{Value of work done per day during period of delay on contract}}{\text{Average value of work done per day during total contract period}}$$

Amount of overheads (and profit) = Normal formula result × F

An alternative would be to examine total costs of all projects, the cost of the delayed project and actual overheads during the period of delay (similar to *Eichleay*). This could be ascertained by monthly records. For an example (see also Figure 5.19):

Total cost of all projects, March and April = £160000
Total head office overheads, March and April = £12000
Cost of delayed project, March and April = £30000

$$\text{Overheads percentage} = \frac{£12000}{£160000} \times 100 = 7.50\%$$

Overheads allocated to delayed project during March and April = £30000 × 7.5% = £2250

$$\text{Overheads during 45 days' delay} = £2250 \times \frac{45}{61} = £1660$$

As stated earlier, head office overheads were considered in the case of *Property and Land Contractors Ltd* v. *Alfred McAlpine Homes North Ltd* (1996) 76 BLR 59. JCT80 conditions applied with some amendments. The contractor was instructed to suspend the works which led to a claim being submitted in the alternative for head office overheads. The matter was referred to arbitration.

The claim was based upon the application of *Emden's* formula. The contractor usually undertook only one major project at any one time. A second project at Tollerton was planned and it was agreed that the contractor intended to carry out this development for its parent company after completing the current project (the subject of the claim). It was claimed that, due to the postponement,

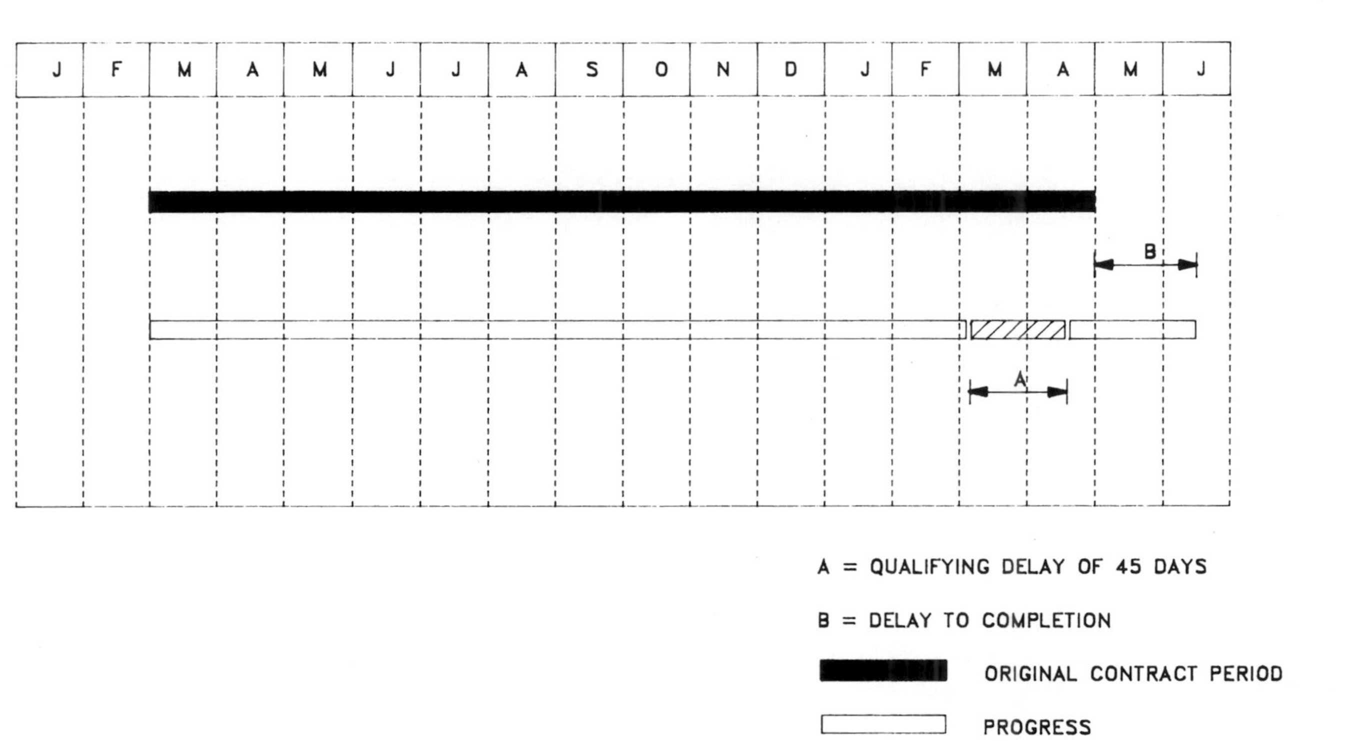

Figure 5.19 *Overheads and profit based on monthly accounts during period of delay*

completion of the work was delayed from 20 May 1990 until 25 November 1990, and that the delay prevented the contractor from carrying out the second project at Tollerton. The contractor claimed that, due to the overrun, he lost an opportunity of carrying out the second project which would have contributed to overheads. *Emden's* formula was employed as a means of calculating the head office overheads. This argument was rejected by the arbitrator, who was not convinced that the suspension resulted in the contractor being unable to work at the second project or elsewhere.

The contractor's alternative claim was for the recovery of head office overheads actually expended. The arbitrator was satisfied that the head office costs were related to the works for the delay period. The contractor's method of calculation was

> 'to extract from the company's account the overhead costs excluding fixed costs not related specifically to progress on the site (i.e. directors' remuneration, telephone, staff salaries, general administration, private pension plan, rent, rates, light, heat and cleaning and insurance to express such annual costs as weekly averages for both 1990 and 1991, and multiply the resulting weekly averages by the period of overrun in each year and thus produce a figure referred to as 'C').'

The total overheads for the period of delay and to be allocated between the delayed project and other work being undertaken at the same time was calculated as follows:

$$\frac{\text{Value of work at Shipton}}{\text{Total value of work}} \times \text{Total overheads (C)} = \text{Amount claimed}$$

The above formula contains a variant of the *Eichleay* formula and the method described using Figure 5.19.

The employer argued that the arbitrator had erred in law because he had awarded costs which would have been incurred by the contractor in any event and could not therefore be classed as direct loss and expense.

The court found in favour of the contractor with the following observations (at page 98 of the judgement):

> 'All these observations like those of Lord Lloyd in *Ruxley*, of Forbes J in *Tate and Lyle*, and of Sir Anthony May in *Keating* all suppose, either expressly or implicitly, that there may be some loss as a result of the event complained of, so that in the case of delay to the completion of a construction contract there will be some "under recovery" towards the cost of fixed overheads as a result of the reduced volume of work occasioned by the delay, but this state of affairs must of course be established as a matter of fact. If the contractors overall business is not diminishing during the period of delay, so that where for example, as a result of an increase in the volume of work on the contract in question arising from variation etc., or for other reasons, there will be a commensurate contribution towards the overheads which offsets any supposed loss, or if, as a result of other work, there is no reduction in overall turnover so that the cost of the fixed overheads continues to be met from other sources, there will be no loss attributable to the delay.'

It will be seen from Figure 5.17 (*supra*) that by comparing anticipated turnover (b–b) with actual turnover (e–e) on a delayed project, the volume of work ought to fall below the anticipated turnover. That is precisely what the court was saying in the above observation.

Problems occur when the cause of delay is a suspension order which applies to the whole, or a substantial part of the works. It is self-evident that the above method would produce a result of zero if all of the works were suspended and no costs were allocated to the project. Nevertheless, *fixed* head office resources would have to be covered by a contribution from the delayed project. It is possible that no management time would in fact be spent on the delayed project. However, this does not mean that more *effective* management time is spent on other projects. Management resources would not be expended on the delayed project (so, in theory, there would be no cost which could be allocated to the delayed projects) thereby making it impossible to justify a claim based on costs as required in *Tate & Lyle* v. *GLC* (*supra*). It must be reasonable to argue that the loss of contribution to overheads should be recovered from the delayed project on the grounds that the contractor's head office resources could not earn the shortfall in contribution on any other project.

Numerous variations to the recognised formulae may be appropriate. In *Finnegan* v. *Sheffield City Council* (*supra*), the contractor argued (unsuccessfully) that the percentage to be used in the formula should be based on a notional contract and the contractor's direct labour cost (excluding subcontractors).

In summary, it is suggested that, unless there are compelling reasons to modify one of the formulae, no adjustment should be necessary when calculating the loss of contribution to overheads (and profit). In most cases, *Emden's* formula, or *Eichleay's* formula, are preferable to *Hudson's* formula.

Adjustment for overheads and profit in variations

Many practitioners argue that any recovery of overheads and profit in variations should be deducted from the overheads and profit included in a claim for prolongation. This may be the case in the event of all of the variations being the cause of all of the period of delay. It may not be the case where some (or all) of the variations can be executed within the contract period or they do not cause delay. (See also *The Presentation and Settlement of Contractors' Claims* by Geoffrey Trickey and Mark Hacket.)

For example, if variations were executed during a period when there was no delay, the contractor would be paid for them at rates which would include additional overheads and profit. If the contract was to complete on time, no adjustment would be made (but see Variations, *infra*). Therefore, if (after completion of all varied work) there should be delay for another reason (such as suspension), the overheads and profit recovered for this delay (using a formula) would be the appropriate measure of damages for the period of suspension and should stand on its own without adjustment for the overheads and profit recovered in the variations. Similarly, if variations are executed concurrently with other recoverable delays, if it can be shown that they could have been incorporated within the contractor's programme (in the event that the other recoverable delays did not occur), then they may also be discounted and no adjustment made.

In short, any variations which do not cause the delay which is the subject of the prolongation claim may be ignored when making any adjustment for overheads and profit. Conversely, if a variation is the cause of a claim for prolongation, an adjustment should be made.

However, if *Emden's* formula has been used to calculate the overheads and profit during the period of prolongation, the percentage to be used in the adjustment may not be the same as that used in the formula. It should be that percentage which was included in the contractor's tender.

Adjustment for non-recoverable delays

Some delays, such as exceptionally adverse weather conditions, do not qualify for additional payment. Where such delays occur in isolation, it is a simple matter to ignore the period of delay in any calculation of prolongation costs (see Figure 5.20). Where such delays occur in parallel with recoverable delays, reimbursement will depend on the particular circumstances of the case (see Concurrent delays, *infra*).

It should be remembered that where a contractor has been forced into a period of adverse weather by a variation, or other qualifying recoverable delay, it may be entitled to reimbursement (*Fairweather* v. *London Borough of Wandsworth, supra*). In these circumstances the adverse weather conditions need not be exceptional in order to qualify for an extension of time and additional payment.

Concurrent delays

A single cause of delay often presents no problem when dealing with prolongation claims. However, in practice, many delays occur at the same time. Previous examples have illustrated the difficulties which arise when considering extensions of time in such circumstances. The situation is far more complicated when deciding whether, or not, the contractor is entitled to additional payment. There are no easy solutions to the wide variety of practical problems which arise when more than one cause of delay is affecting the progress of the works at the same time. Some delays will qualify for additional payment, while others, such as adverse weather conditions (which may qualify for an extension of time) and culpable delay by the contractor, will not normally qualify for additional payment.

Contractors are unlikely to offer any concession for concurrent delays when putting forward a claim for prolongation. They cannot be blamed for that (see Negotiation – Chapter 8). The following notes assume that the author of the claim is impartial and is attempting to establish what is reasonable reimbursement in the circumstances.

The law applicable to the rights of the parties to damages in the event of concurrent delay is complex. In *Keating on Building Contracts, eighth edition* (pages 271–6), the authors discuss the various options which may apply, taking the view that while the law appears to be unclear, in the majority of cases, the dominant cause of delay should be the deciding factor. This has been established in cases of exception clauses used in policies of insurance: *Leyland Shipping Company* v. *Norwich Union Fire Insurance Society* [1918] AC 350. It does not appear to be applicable to contracts generally. However, this may sometimes be the case where the facts are clear and the interaction of the various delays are relatively simple to determine. At paragraph 8.021 of *Keating* the authors state that it is now generally accepted that the dominant cause principle should not be applied to extensions of time if the contract makes provision for an extension of time for an event which has caused delay and the event has at least 'causative potency'. The authors go on to state that there is no enthusiasm for extending the principle to the recovery of losses despite the fact that the same arguments could be said to apply.

It is submitted that the 'dominant delay' principle is generally inappropriate for the majority of construction delay claims (with some exceptions). This appears to be supported by the judgement in the *Fairweather* case. If the responsibility for delays

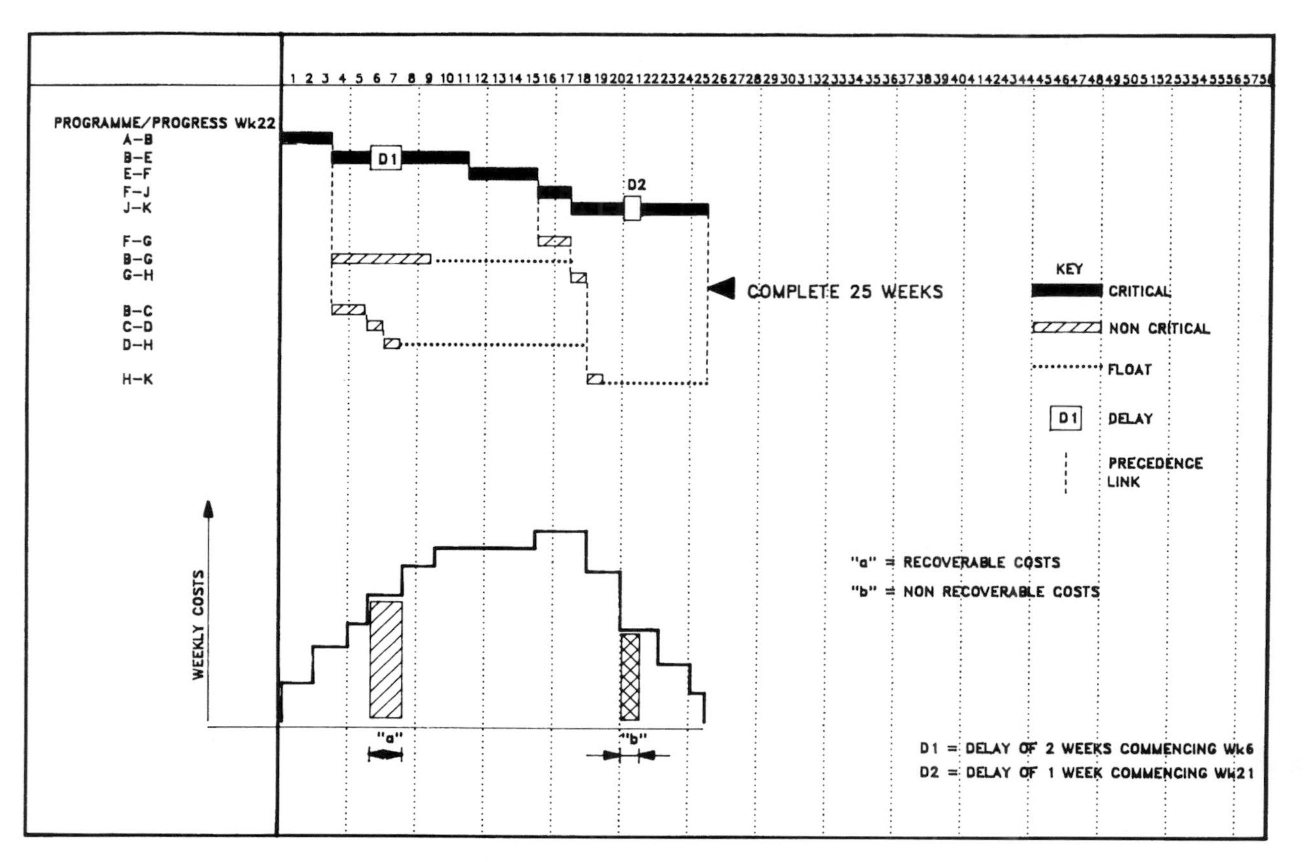

Figure 5.20 *Recoverable and non-recoverable delays*

can be divided according to the circumstances, apportionment may be appropriate. If it is impossible to disentangle the causes and effects of the delays, the claim may fail entirely: *Government of Ceylon* v. *Chandris* [1965] 3 All ER 48. If the competing causes of delay are in parallel, only nominal damages may be appropriate: *Carslogie S.S. Co.* v. *Norwegian Government* [1952] AC 292.

The following guidelines may be applicable in circumstances where more than one delay is affecting the progress of the works during the same period of time:

- where the non-recoverable delay is on the critical path and the qualifying recoverable delay is non-critical, no reimbursement should be permitted;
- where the non-recoverable delay is non-critical and the qualifying recoverable delay is on the critical path, reimbursement should normally be permitted;
- where both (qualifying and non-qualifying) delays are critical, then so far as they are of the same duration, no reimbursement should normally be permitted;
- where a qualifying recoverable delay occurs first, followed by a non-qualifying delay (both delays being on the same or parallel critical paths – see Figure 5.21), there is an argument to support the view that reimbursement should be permitted;
- where a non-recoverable delay occurs first, followed by a qualifying recoverable delay (both delays being on the same or parallel critical paths), there are grounds to argue that no reimbursement should be permitted.

There may be circumstances which merit a departure from the above guidelines. For example, the greater part of the contractor's management and supervisory staff may have been retained on site to deal with a complex variation which has caused a delay of lesser duration than a concurrent period of exceptionally inclement weather. If it can be shown that the contractor's staff could have been released at an earlier date (had there been no variation), then reimbursement may be permitted notwithstanding the concurrent non-recoverable delay.

The above guidelines should not affect the contractor's rights to recover time-related costs which are exclusively in connection with an activity which has been delayed by the employer (such as the cost of supervisory staff wholly employed on the section of work which has been delayed by the employer).

Delayed release of retention

When a project is delayed, the certificates which release the retention held by the employer are also delayed. The delay in issuance of the necessary certificates will give rise to a claim for finance charges on the retentions for the period of delay. Allowance will have to be made for non-recoverable delays.

5.9 Disruption and loss of productivity

The term 'disruption' when used in the context of construction and engineering claims includes any one or a number of the following considerations:

- delays to individual activities (whether, or not, such delay caused completion of the works to be delayed), thereby causing manpower to be retained over a longer period to execute the same amount of work;
- changed sequence of working arising out of delays to individual activities, thereby causing the effective use of manpower to be interrupted and disturbed so that no

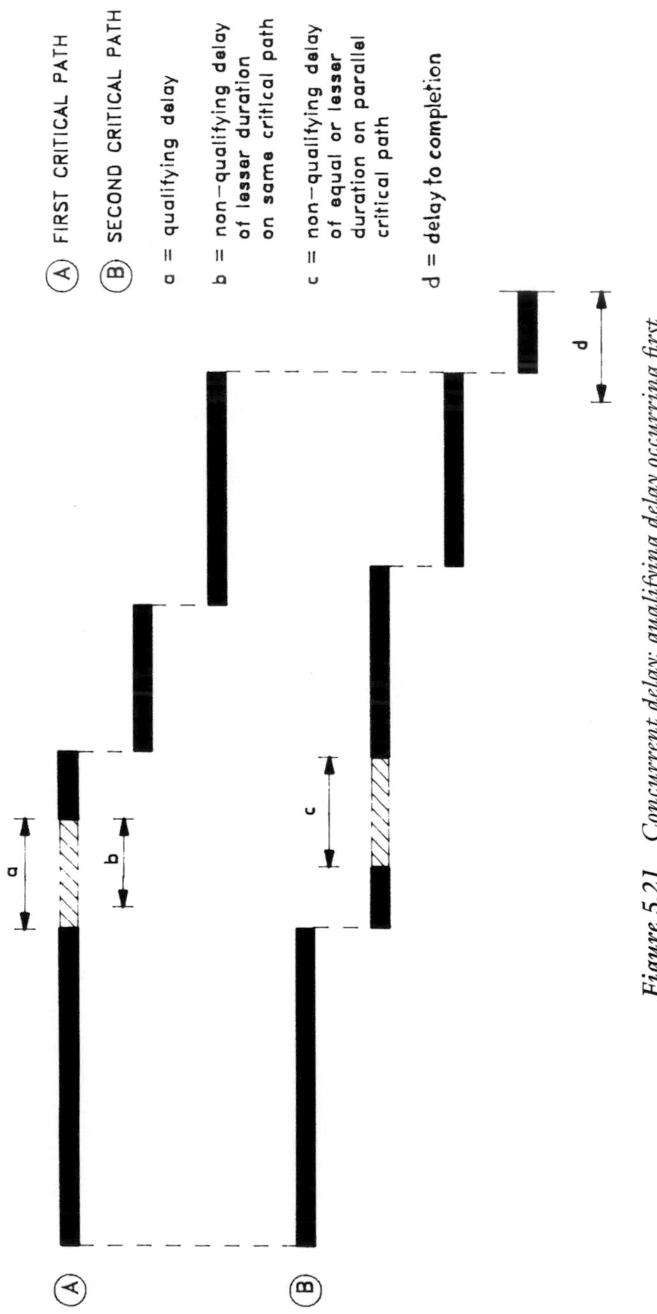

Figure 5.21 *Concurrent delay; qualifying delay occurring first*

production takes place during such interruption and lower production occurs in the initial stages of the activity to which the manpower has redeployed;
- interruption and disturbance to other secondary activities (not directly affected by the cause of disruption) caused by delay to the affected activities or changed sequence of working so that lower production is achieved in carrying out these secondary activities;
- idle (or non-productive) time caused by rescheduling and out-of-sequence working, thereby adversely affecting the progress of the work;
- congestion in sections of the work to which rescheduled manpower is transferred, thereby affecting productivity and progress of the work;
- general loss of productivity due to work being done piecemeal.

The following authorities and references refer to disruption under a number of descriptions:

(a) 'Many serious breaches or substantial variations may involve neither delay nor disturbance beyond their immediate direct cost. They may not be on the critical path of progress, so overall delay will not be involved. They may take place at a time when prompt action and direct expenditure by the contractor can avoid any disturbance of the remaining work. Nevertheless, even where overall delay is not involved, there will often be serious disturbance of the contractor's internal programme. This is particularly true of information or access breaches. Even in the absence of immediate direct costs, labour cannot be suddenly hired or fired, specific tasks cannot be suddenly stopped and restarted, and labour and plant cannot be moved backwards and forwards across the site, without an often substantial general loss of productivity. This will express itself, of course, in a generally heavier labour and plant expenditure, relative to actual work done. This may result from the particular plant and labour force being engaged for a longer period, or the recruitment of additional plant and labour to avoid or recover delay. Theoretically, in reaching a decision as to which course to follow, a contractor able to pre-plan will weigh the effect on his extended time-related costs if there is to be a delay, against the possibly marginal economic advantage of increasing his plant or labour force – it may be reasonably assumed that he will have endeavoured to optimise productivity when planning his original plant and labour force, so that an increase in it may not be economical in terms of production.'
[*Construction Contracts: Principles and Policies in Tort and Contract* by I.N. Duncan Wallace (p. 124, para. 8–23)]

(b) 'Loss of productivity or uneconomic working
This is a head of claim sometimes made where there has been delay in completion or disturbance of the contractor's regular and economic progress even though, on occasion, the ultimate delay in completion is small or does not occur.'
[*Keating on Building Contracts, eighth edition* (para 8.054)]

(c) 'A claim for the effect of an event upon the contract works themselves which does not necessarily involve a delay in completion of the works. This is a disruption claim and can arise even where the works are completed within the contract period.'
[*Problems in Construction Claims* by Vincent Powell-Smith (p. 3)]

(d) 'Delay and disruption can lead to increased expenditure on labour and plant in two ways. It may be necessary to employ additional labour and plant or the

existing labour and plant may stand idle or be under-employed. The latter is sometimes referred to as "loss of productivity"' [Emphasis added].
[*Building Contract Claims*, second edition by Vincent Powell-Smith and John Sims (p. 139) (p. 161 in the third edition)].

The principal elements accompanying and/or causing disruption are

- Rescheduling and out-of-sequence working.
- Causes of disruption which interrupt individual activities (such as late or incomplete information, variations or change orders, design errors and other matters for which the employer is responsible) may sometimes be absorbed within the original programme or schedule of work. This is particularly the case where the affected activities are not on the critical path and/or the number of alternative 'work-faces' is sufficient to facilitate relocation of resources from the affected activity to a location where there is other work capable of being done (by the relocated resources) without affecting other trades or the overall programme or schedule due to the extensive numbers of alternative work-faces becoming available.

However, as the installations become progressively completed, the number of alternative work-faces decreases, bringing about an increase in lost or idle time, additional supervision and consequential effects on other trades, disciplines and activities. As the available alternative work-faces decrease and the consequential effects on other trades, disciplines and activities intensify, the result may be to cause loss of productivity and actual delay to the programme or schedule of work, whether or not the original causes of disruption are on the critical path.

Loss of productivity

The authorities and references cited above confirm the view that disruption or dislocation invariably brings about a loss of output or loss of productivity.

A claim for loss of productivity will usually arise out of

- the employer's default or breach of contract;
- matters expressly permitted under the contract (such as variations or change orders and suspension orders);
- matters for which the employer has given an indemnity or has agreed to reimburse the contractor therefor.

'Loss of productivity' is recognised as a valid head of claim:

> 'While this [loss of productivity] is clearly an allowable head of claim, it can be difficult if not impossible to establish the amount of the actual additional expenditure involved.'
>
> [*Building Contract Claims*, second edition by Vincent Powell-Smith and John Sims (p. 139) (p. 161 in the third edition)]

In order to illustrate the effects of disruption and/or loss of productivity it may be necessary to establish that a planned orderly timing and sequence of events was affected by causes within the employer's control to the extent that the contractor was prevented from carrying out the work in the planned orderly timing and sequence. The planned sequence may not be that which was envisaged at tender stage. The project manager may have planned an alternative sequence and this should be the basis of comparison. It may not be necessary to show that there was delay to any activity or that the completion date has been delayed.

Much has been written about the contractor's rights to additional payment in the event of delay when the contractor's programme shows early completion: *Glenlion* v. *Guinness Trust* (*supra*). While this issue was not decided, the judge referred to two authorities of importance:

> 'In regard to claims based on delay, litigious contractors frequently supplied to architects or engineers at an early stage in the work highly optimistic programmes showing completion a considerable time ahead of the contract date. These documents are then used (a) to justify allegations that the information or possession has been supplied late and (b) to increase the alleged period of delay, or to make a delay claim possible where the contract completion date has not in the event been extended.'
>
> [*Hudson's Building and Engineering Contractss, tenth edition*, p. 603]
>
> *and*
>
> 'Sometimes contractors at the commencement of or early in the course of a contract prepare and submit to the architect a programme of works showing completion at a date materially before the contract date. The architect approves the programme. It is then argued that the contractor has a claim for damages for failure by the architects to issue instructions at times necessary to comply with the programme. *While every case must depend upon the particular express terms and circumstances*, it is thought that the contractors' argument is bad' [Emphasis added]
>
> [*Keating on Building Contracts, fourth edition, First Supplement*]

Example

If, for example, the delay of five weeks on bar D (see Figure 5.5) was caused by a suspension order issued immediately upon commencement of the works, the contractor would be entitled to claim the non-productive costs of its site establishment and overheads during the period of delay. These costs would not have been incurred (or they would have been productive costs) if the suspension order had not been issued. Similarly, if the delay of four weeks on bar E (see Figure 5.5) was caused by a variation, the time-related costs and any disruptive element of cost would be recoverable as part of the value of the variation. These arguments are valid whether, or not, the delays caused the completion date to be extended. These problems appear to have been contemplated by the judge at page 104 of the report: 'It is unclear how the variation provisions would have applied.'

While the majority of costs claimed are likely to be time-related, they are claimed for disruption rather than prolongation. The *Glenlion* case does not appear to affect the contractor's rights to claim in the appropriate circumstances.

The *Glenlion* case prompted numerous articles and exchanges of correspondence in the technical and legal press on the subject of delays when the contractor's programme showed early completion. There appeared to be two equal schools of thought, the first supporting the judgement (some adamant that it was also the death of similar claims for recovery of additional costs due to the delay) and the second being critical of the decision, especially with regard to the recovery of additional costs (which *Glenlion* did not decide). The following commentary may put the debate to rest.

In *Ovcon (Pty) Ltd* v. *Administrator Natal* 1991 (4) SA 71, the contract provided for completion of the work within fifteen months. The contractor, however, contemplated completion of work within eleven months. The contractor had calculated its tender on that basis and prepared a progress chart showing completion in eleven months. The progress chart was approved by the employer as required in terms of the Bills of Quantities. Completion of the work was delayed by the employer through the issue of variations. Despite the delay, the work was completed within the fifteen months but not within the eleven months contemplated by the contractor (see Figure 5.6, *supra*). The contractor's prolongation claim for recovery of additional expense or loss caused by the delay (additional P & Gs) was rejected by the court. It was held that acceptance by the employer of the progress chart did not impose any obligations on the employer and the contractor was not entitled to claim for delays. The contract provided for completion within fifteen months and, had the contract taken the full fifteen months (assuming no variations had been issued), it must be presumed that the contractor had included all the expenses associated with the period.

The arguments put forward on behalf of *Ovcon* for a prolongation claim appeared to miss the point entirely. The contract had not been prolonged as *Ovcon* had completed within the contract period. Based on the law in South Africa (and in the UK), the decision appears to be at odds with the principles of assessing damages for breach of contract:

> 'The sufferer by such breach [of contract] should be placed in the position he would have occupied had the contract been performed, so far as that can be done by the payment of money and without due hardship to the defaulting party.'
>
> [*Victoria Falls and Transvaal Power Co Ltd* v. *Consolidated Langlaagke Mines Ltd* (1915) AD at p. 22]

The presentation of *Ovcon's* case by way of a general prolongation claim possibly took the judge's eye off the ball with respect to the cause and effect of the delays which occurred. For example, if the employer failed to give possession of the site for several weeks, the contractor would have incurred loss and expense which it would not otherwise have incurred save for the failure to give possession. The payment of loss and expense to *Ovcon* would only have put *Ovcon* back in the position in which it would have been had there been no default by the employer. If each delay had been looked at individually in this way, perhaps the force of the argument would have persuaded the court to adopt a different view.

Further, although reference was made to various authorities, counsel for *Ovcon* informed the judge that no case law on the topics could be found. However, various cases and authorities addressed this topic, and reference to those cases and authorities may have assisted in obtaining a decision which would be consistent with the principles for assessing damages for breach of contract (*supra*).

Firstly, the English case of *Glenlion Construction Ltd* v. *The Guinness Trust* (*supra*) only dealt with extensions of time and it is no surprise (in that case) it was decided that extensions of time could only be granted if the delay caused the completion date to be delayed. That is to say, the extension should not be granted merely because the planned (earlier) date had been delayed. However, the *Glenlion* case did not address the matter of loss and/or expense caused by the delay. The judge did venture to say: 'It is unclear how the variation provisions would have applied.' In both the *Ovcon* case and the *Glenlion* case, reference was made to similar authorities and, in addition, to *Keating on Building Contracts*. In the edition referred to in these cases (the *Supplement to the fourth edition*), *Keating* states:

> 'While every case must depend upon the particular express terms and circumstances, it is thought that, upon the facts set out [in *Wells* v. *Army and Navy Co-operative Society* (1902) 86 LT 764] the contractor's argument is bad; and that is the case even though the contractor is required to complete "on or before" the contract date ... There is no authority on this point.'

However, in the eighth edition of *Keating* the author, referring to JCT 2005, goes on to say, at paragraph 19.276:

> 'Where the programme date is earlier than the Date for Completion stated in the Contract, it may be that some direct loss and/or expense may be recoverable on the grounds of disruption. However, provided that the contractor can still complete by the Completion Date, he cannot recover prolongation costs.'
>
> [*Glenlion Construction Ltd* v. *The Guinness Trust*]

It is important, therefore, to distinguish between prolongation costs (costs of overrun beyond the contract completion date) and disruption costs (costs arising as a result of delays and/or disruption caused by the employer whether, or not, such delays caused completion to be delayed beyond the contract completion date). Counsel for *Ovcon* did not appear to make this distinction on a case-by-case basis.

It appears, therefore, that in the appropriate circumstances, the door is open to claim direct loss and/or expense if delays occur but do not necessarily endanger the contract completion date, and that may include time-related costs which would not have been incurred save for the delay.

Secondly, as to there being no authority on the point (quoted both in the *Ovcon* case and referred to in *Keating* in the *Glenlion* case), this topic has been addressed on several occasions in the United States:

> 'Costs are no less damaging merely because they occur fortuitously before a contract deadline rather than after.'
>
> [*Sun Shipbuilding & Dry Dock Co* v. *United States Lines Inc.* 76 US C.Cls 154 (1932)]

> 'The Government may not hinder or prevent earlier completion without incurring liability.'
>
> [*John F. Burke Engineering and Construction*, ASBCA No 8182, 1963 BCA]

> 'While it is true that there is not an "obligation" or "duty" of defendant [owner] to aid a contractor to complete prior to the completion date, from this it does not follow that the defendant may hinder and prevent a contractor's early completion without incurring liability. It would seem to make little difference

whether the parties contemplated early completion, or even whether the contractor contemplated an early completion. Where the defendant [owner] is guilty of "deliberate harassment and dilatory tactics" and a contractor suffers loss as a result of such action, we think that the defendant is liable.'

[*Housing Authority* v. *E W Johnson Construction Co* 573 S W 2d at 323]

Some US cases address other relevant matters:

'The contractor must demonstrate that its planned schedule for the early completion of its work was both reasonable and attainable.'

[*Owen L. Schwam Construction Co* ASBCA No 22407, 79-2 BCA (CCH)]

'It is not necessary for the contractor to communicate its intent to finish early to the owner.'

[*Sydney Constructions Co* No 21377, 77-2 BCA (CCH)]

In most situations, it is not the programme which is relevant. The contractor must show that his progress was affected and that he suffered loss and/or expense thereby.

It is submitted that the *Ovcon* decision was wrong in the light of the arguments set out above. A contractor is entitled to loss and/or expense if the employer causes delay or disruption to the contractor's progress, whether or not the programme showed early completion and whether or not the contractor finished after the contract completion date. However, it is important to consider the facts of each case very carefully as there may be some compelling reasons, in some circumstances, to take a different view.

Evaluation of loss of productivity

It is universally recognised that the evaluation of the additional costs arising out of loss of productivity is difficult, if not impossible, but that this should not be a bar to a claim for reimbursement of these additional costs where loss of productivity can be demonstrated. Leading authorities have said:

'however, the classic element in a contractor's claim which gives rise to most difficulty arises where delay in completion or disturbance of economic working has been caused, whether by the owner's breaches of contract, or by late or numerous variations. Either of these can be present by themselves, though often they will be present together.'

[*Construction Contracts: Principles and Policies in Tort and Contract* by I. N. Duncan Wallace (para 8–10 at p. 115)]

'A reasonably efficient contractor should be able to establish actual costs incurred, but it will clearly be impossible to prove as a matter of fact what the costs would have been had the delay or disruption not occurred …

All that can be said is that the architect or quantity surveyor must do his best to arrive at a reasonable conclusion from whatever evidence is available. In our view, it must be a reasonable assumption that some loss will have been suffered in these respects where delay or disruption has occurred and the architect or quantity surveyor cannot resist making some reasonable assessment simply on the grounds that the contractor cannot prove in every detail the loss he has suffered.'

[*Building Contract Claims*, second edition by Vincent Powell-Smith and John Sims (pp. 139–140) (p. 162 in the third edition)]

See also *Wood* v. *Grand Valley Railway Co* (*infra*).

A number of methods of assessing or estimating the cost of lost production (loss of productivity) have been used with varying degrees of success.

Comparison of actual costs with allowance in the tender

This method is based on the difference in actual expenditure on manpower, according to the contractor's labour records, with the manpower allowed in the tender, after making adjustments for variations and inefficiency. This method is put forward as a possible means of assessment by a number of authorities:

> 'There can be no custom or general rule because the loss will vary in each case. A better starting point is to compare actual labour costs with those contemplated.[58] Thus a particular activity or part of the works is taken and, where the contract price can be ascertained, as by reference to the priced bills, the labour element is extracted. This is a matter for experienced surveyors and is done by taking the unit price and applying constants which are generally accepted in the trade. From the contractor's records the actual labour content for the activity or part is extracted. From the difference must be deducted any expenditure upon labour which was not caused by the breach, e.g. delay or disturbance caused by bad weather, strikes, nominated sub-contractors or the contractor's own inefficiency. If the original contract price was arrived at in a properly organised competition or as the result of negotiation with a skilled surveyor acting on behalf of the employer, the adjusted figure for the difference is some evidence of loss of productivity.
>
> [58]Such an approach was adopted in *Whittall Builders* v. *Chester-le-Street District Council* (unreported).'
>
> [*Keating on Building Contracts, eighth edition, paragraph 8.054*]

The case cited in *Keating* – *Whittall Builders* v. *Chester-Le-Street District Council* (unreported) – is misleading, as it suggests that the method of comparing actual costs with the tender was used and accepted in this case. However, that is not so (see commentary on this case, *infra*).

Legal acceptance of this approach has been mixed. In *London Borough of Merton* v. *Stanley Hugh Leach Ltd* (1985) 32 BLR 51 it was held that no evidence was available to support such a contention and that the result was too speculative.

However, in *Penvidic Contracting Co. Ltd* v. *International Nickel Co. of Canada* (1975) 53 DLR (3d) 748, the Supreme Court of Canada upheld the lower court's decision to accept the difference between the contractual sum per ton of ballast (in a track for a railroad) and the larger sum which was attributable to the adverse conditions caused by the employer's breach of contract. The court was impressed by the decision in *Wood* v. *Grand Valley Railway Co* (1916) 51 SCR 283, where Davies, J. said:

> 'It was clearly impossible under the facts of that case to estimate with anything approaching to mathematical accuracy the damages sustained by the plaintiffs, but it seems to me to be clearly laid down there by the learned Judges that such an impossibility cannot "relieve the wrongdoer of the necessity of paying

> damages for his breach of contract" and that on the other hand the tribunal to estimate them whether jury or Judge must under such circumstances do "the best it can" and *its conclusion will not be set aside even if "the amount of the verdict is a matter of guess work"*.' [Emphasis by the Supreme Court of Canada]

In *Construction Contracts: Principles and Policies in Tort and Contracts* by I. N. Duncan Wallace, the distinguished author respectfully submits that the decision of the Court of Appeal (which rejected the basis of assessing damages accepted by the lower court and ultimately upheld by the Supreme Court) is to be preferred, but the author goes on to say that there is no evidence that the author's reservations were canvassed in evidence or argument.

Acceptance of this method, it is submitted, will depend on

- to what extent the cause and likely effects are supported by evidence to satisfy the requirement to prove the extent of the loss 'on the balance of probability';
- whether the claim arose out of a breach of contract or under one of the provisions of the contract.

Perhaps the courts may be persuaded to accept this method in the case of breach of contract but may be less willing in the case of such additional costs arising out of variations or change orders. Each case must be viewed on its merits.

Assessed percentage addition on disrupted work

The method of adding a percentage on to the direct costs of labour or plant is perhaps the most common in construction and engineering contracts. Arbitrary additions are unacceptable:

> 'Some contractors add an arbitrary percentage to the contemplated labour costs. It is difficult to see how this can be sustained.'
>
> [*Keating on Building Contracts, eighth edition, para 8.054*]

Where no other method is possible, calculations based on sound reasoned assumptions may be acceptable, depending on the circumstances. However, persuading a court that such an approach is acceptable will be a difficult task to perform.

In the United States, the Armed Services Board of Contract Appeals (ASBCA) accepted a 25 per cent inefficiency for winter work: *Appeal of Pathman Construction Co* ASBCA 14285, 71–1 B.C.A. (CCH) 8905 (1971) – *Construction Delay Claims* by Barry B. Bramble and Michael T. Callahan at p.199 (p.3–56 in the third edition).

Comparison of output or productivity with previous or other projects or industry statistics

Where the contractor keeps records of output and productivity on similar projects, comparison of output or productivity on the affected project with that achieved on unaffected projects may be a basis for assessment. Alternatively, published industry statistics may be a guide for comparison.

This method does not take into account different (and sometimes unique) circumstances in any individual project or the difference in managerial supervisory or organisational skills employed on the affected and unaffected projects. Nevertheless, this method may be an acceptable basis in some circumstances and may be used in addition to the other methods described above as a means to support other calculations or assessments.

In *Construction Delay Claims* by Barry B. Bramble and Michael T. Callahan at p. 201 (pp. 12–69 to 12–70 in the third edition), the authors cite 'Effects of Job Schedule Delays on Construction Costs issued by the Mechanical Contractors Association of America, at 7 n.124':

> 'Successful contractors have learned to predict with considerable accuracy the number of man-hours that would normally be expended by their production workers to accomplish the tasks to be performed if conditions remain as expected at the time estimates are prepared. Most contractors have performed similar work many times in the past and have kept records of man-hours expended to accomplish various tasks. In addition, reference manuals indicating average times consumed for a wide variety of tasks are used as estimating guides. Individual contractors can add or subcontract percentage factors to the average times to allow for circumstances they expect to encounter on a given project which differ from those encountered on previous projects or those on which industry averages are based.'

Comparison of output or productivity during known disruption with output or productivity when little or no disruption occurred

This method takes into account what productivity the contractor could (and did) actually achieve when allowed to execute the work normally, and the actual productivity when the work was affected (disrupted or dislocated) by the causes relied upon by the contractor to justify his claim.

Apart from simultaneous contractor defaults during the period of disruption or dislocation (which had not also been evident during the period when the work was not affected), this method overcomes all of the problems associated with any of the other methods mentioned (including general contractor inefficiency). That is to say, if the contractor is generally inefficient over the duration of the project, this factor is taken into account in the direct comparison of productivity, but if a new element of contractor inefficiency is introduced during the affected period (such as changes in supervision and/or labour force), then this new inefficiency must be addressed by making appropriate adjustments to the results obtained by direct comparison of productivity.

This method is put forward in *Emden's Building Contracts and Practice, eighth edition, Volume 2* by S. Bickford-Smith (p.N/45):

> 'Initially, a period is examined when the contract was running normally, and the value of work done during that period is assessed and then divided by the number of operatives and/or items of plant on site. The figure thus arrived at is compared with the same figure calculated for the period of delay or disruption, and the comparative figures are then used to calculate the amount of loss.'

In *Problems in Construction Claims* by Vincent Powell-Smith (p. 112) the distinguished author expresses doubt about the legal basis of this method. The author does not, however, make reference to the case of *Whittall Builders Company Ltd* v. *Chester-le-Street District Council* (*infra*), in which this method was clearly accepted. It is possible that the author missed the fact that the *Whittall* case dealt with this issue (as it was more widely referred to in connection with head office overheads to which the author referred elsewhere), since the lack of any reference to it with respect to disruption claims (to criticise or support the decision) is inconsistent with the otherwise meticulous reference to the latest cases throughout the author's publication.

This method was approved in *Whittall Builders Company Ltd* v. *Chester-le-Street District Council* (1985) unreported. Mr. Recorder Percival QC said:

> 'Therefore I take the view that the total paid to the men employed, whether by wages or bonus, should be taken as the cost actually and properly incurred by the plaintiff for labour in pursuance of the contract up to the end of November 1974. Clearly the consequence of the defendant's breaches was that the plaintiff received much less value for that expenditure than he would have done if there had been no breaches ...
>
> Several different approaches were presented and argued. Most of them are highly complicated, but there was one simple one – that was to compare the value to the contractor to the work done per man in the period up to November 1974 with that from November 1974 to the completion of the contract. The figures for this comparison, agreed by the experts for both sides, were £108 per man week while the breaches continued, £161 per man week after they ceased.
>
> It seemed to me that the most practical way of estimating the loss of productivity, and the one most in accordance with common sense and having the best chance of producing a real answer was to take the total cost of labour and reduce it in the proportions which those actual production figures bear to one another – i.e. by taking one-third of the total as the value lost by the contractor.
>
> I asked both Mr. Blackburn and Mr. Simms if they considered that any of the other methods met those same tests as well as that method or whether they could think of any other approach which was better than that method. In each case the answer was no. Indeed, I think that both agreed with me that that was the most realistic and accurate approach of all those discussed.'

The above case is illustrated in Figure 5.22.

In *General Insurance Co of America* v. *Hercules Construction*, 385 F.2d 13 (8th Cir 1967), productivity and costs during the period when there were difficulties in delivery of pre-cast units (February 12 until May 6) were compared with productivity and costs during the period when pre-cast units were delivered in substantially proper sequence with minimal fabrication deficiencies (after 6 May). The increase per unit that it cost *Hercules* was then multiplied by the number of units erected during the period from February 12 to May 6 in order to determine the amount of damages.

The court found in favour of *Hercules* and awarded damages of US$21900. *General Insurance Company* appealed on the grounds that the proof of damages put forward by *Hercules* was illogical and not in accordance with law. It was held that *Hercules's* method of computing damages was not unreasonable as a matter of law.

The above case is illustrated in Figure 5.23.

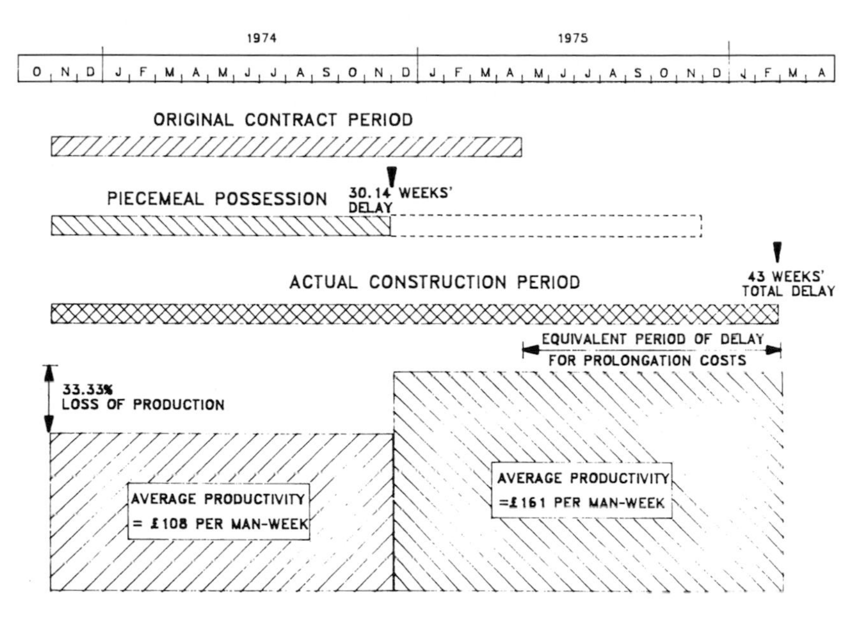

Figure 5.22 *Whittall Builders Company Ltd v. Chester-le-Street District Council 16.5.1985*

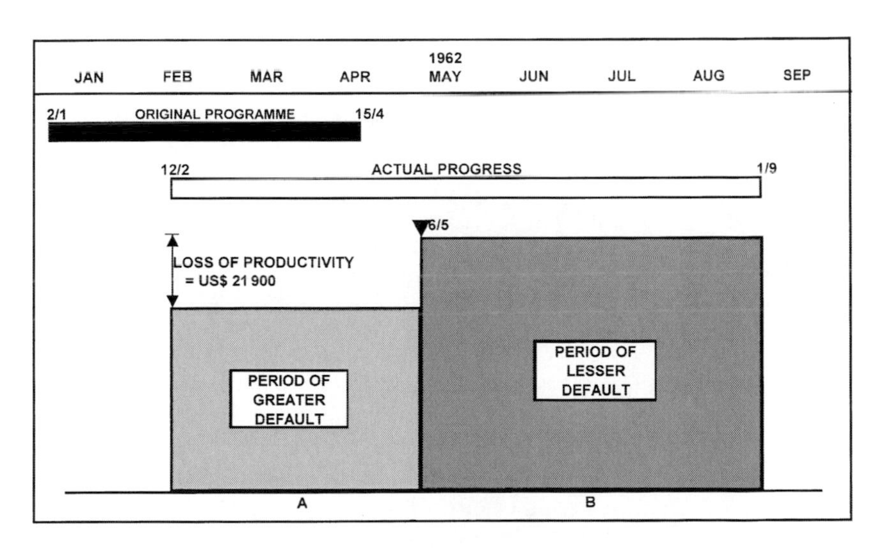

Figure 5.23 *General Insurance Co of America v. Hercules Construction 385 F.2d (8th Cir 1967)*

In *Natkin & Co* v. *George A Fuller Co* 347 F.Supp.17 (WD Mo 1972), reconsidered 626 F.2d 324 (8th Cir 1980), the court accepted comparison of productivity as a basis of assessment of damages (p. 34, para XII D):

> 'As of 11/25/66, on which date all parties accepted *Natkin*'s performance of the original contract as 43% complete, *Natkin*'s cost experience on that work

which was comparable to the work remaining to be performed......was 0.181 manhours for each standard piping unit, as contrasted with *Natkin*'s original estimate of 0.20 for each such unit.'

The *General Insurance* and *Natkin* cases are cited in *Construction Delay Claims* by Barry B. Bramble and Michael T. Callahan (pp. 201–204) (pp.12–70 to 12–73 in the third edition) where, with respect to *Natkin* v. *Fuller*, the authors write:

'Costs for performing *Natkin*'s work prior to November 25, 1966 were 0.181 man-hours for each standard piping unit compared to 0.20 man-hours after November 25, 1966......The court awarded *Natkin* $715,567 for its lost productivity claim. The court stated that comparing actual costs before and after the point in time defendant's failures caused damage to plaintiff was a reasonable method for computation of damages. The court also said *Natkin*'s evidence of comparing the man-hour cost for a standard piping unit before November 25, 1966, with the cost after that date was a logical basis for computing *Natkin*'s damages.'

There is an important difference between the extract from the judgement (which, in paragraph XII D, compares actual productivity with the tender productivity) and Bramble and Callahan's interpretation (which appears to compare actual productivity before the disruption with productivity during disruption). However, the authors' interpretation of the court's findings are otherwise consistent with the judgement which states at page 34, paras XIII A and B:

'A. Plaintiff's cost for performing each unit of its work under the contract after November 25, 1996 were greater than they were prior to November 25, 1996.

B. Plaintiff's costs were greater after November 25, 1996 because it was compelled to accelerate when the defendants failed and refused to grant extensions of time, and there was a resulting impact.'

And in its Conclusions of Law at page 35, Appendix B, Conclusion IX:

'Plaintiff's evidence of comparing the manhour cost for a standard piping unit before November 25, 1996 with the manhour cost for a standard piping after said date, is a logical basis for computing plaintiff's damages pertaining to additional labor costs.'

It should be noted that the court accepted that *Natkin*'s actual productivity before the disruption commenced (0.181 man-hours per piping unit) was the starting point (baseline productivity) from which to calculate loss of productivity. That is to say, even if (as the figures quoted suggest) *Natkin*'s productivity fell to the same level as its tender allowance during the period of disruption (0.20 man-hours per piping unit), it was right to compensate *Natkin* if his productivity during the disrupted period was no lower than its tender. Conversely, if a contractor's achieved productivity before disruption was less than the tender, then that would be the baseline from which to measure loss of productivity.

The above case is illustrated in Figure 5.24 based upon the assumption that the quoted productivity figures before and during disruption were as stated by Bramble and Callahan.

Which method of calculating loss of productivity should be adopted?

Any of the methods described above may be a reasonable method of evaluation in the appropriate circumstances, but the various methods are subject to varying degrees

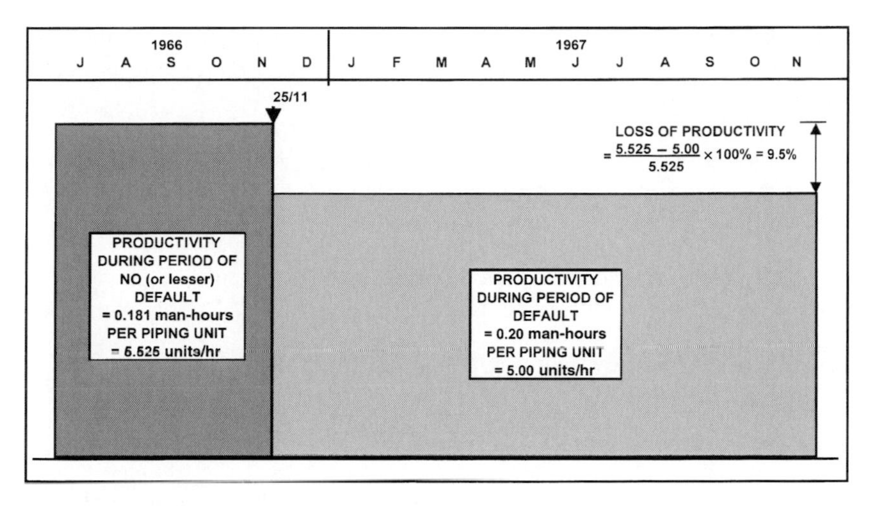

Figure 5.24 *Natkin & Co v. George A Fuller Co 347 F. Supp 17 (WD Mo 1972), reconsidered, 626 F.2d 324 (8th Cir 1980)*

of certainty and accuracy. The situation is best summed up by I.N. Duncan Wallace in *Construction Contracts: Principles and Policies in Tort and Contract* (para 8.24, pp. 124–5):

> 'The computation of loss of productivity claims is one of the more difficult problems in this field. An arbitrary guess or assertion of some percentage of the total affected labour or plant costs of the trades in question is not convincing. Another highly unconvincing method would be to compare actual total costs of the trades affected against alleged pre-contract estimates of those costs[23]...... More helpful will be a close analysis of any contract programme required to be supplied by the contractor, and a close correlation of it to the contractor's recorded labour and plant and work output on site, together with the chronology and contemporary evidence of the breaches or variations in question. In addition, expert evidence coupled with available publications showing the plant, labour and material elements of the better known construction processes, with various factors for the special conditions of particular contracts, are available in the civil engineering as well as the building industries. *But the most convincing of all* will be comparisons of actual hours and output, during a period known to be unaffected, with those in the affected period. In addition, of course, there will frequently be found to be contemporary site records kept of standing time of men or plant on well-organised contracts. In practice, good quantity surveyors in both industries, on each side of the negotiating table, can always do much better than asserting arbitrary percentages on affected turnover, or comparing contract with actual total cost. As will be seen, there are very powerful legal as well as logical objections to the use of this latter "total cost" method.' [Emphasis added]

The reference cited by Duncan Wallace at 23 is *E.C. Ernst, Inc* v. *Koppers Co* 476 F. Supp.729 (WD Pa 1979).

The most convincing method, that is comparing productivity during a period when there was no disruption with productivity during a disrupted period, is not without its problems.

In *Whittall Builders Company Ltd* v. *Chester-le-Street District Council* (*supra*) the method accepted by the court was based on a comparison of productivity over all trades for the duration of the project by expressing the output per man-week in pounds sterling, that is:

> Average productivity during period of default £108 per man-week
> Average productivity during period of normal working £161 per man-week

Therefore, loss of productivity during period of default was:

$$\frac{£161 - £108}{£161} \times 100 \text{ per cent} = 33 \text{ per cent}$$

This percentage was then applied to the total cost of labour during the period of default resulting in 33 per cent of the cost of labour (representing the loss of productivity), being a total of £21479.35.

Because this project was for the refurbishment of 108 dwellings, the proportions of each trade and the type of work being undertaken in each week were probably similar (save for the beginning and end of the period). These circumstances lend themselves to comparison in the manner used in this case.

In *General Insurance Co of America* v. *Hercules Construction* (*supra*), the comparison was made between productivity on the particular sections of the work affected (in this case erection of pre-cast units). These circumstances also lend themselves to comparison in this manner because of the repetitive nature of the delayed and disrupted work.

Similarly in *Natkin & Co* v. *George A. Fuller Co* (*supra*), installation of piping units were the subject of delay and disruption, thereby making it suitable for comparison purposes. In this case the loss of productivity may have been calculated as follows (assuming Bramble and Callahan are correct – *supra*):

> Productivity during period of no default 0.181 man-hours per unit
> (or 5.525 units per hour)
> Productivity during period of default 0.20 man-hours per unit
> (or 5.00 units per hour)

Therefore, the loss of productivity during the period of default was:

$$\frac{5.525 - 5.00}{5.525} \times 100 \text{ per cent} = 9.5 \text{ per cent}$$

The above method (unamended) may not be appropriate where the proportions of the various trades, disciplines and activities are substantially different during the period of disruption or dislocation when compared with the period when there was no disruption or dislocation. Significant errors can occur if it is not recognised that the man-hour content may be very different during the following phases of the project:

- *Phase 1*: Superstructure – a comparatively low labour content may be involved in this stage because of the high proportions of mechanised plant and large material sections, such as steel and prefabricated units, involved.

- *Phase 2*: 1st and 2nd fix carcassing and service installations – a higher labour content is invariably involved during this section of work.
- *Phase 3*: Final fitting-out and installation of equipment – during this period the manpower element is likely to be a lower proportion of the total cost because of the high value of fittings, finishings and hi-tech equipment.

Further changes in the proportions occur as the three phases overlap, so that the labour content as a proportion of the total may be constantly changing.

This difficulty may not be overcome simply by comparing the productivity of each individual trade, discipline or activity, as there may be problems in showing that periods of lower productivity in any single discipline are due to causes of disruption directly linked to that discipline. For example, substantial causes of disruption to pipe fitting may cause wholesale disruption to electrical installations, HVAC installations and fitting-out (even where there may have been no changes to those disciplines).

It is also essential to take account of all of the following:

- variations and change orders;
- other claims and additional work;
- growth (or re-measurement of contract work);
- any other contract adjustments.

One solution is to employ a method based on *Earned Value Costing*. Earned Value Costing has developed from the US DoD Cost/Schedule Control Systems. In its basic form, Earned Value Costing measures performance by monitoring total cost or value against the planned budget. However, the Earned Value Costing method deals with costs and not man-hours. The principal objectives of and results from Earned Value Costing (that is to measure performance), and comparison of performance during affected and unaffected periods, are valid and admirably suited to satisfy the criteria which were the basis of assessment in the UK and US cases cited above.

In order to utilise the basic techniques of Earned Value Costing to calculate the loss of productivity of labour (or plant), the following process takes into account most of the shortcomings which would otherwise be inherent in this method of calculation:

(1) Determine the actual man-hours (cumulative and monthly) from labour records.
(2) Determine the planned man-hours at the same dates (as 1), based on measurement or by reference to the schedule or programme and planned resource allocation. (If the planned man-hours have been based on the schedule or programme, it is essential that these should be adjusted to account for any delay or 'slippage'.)
(3) Add the man-hour content in all variations, change orders, additional work and other claims to the man-hours determined in 2 above.

The performance index or productivity factor (PF) of labour is then calculated in the same way as in the Earned Value Costing method:

$$PF = \frac{\text{Achieved man-hours during the period}}{\text{Actual man-hours expended during the period}}$$

where Achieved man-hours is the sum of the man-hours included in the tender plan for the original contract work plus the man-hours in any additional work (variations, change orders and other claims etc.).

Apportionment or allocation of the man-hours in the additional works should be done as accurately as possible. Day-works are the easiest to allocate to the time when the work was carried out. Variations and change orders may be allocated to periods of time according to the nature of the work and the schedule of work or programme. Subcontract work or 'work packages' may be allocated according to known or assessed periods of execution. Growth or changes due to re-measurement may be identified according to the individual disciplines or activities affected and allocated based on the progress of the changed work.

Comparison of productivity may then be done as shown in the following example (assumed data):

Analysis of man-hours and productivity in affected or disrupted period A

Actual man-hours expended during the period = 905
Planned (or achieved) man-hours during the period = 825

$$PF = \frac{\text{Achieved man-hours}}{\text{Actual man-hours}} = \frac{825}{905} = 0.912$$

that is, for every 1.0 man-hour worked, 0.912 man-hour's value of work was produced.

Analysis of man-hours and productivity in unaffected or normal period B

Actual man-hours expended during the period = 601
Planned (achieved) man-hours during the period = 623

$$PF = \frac{\text{Achieved man-hours}}{\text{Actual man-hours}} = \frac{623}{601} = 1.031$$

that is, for every 1.0 man-hour worked, 1.031 man-hours' value of work was produced.

Loss of productivity in affected period A (compared with unaffected period B)

$$= \frac{1.031 - 0.912}{1.031} \times 100 = 11.54 \text{ per cent}$$

It should be noted that the data used for this example does not, in itself, indicate separate periods for which a loss of productivity claim may arise. In order for the Earned Value method to succeed in a loss of productivity claim it is also necessary to be able to show distinct periods for comparison purposes, and that the period for which loss of productivity is claimed is affected by a significantly higher incidence and/or volume of defaults or disruptive matters relied upon as causes of the loss of productivity. This process will need considerable research but is essential to illustrate cause and effect.

Where the productivity factor (PF) departs significantly from 1.0, the figures may be distorted if a substantial amount of the additional work is based on cost (for example day-work), as this work will always be executed with a productivity factor (PF) of 1.00. That is to say, there is no loss of productivity on the actual work carried out on a day-work basis or on work which is priced from hours actually worked. This distortion may be overcome by the following modification to the formula:

$$\text{PF} = \frac{\text{Achieved man-hours during the period less the man-hours expended at cost}}{\text{Actual man-hours expended less the man-hours recovered in additional work at cost}}$$

In the modified formula, the achieved man-hours includes the total value (in man-hours) in the original contract work and in additional work executed during a given period based on rates or prices applicable to the work executed (that is all man-hours at cost, such as day-work, have been excluded from the calculation).

If, during the periods in the above example, a significant amount of work had been done at cost (that is for every hour worked, one hour's value of work had been achieved, or PF = 1.0), then the calculation of loss of productivity may be as follows:

Analysis of man-hours and productivity in affected or disrupted period A

Man-hours expended at cost during period = 125
Actual man-hours expended during the period = 905
Planned (or achieved) man-hours during the period = 825

$$\text{PF} = \frac{\text{Achieved man-hours}}{\text{Actual man-hours}} = \frac{825-125}{905-125} = 0.898$$

that is for every 1.0 man-hour worked, 0.898 man-hour's value of work was produced.

Analysis of man-hours and productivity in unaffected or normal period B

Man-hours expended at cost during period = 75
Actual man-hours expended during the period = 601
Planned (achieved) man-hours during the period = 623

$$\text{PF} = \frac{\text{Achieved man-hours}}{\text{Actual man-hours}} = \frac{623-75}{601-75} = 1.042$$

that is for every 1.0 man-hour worked, 1.042 man-hours' value of work was produced.

Loss of productivity in affected period A (compared with unaffected period B)

$$= \frac{1.042-0.898}{1.042} \times 100 = 13.82 \text{ per cent}$$

Similar calculations may be done to determine the loss of productivity of mechanical plant.

Unfortunately, when projects go wrong from the outset, it may be impossible to identify any period when the progress of the works was relatively free from disruption. Alternatives such as comparing parts of the works which were not disrupted with parts of the works which suffered from disruption may be applied. If neither of these methods can be adopted, one of the other alternative methods mentioned above may be the only solution.

In many circumstances, it is difficult or impossible to calculate the cost of disruption of each individual element. A global approach may be the only solution, *J. Crosby & Sons Ltd* v. *Portland Urban District Council* (*supra* – Chapter 1). This method may be appropriate where the evidence of delay and disruption is overwhelming and there is no significant default on the part of the contractor. If it can be shown that the contractor was partly responsible for the disruption, this type of claim may fail entirely, or the additional costs may have to be borne, in part, by the contractor.

5.10 Claims for acceleration

In the event of delay to the progress of the works, the employer, or the contractor, may be faced with deciding whether, or not, there are good grounds to accelerate the progress of the works to bring about earlier completion (to the whole, or part of the works).

From the employer's point of view, acceleration may be advantageous in the following circumstances:

- where it is essential to achieve completion by an earlier date for commercial reasons;
- where the delays qualify for additional payment, there is a real probability that the cost of acceleration will be less than the cost of prolongation for the period, which can be reduced by acceleration;
- where there may be substantial savings in escalation costs as a result of earlier completion;
- where the actual loss to the employer for late completion is greater than the liquidated damages which may be recovered from the contractor.

Some forms of contract (for example GC/Works/1 (1998), clause 38 and NEC 3 clause 36) provide for acceleration. However, the contractor's consent is usually required and the acceleration cost is normally agreed beforehand. Where there are no contractual provisions, a separate agreement will be required. In any event, the terms of an acceleration agreement (including matters required to be dealt with pursuant to clause 38(2)(e) of GC/Works/1) should contain provisions in the event of

- subsequent delay by qualifying events which would entitle the contractor to an extension of time for completion (thereby delaying the earlier date for completion);
- failure to complete by the earlier completion date for reasons which do not qualify for extensions of time (the employer may wish to increase the rate of liquidated damages in the light of his revised anticipated loss).

Whatever the reason for acceleration (even if the contractor is partly responsible for delay and is already liable for liquidated damages), the contractor is likely to be in a strong bargaining position when terms are agreed. The employer should be reasonably confident that the objectives of an acceleration agreement will be met before concluding any deal.

From the contractor's point of view, acceleration may be advantageous if he is in culpable delay and the cost of acceleration is less than the cost of prolongation.

However, when a contract is delayed and no (or insufficient) extensions of time have been made, the contractor may be faced with a dilemma. Should the contractor proceed to complete later than the completion date and run the risk of liquidated damages or should he accelerate the progress of the works to eliminate or reduce that risk?

Very often, pressure is brought to bear on the contractor to improve progress. The language used in these circumstances usually avoids the term 'accelerate', but the contractor is intended to be left in no doubt that he is being pressed to take measures to improve the progress of the works. Veiled, or patently open, threats of deducting liquidated damages may sometimes be used. The contractor's options are

- to keep his nerve in the belief that the extensions of time will eventually follow (or be awarded in adjudication or arbitration), *or*
- to take all of the necessary measures to improve progress and bring about earlier completion, *or*
- to take some measures to improve progress in the hope that some extension may subsequently be made to the actual completion date.

The decision to accelerate in such circumstances is not easy. If the contractor has a 'cast iron' case for extensions of time, then the first option is probably the best. In these circumstances, the right to recovery of acceleration costs may be in doubt. If the architect, or engineer, has responded to all requests for an extension of time, giving reasons for not making an extension, or explaining why an extension was for a lesser period than the contractor's estimate, the contractor is better placed to judge whether, or not, the extension is reasonable or capable of being reviewed. However, if there is no response, or if the response is an unreasoned rejection of the contractor's application for an extension of time, the contractor has no means by which to judge the eventual outcome which may result from further representations. All of these circumstances, including the pressure which may be brought to bear to improve progress, will influence the contractor's decision to accelerate.

Where it can be shown that the contractor was entitled to an extension of time when he took the decision to accelerate, and that the architect, or engineer, ought reasonably to have made the extension of time promptly, there are grounds to argue that the contractor is entitled to reimbursement of reasonable acceleration costs. The claim will be based on the premise that there was a breach of contract (that is, failure to operate the extension of time provisions). The success of such an argument will depend on

- whether the contractor had complied with the contractual provisions to give notice and particulars of the delay in accordance with the contract;
- whether the architect, or engineer, had properly considered all of the circumstances and events for each delay before making, or rejecting, an application for an extension of time (there may be a considerable difference between a genuine attempt to make an extension where the conclusion was merely wrong, and a rejection out of hand without proper, or any, consideration being given to the matter);
- to what extent the contractor had communicated his intention to accelerate and the circumstances at the time of making the decision;
- whether, or not, the contractor's decision was a sensible commercial decision in the circumstances;
- whether, or not, the contractor's claim for the costs of acceleration was less than the probable cost of prolongation (often referred to as mitigation of damage). Further, it may be equitable to reimburse the contractor for the costs of acceleration if the employer was ultimately going to benefit by a saving in the

amount of the contractor's probable claim for prolongation – that is to say, the employer should not benefit from his own default: *Alghussein Establishment* v. *Eton College* – Chapter 1, *supra*.

Invariably, it can be shown that the reason for failing to make extensions of time was a result of pressure from the employer on the architect, or engineer. Sometimes this is evident from the conduct of the employer's representatives and the professional team at meetings (or even in correspondence). Where this is not evident, it may come to light during discovery of documents or upon cross-examination in arbitration or litigation. Unfortunately, it is becoming increasingly common for some powerful employers to use the threat of termination of services (or the promise of future work) as a lever to put pressure on, or influence the architect or engineer.

If such pressure or influence was present, the contractor would have a *prima facie* claim for reimbursement (see *Morrison-Knudsen* v. *B.C. Hydro & Power* and *Nash Dredging Ltd* v. *Kestrell Marine Ltd* – Chapter 1, *supra*).

If it should be established that there is a case for reimbursement of acceleration costs, there is the difficult task of proving the actual amount of the claim. Costs which need to be considered are

- *Non-productive overtime* – That is, the premium rates paid to operatives for working outside normal hours. Not all of the overtime hours are recoverable. Only those hours in addition to the allowance in the contractor's tender should be claimed (if the contractor had always planned to work nine hours per day and Saturday mornings in order to complete within the original contract period, he could only claim the additional hours in a claim for acceleration).
- *Additional cost of employing extra staff and operatives* – Higher rates of pay, incentives, travelling time, subsistence and transportation costs of importing labour.
- *Loss of productivity* – An increase in the number of staff and operatives does not necessarily bring with it a proportional increase in production. On a congested site, labour cannot be utilised as efficiently. The co-ordination of various activities and trades becomes more demanding and there is likely to be a greater incidence of waiting time between activities.
- *Increase in the use of lighting and power* – Inevitable in winter and in large buildings and basements.
- *Increase in the hire of equipment and plant* (sometimes fuel only).

Whatever the reasons for acceleration, the contractor ought to be aware, before incurring the additional costs, that care should be taken to keep good records to enable the above costs to be substantiated. It should also be borne in mind that, whatever the moral grounds justifying acceleration, in practice this head of claim is one of the most difficult to justify on legal grounds.

5.11 Variations

Variations to the works are almost inevitable. Therefore, all standard forms of contract contain provisions to deal with them. Some variations can be made without affecting the progress of the work and with no change in the method, sequence and cost of the work to be done in the variation. In such circumstances, the rates applicable to the contract can be applied to the measured quantity of work in order to arrive at the value of the variation. However, even when these simple rules are applied, there may be some indirect costs which need to be addressed.

For example, if the costs of insurance premiums have been included in the 'Preliminaries' sections of the bills of quantities, there may have to be an adjustment made to the 'value related' element of the insurance premiums in the bills to reflect any change caused by variations. Where there is a decrease in the contract price as a result of variations, there may be no adjustment to the cost of insuring the works (depending upon the insurer's practice in this regard). However, a decrease in the contract price may justify a reduction in the allowance for employer's liability insurance. Likewise, if small tools and equipment are priced in the preliminaries section of the bills, an increase may be justified if the contract price is increased by variations. Where there is a decrease in the contract price, the likelihood of the contractor being able to save on the amount of tools and equipment is remote (unless the reduction in work was known well in advance of the need for the necessary tools and equipment).

In practice, most variations have some effect on the progress of the works and the method of executing the work. Where it is possible, each variation should be valued taking into account all of the delaying and disruptive elements which are directly related to the variation. Common factors which affect the valuation of variations are

- *Changed conditions or circumstances* – The varied work may be carried out in different circumstances than those contemplated at tender stage for reasons which are entirely related to the nature of the variation itself. For example, the contractor may have allowed for excavation to reduced levels using scrapers to deposit spoil in a temporary spoil heap for future disposal. Following a variation to add a length of surface water drain across the site in the location of the spoil heap, the contractor is forced to excavate and load into lorries and cart away most of the spoil in one operation. The revised method takes longer so that more work is done in wet weather and the operation is more costly. There is no delay or disruption to the works as a whole. This change could, and should, be dealt with by valuation under the variation provisions in the contract. There is express provision for such an eventuality in clause 5.9 of the JCT Standard Building Contract 2011 provided there has been a '*substantial change*'.
- *Changed quantities* – Some changes in quantities have a significant effect on cost, even when the nature of the work and the method of executing the work are unchanged. For example, an increase in the volume of concrete may require working overtime in order to complete a floor slab which may be critical to the activity planned to commence the following day. Another example is where an increase in quantities causes some of the work to be carried out later. If the quantity of brickwork increased by twenty per cent, and using the same resources, the time to execute the work (but not any other activities or the contract as a whole) was extended into another pay increase, then the extra costs resulting from the pay increase should be reflected in the value of the variation (assuming a fixed price contract).
- *Changed timing* – Work of a similar nature to that contained in the contract may be ordered at different times so that material and labour costs are not the same as those for the original work.
- *Small quantities* – Variations requiring ordering and execution of similar work in small quantities may involve loss of purchasing discounts and increased prices

payable to subcontractors who may have to return to site after completion of the original subcontract work.

- *Time-related costs* – Where it is possible to isolate a period of delay to part, or the whole, of the works to a single variation (or group of variations), the time-related costs may be reflected in the value of the variation. For example, a major variation to the ground floor structure may cause the time taken to reach completion of the first floor slab to be delayed by one week. If may be appropriate to include the costs of the entire concrete, steelwork and carpenter resources, including concrete mixers, pumps, dumpers, tower-crane, supervision and other preliminary items in the value of the variation. Additional time may be required as a result of actual remeasured quantities exceeding the quantities in the contract bills.

Time-related costs were the subject of a dispute under conditions of contract which were similar to those contained in clause 52 of the FIDIC and ICE conditions of contract. In *Mitsui Construction Co Ltd* v. *Attorney General of Hong Kong* (1986) 33 BLR 1, the executed work in a tunnelling contract was significantly different from that measured in the bills of quantities. The changes in quantity were not a result of a variation order given by the engineer. The contract period was twenty-four months. The result was that the contractor had taken much longer to complete the works and the engineer had granted an extension of time of 784 days. The contractor argued that he was entitled to compensation for the costs of the extra time taken to complete the works. The employer argued that the contract did not empower the engineer to agree or fix any adjusted rates. The Privy Council ruled that the engineer was empowered to vary the rates, thereby opening the way to take account of the time-related costs in the valuation of the variation. It should be noted that clause 2.14 of the JCT Standard Building Contract 2011 contains provisions which would enable time-related costs to be taken into account in the event of a variation arising out of errors in the quantities in the contract bills (other than errors in the Contractor's Proposals).

Clause 52(3) of the seventh edition of the ICE conditions provides for rates for varied works to be varied from the contract rates if the work is not of a similar character or is not carried out under similar conditions as those of the original contract work. Clause 52(4) provides for the contract rates to be revised for the original contract work if the execution of the original work renders such rates to be unreasonable. That is to say, the method or conditions under which the contract work is executed must be significantly affected by virtue of the varied work so that the contract rate is no longer reasonable.

The 1999 FIDIC Red Book contains completely new provisions:
Sub-clause 12.3 provides for the rates or prices applicable to the measured work (including variations) to be the rates stated in the contract. However, a rate or price for an item of work may be amended if

'(a) (i) the measured quantity of the item is changed by more than 10% from the quantity in the Bill of Quantities or other Schedule,

(ii) this change in quantity multiplied by such specified rate for this item exceeds 0.01% of the Accepted Contract Amount,

(iii) this change in quantity directly changes the Cost per unit quantity of this item by more than 1%, and

(iv) this item is not specified in the Contract as a "fixed rate item";

or

(b) (i) the work is instructed under Clause 13 [Variations and Adjustments],
(ii) no rate or price is specified in the Contract for this item, and
(iii) no specified rate or price is appropriate because the item of work is not of similar character, or is not executed under similar conditions as any item in the Contract.'

The requirement to give notice and particulars etc. is given in sub-clause 20.1 (see 1.7 and 4.9, *supra*).

In some circumstances, there may be arguments as to whether the contractual provisions permit the valuation of disruptive, or time-related, elements as part of the variation. It would appear that some rules governing the valuation of variations are sufficiently flexible to permit a very wide interpretation of them so as to enable the quantity surveyor to adopt a sensible approach according to the circumstances. Contractors should bear in mind that it is in their interests to include as much as possible in the valuation of variations so that an element of profit can be recovered on the extra costs. This is particularly important where the provisions of the contract limit reimbursement to cost, or expense, if the additional payment is claimed under any other provisions.

5.12 Dayworks

Payment for work on daywork is usually reserved for circumstances where there is no other reasonable means of valuing the work to be done. Some contracts provide for the contractor to give advanced notice of any work to be done on daywork. There are usually strict time limits for submission of daywork vouchers. It is important to follow the contractual provisions so that the time and materials can be properly recorded and agreed. Contemporary notes setting out the reasons for recording the work on daywork may be helpful. It is important to include all incidentals, such as small tools and transport. Signatures verifying the times and materials used may not signify that payment will be made in the daywork account. However, proper records of such work can be of assistance as supporting documents for other methods of payment.

5.13 Fluctuations

Most fluctuating price contracts use a recognised formula which is applied to the value of work done each month. The base date is predetermined at tender stage and fluctuations are calculated by reference to the published indices each month and the base index. Some contracts contain a 'cut-off date' in the event of delayed completion. However, not all of the effects of price increases may be recovered under the fluctuations clause. If there is a qualifying recoverable delay, any shortfall in recovery which can be substantiated may be included in the contractor's claim for additional payment under the appropriate contract provisions.

In the event of delay during a fixed price contract, work is progressively carried out at later times than allowed for in the tender. The estimator ought to have allowed for the anticipated increases in cost during the contract period in accordance with the tender programme. By comparing actual progress and the value (or cost) of work done each month with anticipated progress and value (or cost) of work in accordance with the programme, it is possible to determine the probable effects of inflation as a result of the delay. The actual monthly value and relevant monthly index can be used to compare the planned monthly value and index as shown in Figure 5.25.

It should be borne in mind that this method may not be accepted as a means of measuring the additional cost due to the delay. However, provided that suitable adjustments can be made for materials and subcontracts let at fixed prices (which are not changed during the contract), materials on site and other factors which may be applicable, this method is generally recognised as a reasonable means of calculating reimbursement. Other evidence, such as comparison of actual invoices and wage rates paid at different times may be required.

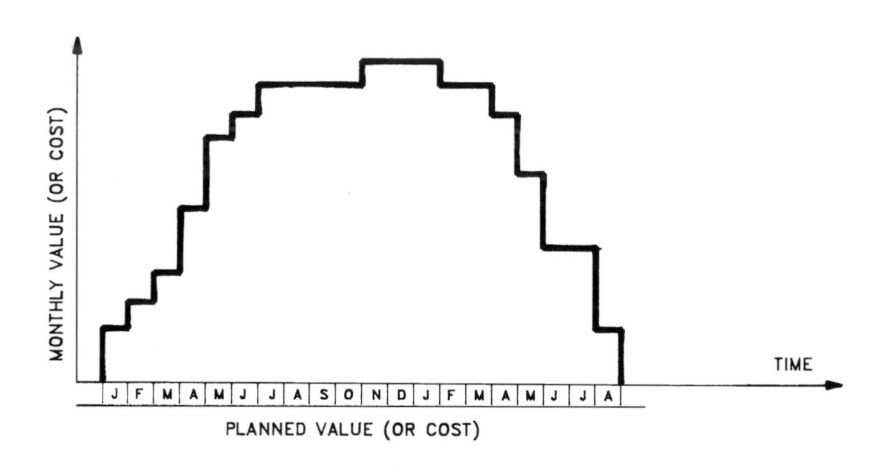

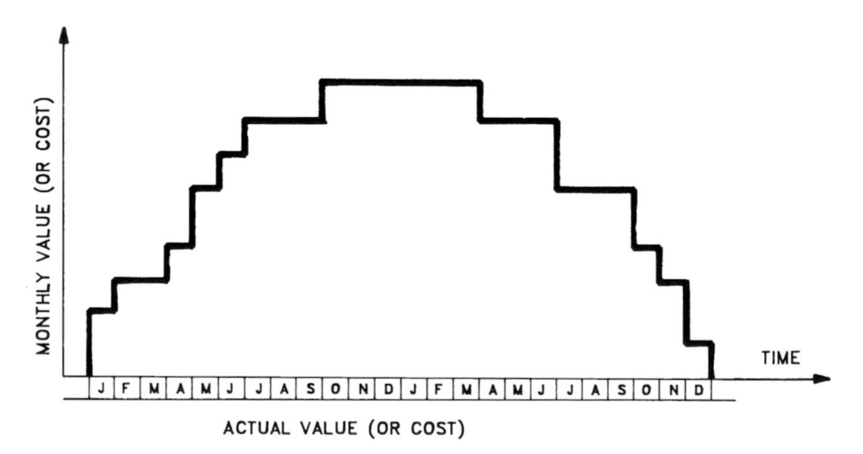

$$\text{INFLATION} = \Sigma\left(\frac{\text{AMV} \times \text{MI}}{\text{BI}}\right) - \Sigma\left(\frac{\text{PMV} \times \text{MI}}{\text{BI}}\right)$$

AMV = ACTUAL MONTHLY VALUE (OR COST)
MI = INDEX FOR RELEVANT MONTH
PMV = PLANNED MONTHLY VALUE (OR COST)
BI = BASE INDEX (AT TENDER)

NOTE: IF MONTHLY VALUE IS USED, RESULT MAY HAVE TO BE ADJUSTED FOR PROFIT ELEMENT

Figure 5.25 Calculation of fluctuations using published indices

5.14 *Quantum Meruit*

A well-drafted variation clause will enable the employer to make substantial changes to the works without invalidating the original contract. Nevertheless, variation clauses do not enable the employer to vary the works without limit. In *Wegan Construction Company Pty. Ltd.* v. *Wodonga Sewerage Authority* (see Chapter 1, *supra*), substantial changes were made and the contractor claimed payment on a *quantum meruit* basis. The variation clause applicable to this case, in part, is almost identical to the pre-1999 FIDIC conditions of contract, and is sufficiently similar to many other forms of contract to justify a detailed analysis of the case.

Clause 40.1 of the contract contained the following terms:

> 'Variations Permitted. At any time prior to practical completion the engineer may order the contractor to
> (a) increase, decrease or omit any portion of the work under the contract;
> (b) change the character or quality of any material, equipment or work;
> (c) change the levels, lines, positions or dimensions of any part of the work under contract;
> (d) execute additional work;
> (e) vary the programme or the order of the work under the contract;
> (f) execute any part of work under the contract outside normal or agreed upon working hours;
>
> and the contractor shall carry out such variation, and be bound by the same conditions, so far as applicable, as if the variation was part of the work under the contract originally included therein.
>
> The extent of all such variations shall not, without the consent of the contractor, be such as to increase the moneys otherwise payable under the contract to the contractor by more than a sum which is the percentage stated in the annexure A of the contract sum, or if not stated, by a reasonable amount.
>
> No variation shall vitiate or invalidate the contract, but the value of all variations shall be taken into account and the moneys otherwise payable under the contract shall be adjusted as provided under cl. 40.4.'

It appears, from the judgement, that no percentage had been inserted in annexure A, and the contract was therefore construed on the basis of the term 'by a reasonable amount'.

In the new plans, excavation was increased by twenty per cent; sewer length was increased from 840 metres to 1181 metres; manholes from nineteen to twenty-seven, requiring a ninety per cent increase in concrete; house connections had increased from forty-seven to ninety-one and the new design included one hundred and sixty metres of excavation below four metres deep which was not shown on the original plans. The contract price was $30867.40 and the revised contract price was $43200.

The contractor argued that the change in design was not a variation permitted by the contract and sought to be released from the contract rates and for payment to be on a *quantum meruit* basis.

Held: In the circumstances the amended plans did not constitute a variation permitted by the original contract.

In practice, where there are very wide variation provisions, and the rules for valuing variations allow for departure from the contract rates, it may be difficult to argue successfully that the works should be valued on a *quantum meruit* basis. There would have to be some compelling reasons which would have made it impossible for the contractor to continue on the basis of the original contract. A substantial increase in the value of work may not, on its own, be sufficient reason to escape from the contract rates.

5.15 Finance charges: Remedies for late payment

In nearly all cases, contractors will allow something in their tender for finance charges on the working capital required to carry out the works. There may not be a positive cash flow until final retention is released. Whatever the contractor's anticipated cash flow, as a general rule, if the value of work increases, the additional financing ought to be recovered in the rates for variations (assuming that the finance costs are allocated throughout the rates for measured work).

However, it is often the case that interim certificates do not reflect the true value of the original contract work including variations. In such circumstances the contractor will be incurring additional finance charges on the under-certified sums. While significant changes have taken place in recent years to compensate contractors for the loss incurred as a result of increased finance charges in cases of default by employers, the commercial reality of the high cost, and potential loss, has not been recognised fully in many modern contracts or in the general law. A claim for finance charges on late, or under-certification, will have to be founded on a contractual provision, or for breach of contract.

In the case of *Morgan Grenfell Ltd* v. *Sunderland Borough Council and Seven Seas Dredging Ltd* (1991) 51 BLR 85, it was held that clause 60(6) of the ICE fifth edition enabled the contractor to claim compound interest on amounts which were included in a statement under clause 60(1) if the engineer failed to certify and it was subsequently found that the amounts ought to have been certified.

However, in *Secretary of State for Transport* v. *Birse–Farr Joint Venture* (1993) 26 BLR 36, Mr Justice Hobhouse said:

> 'The opinion which the engineer is required to form and express in his certificates is a contractual opinion. It must be a *bona fide* opinion arrived at in accordance with the proper discharge of his professional functions under the contract. In sub-clause (3) there is an express reference to "the amount which in his opinion is finally due under contract." It is implicit in sub-clause (2) that the sum certified is that which, in his opinion, he considered to be due under the contract as an interim payment under that month. If it should be the case that the engineer's opinion is based on a wrong view of the contract then it can be said that he has failed to issue a certificate in accordance with the provisions of the contract. This was the case in the *Farr* case [*Farr* v. *Ministry of Transport* [1960] 1 WLR 956]. Therefore, leaving on one side all question of bad faith or improper motive – and none is suggested in the present case – a contractor who is asserting that there has been a failure to certify must demonstrate some misapplication or misunderstanding of the contract by the engineer. For example, it certainly does not suffice that the contractor should merely point to a later certification by the engineer of a sum which had been earlier claimed but not then certified.'

Where the engineer has certified and the employer fails to pay on time, clause 60(7) of the ICE seventh edition and clause 14.8 of the 1999 FIDIC contracts expressly provide for finance charges to be paid.

The case of *Borough of Kingston-upon-Thames* v. *Amec Civil Engineering* [1993] 35 ConLR 39 almost got to grips with the issue as to whether, or not, finance charges could be considered as part of the cost. *Amec*'s claim for finance charges had been rejected on the same grounds as those given in *Secretary of State for Transport* v. *Birse–Farr Joint Venture*. *Amec* argued alternatively that finance charges were part of the cost. His Honour Judge Richard Havey QC stated:

> 'Two questions arise: first, whether interest on any balance found due to the contractor, calculated from the date when that balance could or ought to have been certified, is recoverable as a financing charge representing a cost, or part of a cost, recoverable under a relevant clause of the contract; and, second, whether any interest claimed as a financing charge representing a cost, or part of a cost, recoverable under a relevant clause of the contract continues (whether compounded or not) beyond the date when certification or payment could or ought to have been made.
>
> The first question covers the whole of the amount of interest claimed. My answer to that question is that such interest is not recoverable, since no clause of the contract provides for its recovery. The second question seems to me to be academic, since the amount of such interest, if any, is indeterminate having regard to the terms of the Commercial Settlement. Moreover, interest on that basis is not claimed in the points of claim. Mr Stimpson [for the plaintiffs] submitted that there was enough material before the arbitrator for him to award an appropriate sum under this head, and that, if necessary, the case should be remitted to him for determination of that sum. I reject that argument. Such determination would involve re-opening the Commercial Settlement.'

If there had been no commercial settlement and the argument had been included in the points of claim, perhaps a definitive answer would have been forthcoming. However, this case did not appear to deal with the finance charges on the 'prime cost' from the date when the cost was incurred until the date when it ought to have been certified. This is part of the contractor's 'secondary cost' whether, or not, the engineer certifies promptly (see *Rees and Kirby Ltd* v. *Swansea City Council* (1985), *infra*).

In any event, the form of contract in this case was the ICE fifth edition where the definition of 'cost' is not so widely defined as in the seventh edition and the FIDIC contracts.

In the case of *Amec Building Ltd* v. *Cadmus Investments Co Ltd* [1996] 51 ConLR 105, the court held that under a JCT contract it was proper for simple interest to be awarded from the date of under-certification.

Where delay and disruption occur, the interest on the cost, or on the loss and/or expense, may be claimed as part of the cost or expense. This was held to be the case in *Rees and Kirby Ltd* v. *Swansea City Council* (1985) 30 BLR 1.

A diagram illustrating interest or finance charges from the date of expending the 'primary cost' until payment is received in given in Figure 5.26. The first element [F1] represents the finance charges occurring from the date of incurring the cost until the date of certification (the sums approved in *Rees and Kirby Ltd* v. *Swansea City Council* (1985)). The second element [F2] represents the finance charges due to late payment of certified sums under a provision in the contract (such as ICE or FIDIC) or for breach of contract (*infra*).

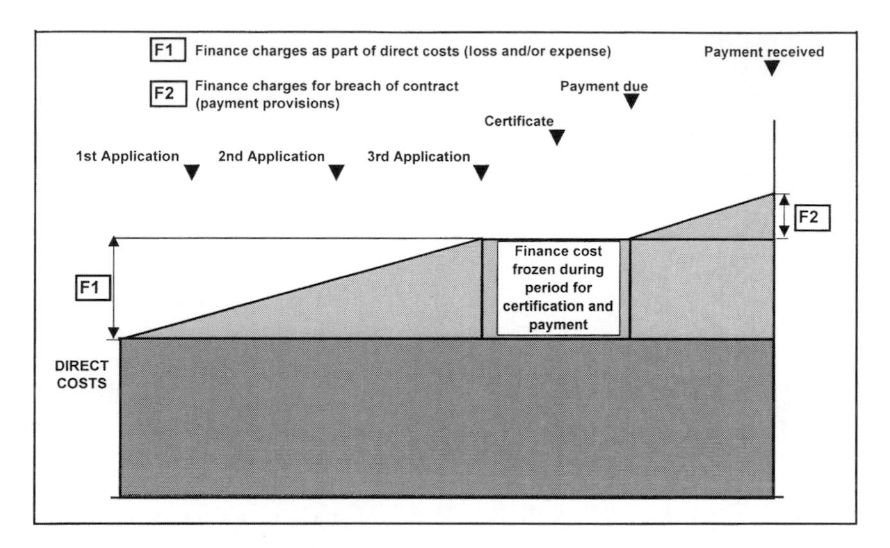

Figure 5.26 Finance charges

The Late Payment of Commercial Debt (Interest) Act of 1998 may be of assistance with respect to late payment of certificates in the UK. Many other jurisdictions have provisions for payment of interest on late payment.

While it is not always essential to include a statement showing the amount of interest on delay and disruption claims, it is a practice which should be encouraged, if only to prompt the architect or engineer to deal with the matters in the earliest possible interim certificate.

Remedies for late payment

Many contractors suffer from late payment of not just one certificate but of several or even all certificates. In international contracting and in domestic contracts overseas, it is not uncommon to experience several unpaid certificates at one time involving several million pounds. Apart from the extreme course of action to terminate the contractor's employment (which contractors are usually reluctant to do), what other redress is available to contractors in these circumstances?

Under section 112 of the Housing Grants, Construction and Regeneration Act 1996 (as amended), a contractor has a right to suspend performance of his obligations under the contract in consequence of non payment provided the contract is a 'construction contract' as defined by the Act. The right is expressly included in NEC 3, the JCT contracts and the ICE contracts. The right is implied into all other contracts which fall under the provisions of the Act. By amendment 3A to the section, a contractor is entitled to recover reasonable costs and expenses incurred in consequence of the exercise of the right to suspend given in section 112.

In most countries, there are no legal rights to suspend work or slow down the progress of work. FIDIC, in its 1999 Red, Yellow and Silver Books, has introduced provisions to enable the contractor to suspend work or slow down his progress (sub-clause 16.1 of the 1999 Red, Yellow and Silver Books). Subject to the contractor giving twenty-one days' notice of his intention to suspend or slow down the progress of the works, if the employer fails to pay by the expiry of the notice period, the contractor may then suspend or slow down the progress of the work. Following such suspension or slowing down, the contractor is entitled to

- an extension of time;
- additional costs;
- a reasonable profit.

These rights and remedies are without prejudice to any other rights (finance charges and/or termination).

The 1999 FIDIC Green Book contains similar but much simplified provisions (sub-clauses 7.3, 10.4 and 12.2).

The NEC 3 forms of contract provide for interest to be paid on late certification of payment and late payment of a certified sum (clause 51.2). The JCT forms of contract also make provision for interest to be paid on late payment of certified sums.

5.16 Cost of preparing the claim

In the vast majority of cases, the cost of preparing the claim is not a recoverable cost. However, there are circumstances in which the cost of preparing claims may be recovered:

- If each claim is prepared by the contractor's staff, as and when they arise during the contract, the salaries and other costs of the staff will usually be included in the site or head office overheads and may therefore be included in the general claim for prolongation.
- If, in spite of all requests for an assessment of the amount of the claim (and provided that the contractor has given all particulars in accordance with the contract) no assessment is made within a reasonable time (and particularly if it has not been made within the period of final measurement or other specified contractual time frame), the contractor would be justified in preparing his own claim and may be entitled to *reimbursement – see James Longley & Co Ltd* v. *South West Regional Health Authority* (1985) 25 BLR 56 at page 57: 'The costs of preparing a final account may be recovered as damages in a suitable cases, e.g. for breach of an obligation on the part of an employer to provide a final account....' This may include the contractor's own managerial time (provided that it is not included in overheads): *Tate & Lyle Food Distribution Ltd and Another* v. *Greater London Council* (*supra*).
- Where certain work is done in connection with preparing a case for arbitration: *James Longley & Co Ltd* v. *South West Regional Health Authority* (*supra*). The cost of preparing unnecessary evidence may not be allowed.

5.17 Assessment and evaluation

Assessment and evaluation of delay and disruption claims will depend on the pricing and accounting policy of the contractor. The following should be established:

The tender

How are the overheads and profit distributed in the tender? Loading rates or preliminaries may merit adjustments to any sums calculated using a formula.

Are all of the site overheads (preliminaries) priced in the preliminaries sections of the bills of quantities? If part, or all, of the preliminaries are included in the rates for measured work, some analysis may have to be done to ascertain the sums to be used as a basis of calculating time-related elements (if it is appropriate to use the contract rates for variation delays). An adjustment may have to be made to account for additional preliminaries recovered in the rates for variations (while there are circumstances where no adjustment should be made for overheads and profit recovered in variations, an adjustment will usually be justified for any preliminaries recovered in variations).

Accounting practice

Are head office overheads charged to the project? If so, on what basis? Time records? Percentage allocation? *Ad hoc*? Unusually high allocation of costs may have to be justified.

Are finance charges included in general overheads? If so there may be duplication with separate claims for finance charges. This may be overcome by deducting interest and finance charges from the general overheads and making a separate assessment of the finance costs on the average working capital required for the delayed project (excluding claims).

Having established the above, the assessment and evaluation of the claim can proceed without fear of unnecessary duplication or omission.

It is important that all facts, evidence and data upon which any calculations are based are collected and bound in an annotated appendix to the claim. In the narrative of the claim, the author should have set out the basis of the claim, giving reasons for any particular method which has been adopted (such as an explanation as to why a particular formula has been used to calculate overheads and profit and any adjustments which have been made).

It is sometimes helpful, and persuasive, to give financial information in tabular and graphical form. This will facilitate a better understanding of the nature of the contractor's claim and may assist in obtaining an early settlement.

Each head of claim should state the source documents used (referring to the appropriate appendix) and any assumptions made for the purposes of calculation or assessment.

5.18 Summary on presentation of claims for additional payment

Similar guidelines to those given for extensions of time are applicable to claims for additional payment. In spite of the fact that contractors may not be reimbursed for preparing a claim, it is usually in the contractor's interest to do so at the earliest opportunity. The temptation to wait until extensions of time are made before submitting a claim should be resisted unless there is real possibility that this will sour relationships beyond repair. However, where there is a contractual requirement to submit a claim within a specified time frame the claim should be submitted within that time frame irrespective of the effect on the relationship with the client. In any event a claim should be prepared (even if not submitted) so that the magnitude of the loss or additional cost can be made available to management. The sooner the opposition are made aware of the amounts which are likely to be claimed, the better the chances that funds will be put aside to meet them.

In addition to the details and particulars mentioned with regard to extensions of time (*supra*), the following may be necessary:

- details of the effects of any delay or disruption on all activities in parallel and subsequent to the circumstances giving rise to the claim;
- an introduction to the claim giving the contractual provisions under which the claim is being made;
- a summary of notices and particulars given during the contract;
- diagrammatic illustrations where appropriate;
- references to recognised authorities and case law relied upon;
- additional, or alternative claims under the general law (if applicable).
- a statement setting out the amount of the claim.

Presentation will depend on the type of claim. If several individual claims are made during the course of the project, these need not necessarily be couched in legal language which is sometimes seen in formal submissions.

5.19 Formal claim submission

Whatever the size or nature of the claim it must be submitted in accordance with any timing and detailing provisions required by the contract. If individual claims are dealt with and settled promptly during the contract, a formal submission setting out the contractual basis and detailed analysis of the contractor's rights and entitlements will not be necessary. However, if settlement is not reached on these claims, the contractor is faced with preparing a document which, it is hoped, will lead to an amicable settlement at the earliest possible time. This type of claim submission may take a form almost approaching pleadings for arbitration. Some contractors spend considerable time and effort in negotiations which fail because of the lack of a sound, comprehensive and persuasive submission which sets out the contractor's claim and the basis upon which the claim is made. The sooner a formal submission is made, the earlier a settlement can be reached or proceedings can commence. In the absence of mandatory requirements of the contact a formal claim submission should include:

Introduction: contract particulars

Names of the parties; description of the works; details of tender and acceptance; the form of contract and any amendments thereto; the contract sum; dates for commencement and completion; phased completion (if applicable); liquidated damages for delay; the programme.

Summary of facts

Date of commencement and practical completion; dates of sectional or partial completion (if applicable); summary of applications for extensions of time; extensions of time awarded; summary of claims submitted; final account and claims assessed (if any); amount of latest certificate and retention; payments received; liquidated damages deducted (if applicable).

Basis of claim

Contract provisions relied upon; common law provisions; contractual analysis and explanation of the basis of the claim.

Details of claim

Full details of every matter which is the subject of the claim. Each separate issue should be carefully set out in a logical format. Key dates, events, causes and effects, references to relevant documents and the like should form the basis of a narrative which fully describes the history of the project and the effects on progress, cost and completion. It is important to distinguish between the causes and effects of delay (and/or disruption), extensions of time and the financial effects of delay and/or disruption. Wherever possible, diagrams, programmes, tables and the like should be included in the narrative (or in an appendix). The extensive use of schedules can be invaluable.

Evaluation of claim

Each head of claim should be calculated, step by step, with explanations and reasons for the methods adopted. Supporting source documents (from which financial data has been used in the evaluation of the claim) should be given in an appendix, or listed, so that the recipient may examine such documents when considering the claim.

Statement of claim

A brief statement setting out the claimant's alleged entitlements and relief sought, such as extensions of time; sums claimed; repayment of liquidated and ascertained damages (if applicable).

Appendices

Copies of all documents referred to in the claim; programmes; diagrams; schedules; financial data.

6 Subcontractors

6.1 Subcontracting generally

An increasing number of contractors do less work by direct labour and they rely to a great extent on subcontractors for the execution of the work. It is perhaps for this reason (at least in part) that contractors are sometimes unable to provide adequate particulars and substantiation in support of their claims.

At tender stage, contractors may rely on subcontractors' quotations for large sections of the works. The tender may be based on the lowest of all the subcontractors' quotations. Once the contract has been awarded, the contractor may then seek to get better quotations (by negotiation with the original tendering subcontractors or by looking for alternative quotations).

In many cases, the contractor will not award the various subcontracts until it is necessary to do so. For example, the subcontract for painting may not be awarded until a few weeks before the painting is due to commence. The contractor runs the risk of price increases in these circumstances. If there has been delay to the project, prior to placing the order for painting, it will be difficult for the contractor to establish a claim for an increase in the cost of the work. Is the increase in the subcontract price due to the delay to the project, or is the market for painting buoyant at the time of subcontracting (whereas it may have been depressed at the time of tender)? If the painting had been ordered at tender stage, the subcontractor may well have had a claim for increased costs due to executing the work at a later date, but this would have been determined by contractual provisions based on conditions at tender stage.

This practice makes it difficult for the contractor to justify a claim for additional payment. The subcontractor will have no interest in providing particulars (because the extra cost is in his price). The employer will not expect to reimburse the contractor for the extra cost caused by a buoyant market. Nevertheless, the contractor may have grounds for a claim.

If all subcontracts were placed at tender stage, based on the same programme and other contractual provisions, the contractor ought to be able to deal with subcontractors' claims as if they were his own (subject to the practical difficulty of getting subcontractors to give the same notices and particulars to the contractor as the contractor is required to give under the principal contract). In practice, subcontracts are placed progressively during the course of the project. If delays occur throughout the project, as the magnitude of the cumulative delay increases, various subcontracts will be placed on different programmes and base costs. Very often subcontracts will be placed when the contractor's current programme is out of date (sometimes the programme may be obsolete to the extent that the programme shows completion of the subcontract works before the date of placing the order for the subcontract). These problems are not imaginary. They occur regularly in real life and are a constant source of contractual disputes.

It is often a problem to establish the subcontractor's obligations regarding progress and completion of the subcontract works when the order, or subcontract, states that the subcontract works shall be carried out 'in accordance with the contractor's programme'. Which programme? Was it the programme which was in existence at the time of making the subcontract (even if the programme shows the subcontract works to be complete before the time of the subcontract)? Is it to be the next revision of the programme? Is it to be any future revision of the programme? What is the situation if the contractor never produces a revised programme?

The dangers which may arise from the above practices are

- The period for completion of the subcontract works may be impossible to determine from the subcontract documents, in which case the subcontractor may have an obligation to complete within a reasonable time. A reasonable time for the subcontractor may not be within the time allowed for in the principal contract.
- The subcontractor may take on board the obligation to execute the works in accordance with any programme of the contractor.

Even more uncertain and onerous provisions (from the subcontractor's point of view) arise when the terms of the subcontract require the subcontractor to proceed with the subcontract works in accordance with the contractor's reasonable requirements. In the case of *Martin Grant & Co Ltd* v. *Sir Lindsay Parkinson & Co Ltd* (1984) 29 BLR 31, the subcontract contained the following terms;

> '2. The Sub-Contractor will provide all materials labour plant scaffolding in addition to that provided by the Contractor for his own requirements haulage and temporary works and *do and perform all the obligations and agreements imposed upon or undertaken by the Contractor under the Principal Contract* in connection with the said works to the satisfaction of the Contractor and of the Architect or Engineer under the Principal Contract (hereinafter called 'the Architect') *at such time or times and in such manner as the Contractor shall direct or require* and observe and perform the terms and conditions of the Principal Contract so far as the same are applicable to the subject matter of this contract as fully as if the same had been herein set forth at length and as if he were the Contractor under the Principal Contract.
>
> 3. The Sub-Contractor *shall proceed with the said works expeditiously and punctually to the requirements of the Contractor and so as not to hinder hamper or delay the work or the portions of the work at such times as the Contractor shall require having reference to the progress or conditions of the Main Works and shall complete the whole of the said works to the satisfaction of the Contractor* and of the Architect and in accordance with the requirements of the local and other authorities.' [Emphasis added]

The works under the principal contract were delayed and the subcontractor was retained on site for a considerably longer period dictated by the progress of the principal contract. The subcontractor contended that there was an implied term that the contractor would make sufficient work available to enable the subcontractor to maintain reasonable and economic progress and that the contractor would not hinder or prevent the subcontractor in the execution of the subcontract works. The subcontractor's claim failed and he was unable to recover the extra costs arising as a result of working on site for a much longer period.

Unfortunately (from a subcontractor's point of view) terms similar to those in the *Martin Grant* case are still all too common in main contractor proposed subcontracts. Subcontractors would be wise to avoid such terms and conditions.

Some of these problems can be avoided by using one of the standard forms of contract which are tailor-made for use with the appropriate principal contract. Some contractors have their own 'look-alike' forms of contract which resemble the standard forms of subcontract but which contain onerous provisions. Subcontractors should not assume that onerous provisions can be defeated by implied terms.

6.2 Nominated subcontractors

Nominated subcontractors have been used in building contracts for over one hundred years but in recent times they have lost favour. They appeared in the RIBA Model Form of Contract at the beginning of the last century. They have a useful and important function where the employer has a genuine requirement to select a subcontractor to execute specialist work. However, the provisions and procedures surrounding their selection and use have become unnecessarily complicated. PC Sums (Prime Cost Sums) in contracts are intended for work to be done by nominated subcontractors or for materials or goods to be supplied by nominated suppliers. There is no room in the current range of JCT contracts for nominated subcontractors but the JCT Intermediate form makes provision for "named" subcontractors. Nominated subcontractors are provided for in the ICE 7th Edition form of Contract (clause 59(1)) and in the FIDIC 'Red Book' (clause 5). Nominated subcontractors have never had a place in the NEC forms of contract.

In general, it is better to limit nominated subcontractors to a minimum, and then only for work which cannot reasonably be included in the contractor's own scope of work. Some of the reasons used to justify the use of nominated subcontractors were

- where the subcontractor is to undertake design responsibility and the features of the subcontractor's design must be co-ordinated with the principal design of the works;
- where it is essential to appoint a nominated subcontractor before appointment of the contractor for the principal contract (for example, there may be long delivery periods for plant and equipment to be provided by the subcontractor);
- where the subcontract works are an extension of work done previously by a particular subcontractor and the same equipment and standards are required to be used in the new works;
- where the subcontract works are the main requirements of the employer and the building, or civil works, are secondary (for example, in process plants);
- where the employer, or its designers, have a particular preference for a subcontractor based on previous performance and standard of work.

Having regard to the increasing amount of sophisticated mechanical and electrical installations, including lifts, escalators, heating and ventilating and air conditioning (HVAC), building automation systems (BAS), security systems (such as closed circuit television – CCTV) and a host of new additions to the field of building services, it was not surprising to find these in the form of PC sums which, in total, may make up more than fifty per cent of the total building cost. The risk of sophisticated installations of any kind is now more often passed to the contractor in the form of design and build contracts. The advantages to the employer of passing the risk to the contractor are very clear to see.

Quite apart from being contemplated on contractual grounds, it is sound commonsense to completely develop the design of all of the specialist subcontract work alongside the design of the building structure and building envelope. If this is not done, how can the design be co-ordinated to ensure that all of the service pipes, ducts, cable trays and equipment be built into the spaces allocated for them? It is this lack of co-ordination which leads to conflicts in the services during construction on site and in some cases renders it impossible to incorporate them in the space allowed. This may require late variations to re-route some of the services in unsightly bulkheads and lowered ceilings. In extreme cases, valuable floor space may have to be sacrificed or, if it is not too late, storey heights may have to be increased. The 'knock-on effect' may include redesign of curtain walls and substantial changes to lift cables, controls and machinery. The cost of all vertical components and finishes will increase.

The direct costs may be a small proportion of the costs of delay and disruption and may cause substantial loss of revenue for the employer. Consultants who embark upon a design up to tender stage without taking account of these potential problems may find themselves being sued by the employer who has not had his building on time and has paid considerable additional sums of money to the contractor for the privilege. That is little consolation to the employer and the whole problem is removed if the contractor is responsible for the design of the works. There are also clear advantages to contractors. They do not have to subcontract with subcontractors not of their choice.

Traditionally, problems have arisen with nominated subcontract work when the contract contains prime cost sums which are no better than provisional sums. If, for example, the design of kitchen equipment is incomplete or not capable of being adequately defined at tender stage, a provisional sum should be used in preference to a prime cost sum. If prime cost sums are used for work which is really provisional, the employer can expect problems when it becomes clear that what is actually required is not what is provided for in the contract. If the subcontract work (as ordered) is different to that tendered for, the nominated subcontractor and the main contractor will be entitled to treat the ordered work as a variation. Further complications may arise if the subcontractor is aware of the work required by the employer (and has provided a quotation for it) but the main contractor is not.

6.3 Contractor's rights to object to nominees

Clause 59(1) of the ICE seventh edition and clause 5.2 of the FIDIC 'Red Book' contain provisions for the contractor to object to a nominee on limited grounds. In general, the contractor will have a right to object to a nominated subcontractor for the following reasons:

- if the subcontractor will not enter into a subcontract on terms containing provisions which indemnify the contractor against the same liabilities as those for which the contractor is liable to indemnify the employer and which indemnify the contractor against any claims arising out of default or negligence of the subcontractor;
- if the subcontractor will not agree to complete the subcontract works in accordance with the reasonable directions of the contractor and to enable the contractor to discharge its obligations under the principal contract;

- if the subcontractor will not agree to complete the subcontract works within the period specified in the proposed subcontract;
- if there are reasonable grounds for the contractor to believe that the subcontractor is unsuitable or is financially unsound.

The first three reasons are usually catered for in standard forms of subcontract designed to operate alongside the appropriate standard form of principal contract. Any attempt by the contractor to impose more onerous provisions will usually be thwarted by predetermined tender procedures which are known by the contractor. However, if the principal contract contains amendments and more onerous provisions than the standard form of contract, the contractor would be within his rights to insist on similar provisions in the subcontract, so far as they were applicable to the subcontract works.

The third reason may arise if nomination procedures are not followed, or if the nomination is made during a delayed project. If there has been no delay and the period for completion contemplated by the subcontractor is inconsistent with the contractor's original programme, the contractor will have a *prima facie* case to object unless the nominee agrees to comply with the programme. If delay has occurred, various problems may arise.

If the contractor is in delay, but no extension is justified, the contractor may reprogramme the remaining work to allow a shorter period for work to be done by a subcontractor to be nominated at a future date. For example the contractor may cause delay of two weeks to activity B–E (see Figure 6.1). The contractor's revised programme may show a reduction in the period allowed for activity J–K which is for work to be done by a nominated subcontractor (see Figure 6.2) so that the completion date is preserved. Activities A–B, B–E, E–F, F–J and J–K are on the critical path but none of the other activities are critical.

Is it reasonable for the contractor to object if the nominee can complete within the original period allowed, but refuses to agree to a shorter period? Can this be overcome by making an extension of time so that the subcontractor can be accommodated, thereby enabling the contractor to escape liability for liquidated damages for his own delay?

Delays may occur for which extensions of time may be due, but for which no extension has been made. There may be a dispute as to the contractor's entitlement to an extension. If a subsequent nominated subcontractor cannot complete its work by the current completion date, is the contractor justified in objecting to the nominee (even if some of the previous delay was caused by the contractor's own default)? Should an extension be made to accommodate the nominated subcontractor? What is the situation if it should subsequently be found that no extensions of time were justified for delays prior to the date of the nomination? Is the nomination made late (even if the nominee was able and willing to commence work on the day when the contractor would be ready for him to commence work)?

The problems which arise when realistic dates for work to be done by nominated subcontractors are out of synchronisation with the contract completion dates and/or the contractor's programme are common. A commonsense solution may be the only way ahead. Some of these problems have been considered in the courts. The House of Lords heard an appeal in the case of *Percy Bilton Ltd* v. *The Greater London Council* (1982) 20 BLR 1 (HL). A nominated subcontractor withdrew his labour from site on 28 July 1978 and went into liquidation. The subcontractor was behind programme

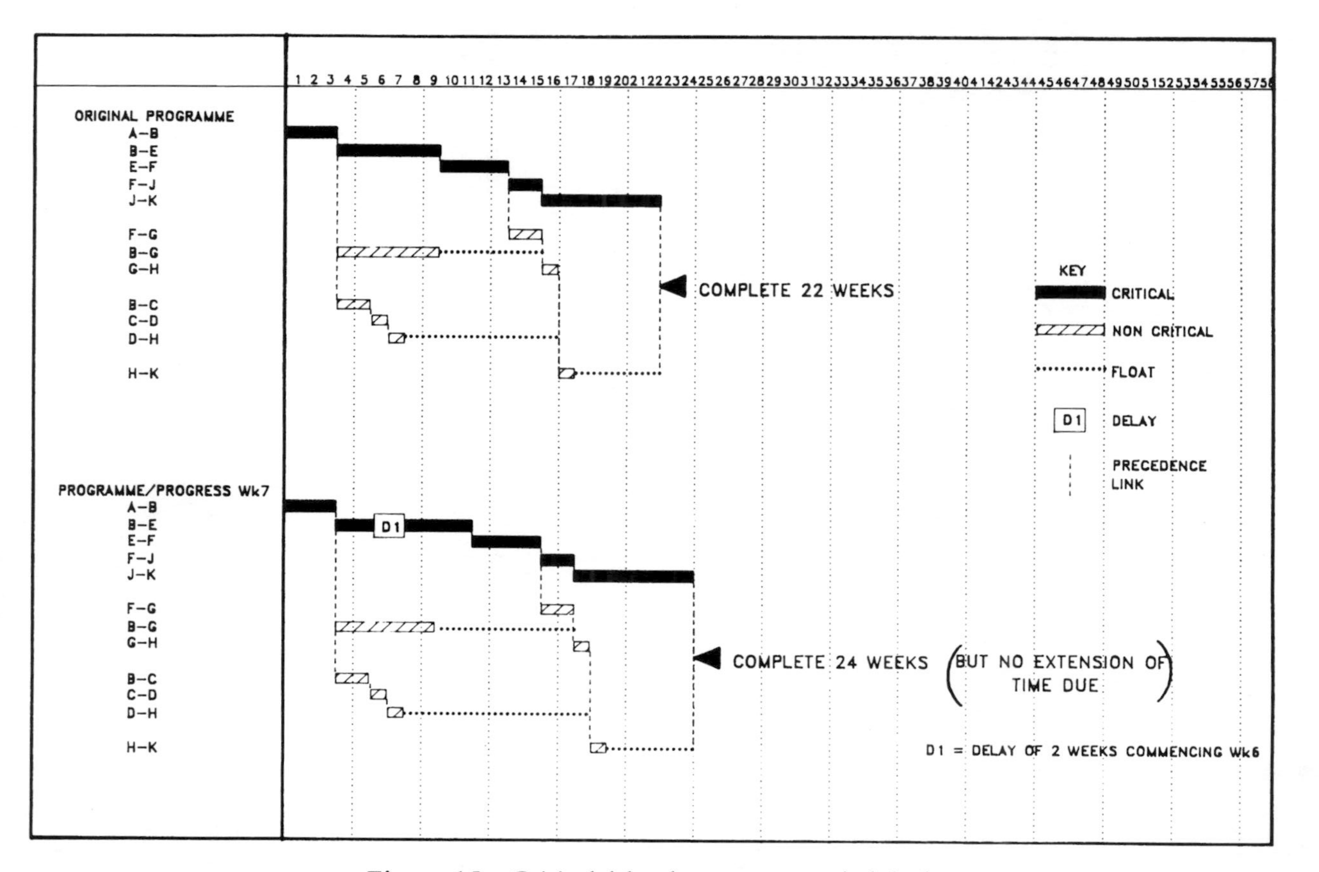

Figure 6.1 *Critical delay due to contractor's default*

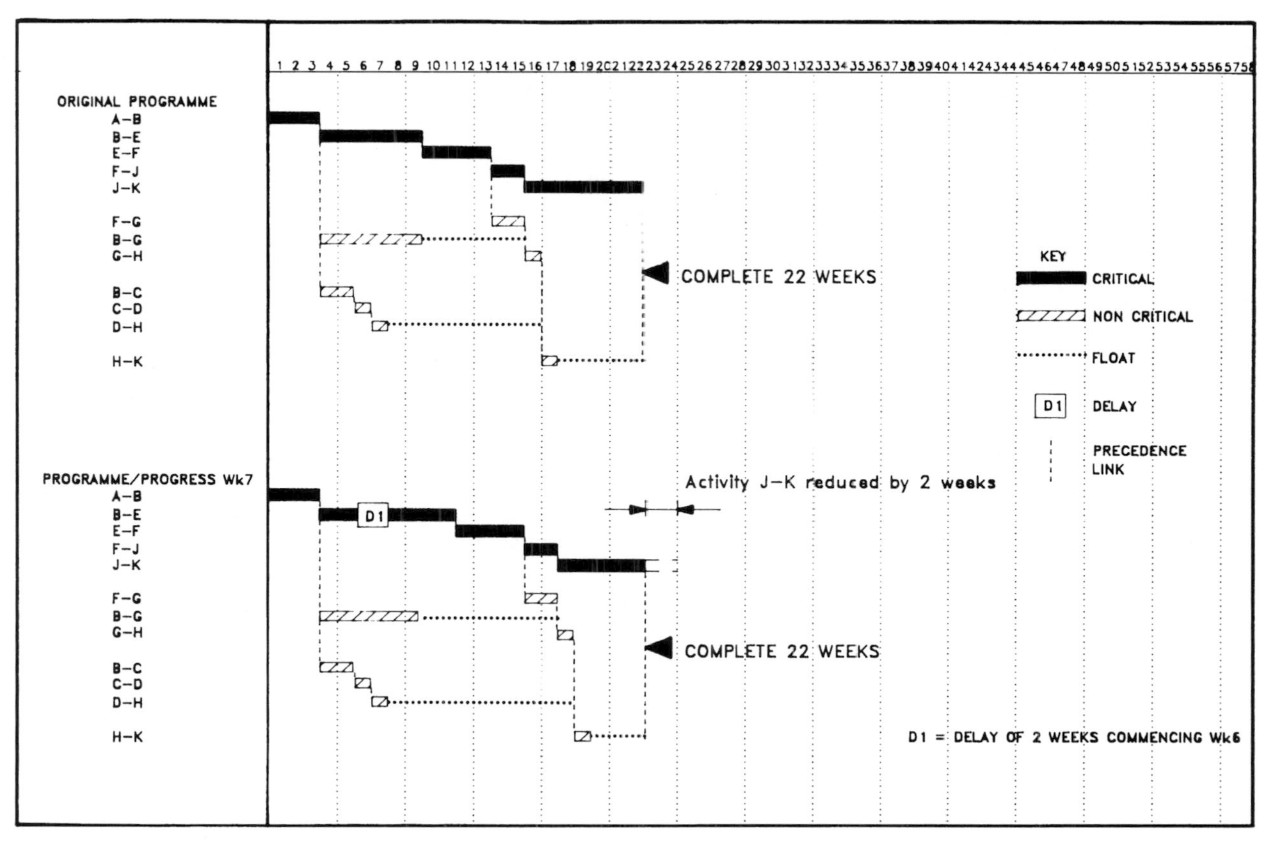

Figure 6.2 *Critical delay due to contractor's default – reduced period for subcontractor to preserve completion date*

at the time of his withdrawal with some forty weeks of the subcontract period remaining. On 31 July 1978, *Bilton* (the contractor) terminated the subcontractor's employment. The (extended) contract completion date at this time was 9 March 1979. Some of the defaulting subcontractor's work was done by a temporary subcontractor (Home Counties Heating & Plumbing Limited) under architect's instructions and on 14 September, *Bilton* was instructed to enter into a nominated subcontract with a new subcontractor (Crown House Engineering Limited). The new subcontractor withdrew his tender on 16 October and on 31 October *Bilton* was instructed to enter into a nominated subcontract with Home Counties. Negotiations between *Bilton* and Home Counties were concluded on 22 December 1978 on the basis that Home Counties would commence work on 22 January 1979 and that the period for completion of the subcontract works would be approximately fifty-three weeks (complete about 23 January 1980). Various extensions of time were granted, but the architect only granted an extension of fourteen weeks (to 14 June 1979) under clause 23(f) of JCT63 for the delay caused by renomination (see Figure 6.3). Further delays occurred; the contractor completed late and the *GLC* deducted liquidated damages. The contractor contended that time was at large and that liquidated damages could not be deducted. It was held that the delay arising out of the renomination fell into two parts. The first part was due to the original subcontractor's default and the second part was due to the unreasonable time taken to engage Home Counties to complete the work. No extension of time was justified for the first part of the delay (however, it appears that the extension of time granted by the architect included the first part of the delay), but the architect was empowered to grant an extension of time for the second part of the delay. As the first part of the delay was not due to the employer's default, time was not at large and liquidated damages could be deducted.

An important aspect of this case was reported in the Court of Appeal *17 BLR 1* (at page 18):

'A quite separate argument by Mr Garland is what is described as his "overshoot" submission; that is to say that, at the time of the application for the re-nomination, the new subcontractor's date for completion was later than the plaintiff's date for completion and that, since this would make it impossible for the plaintiffs both to accept the new subcontractor and to comply with the provision in their own contract as to time for completion, therefore the time provision must go completely, time will be at large and the right to liquidated damages will disappear.

I do not accept this argument. The contractor, faced with a subcontract with such a provision as to completion, would be entitled to refuse to accept the subcontractor under clause 27 [of JCT63]; or what the subcontractor could do would be to say that he would not agree to accept the subcontract unless at the same time the employer would agree to an extension of time for the completion of the main contract.'

The above argument found support in the House of Lords, *20 BLR 1* (at page 15).

It should be noted that this case dealt with *renomination* which was not due to the employer's default. If these circumstances arose with respect to the original nomination of a subcontractor to execute the work covered by a PC sum, the result would probably be very different. The contractor may have a claim for breach of contract and/or a claim arising out of a late instruction pursuant to provisions in the contract.

In a similar case of *Fairclough Building Ltd* v. *Rhuddlan Borough Council* (1985) 30 BLR 26, a nominated subcontractor ceased work in September 1977 and the

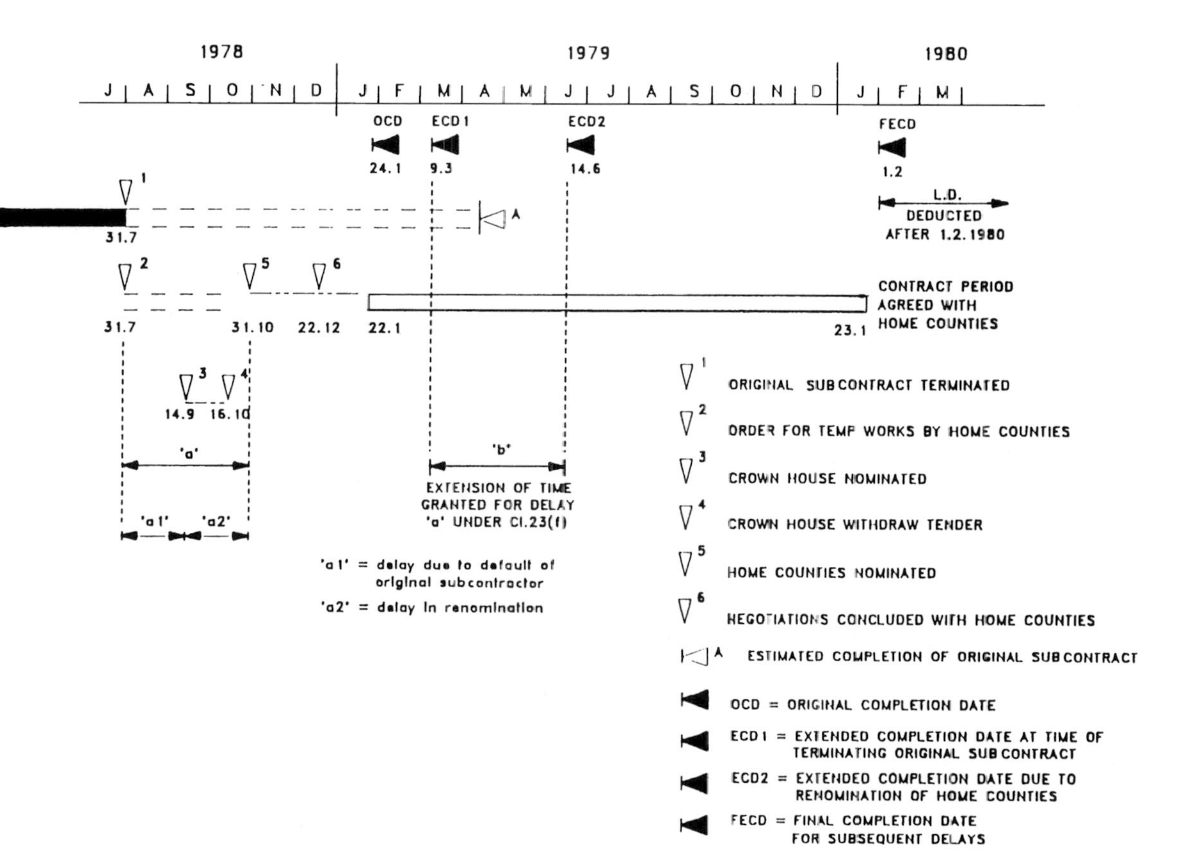

Figure 6.3 *Percy Bilton v. Greater London Council*

subcontractor's employment was terminated. The subcontractor was eight weeks late at the time of termination. The standard conditions of JCT63 had been amended to exclude delay by a nominated subcontractor (unless such delay was due to a reason for which the contractor could obtain an extension). The original date for completion of the principal contract was 2 May 1977 and an extension of time for strikes occurring prior to the subcontractor's withdrawal from site was granted to 10 May 1978. The architect did not issue an instruction to renominate a new subcontractor until 24 February 1978. The contractor objected to the renomination on the grounds that it did not include making good defects in the original subcontract work and that an extension of time would be required to cover the time required by the new subcontractor (twenty-seven weeks from acceptance of tender) which would overrun the date for completion of the main contract (see Figure 6.4). The architect replied (on the latter issue) stating 'I would confirm our intention to grant an extension of time in connection with the re-nominated Sub-contractor's programme time at such time as the effect on your overall programme can be ascertained.'

It was held that the contractor was entitled to refuse the nomination. With respect to extensions of time, the following is of practical importance, *30 BLR 26* (at page 41):

> 'In the present instance delay until 24 February therefore falls on the contractor [following *Bilton* v. *GLC*, but on the grounds that the period taken to renominate by 24 February 1978 was not an unreasonable time]. If, when his contractual completion date is some two and a half months off he is asked to do work which

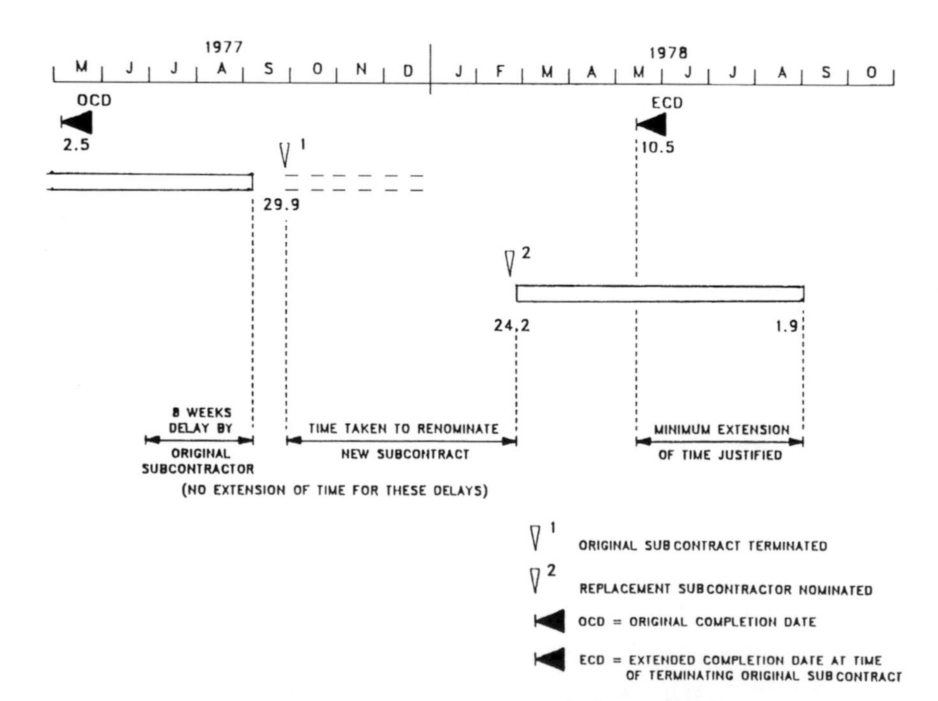

Figure 6.4 Fairclough Building Ltd v. Rhuddlan Borough Council

will take six months to complete we see no reason for saying that the contract must be so construed that he cannot insist on an extension of time under the main contract to bring it in line with the proposed subcontract.'

And at page 42:

> 'It may well be that the doing of such work would not delay *actual* completion of all outstanding work but if the contractor is required on 24 February to do work which cannot be done until September it appears to us at least arguable that he could not be in breach of contract by reason of failure to do that part of the work until September and thus that he is entitled, **if he does not exercise his right to prevent nomination,** to an extension to that date.' [Bold emphasis added]

The main difference between the *Bilton* case and the *Fairclough* case was that *Fairclough* had asked for an extension of time to cover the period to complete the work required by the new nominated subcontractor, and the architect had intimated that he would grant an extension of time, whereas no extension had been requested in the *Bilton* case.

Similar problems arise where the contract contemplates the use of named subcontractors to execute work. However, if the contractor is unable to enter into a nominated or named subcontract for reasons which are justified, appropriately worded provisions could be included in the contract to overcome such difficulties by way of a variation or by omitting the work or by substituting a provisional sum.

6.4 Subcontractors' programmes

In most cases, the contractor's programme will indicate overall periods for work to be done by each subcontractor. The programme may show separately, first, second and final fixing and various sections of the subcontract work. Whatever the level of detail shown on the contractor's programme, many subcontractors will need to subdivide their work into several activities when preparing their own programmes. If the contractor has been given sufficient design information when tendering for the work, he will have been able to prepare his programme taking into account many of the factors which govern the sequence of the subcontractor's work. Assuming that the contractor's programme is still valid (based on progress and the current contractual completion date), the contractor and the subcontractor ought to be able to agree a realistic programme which is consistent with the overall programme. It would be unusual if some minor reprogramming of the principal works and/or the subcontract works was not necessary at the time of subcontracting. A competent contractor, given sufficient information at tender stage, ought to be able to accommodate such reprogramming without raising an objection or subsequent claim.

In some cases, the subcontract works may be on the critical path, in which case the subcontractor's programme and the overall programme need to be given careful attention, preferably before the subcontractors submit their tenders for the subcontract works. This can be facilitated by ensuring that the contractor and all tendering subcontractors have detailed discussions at pre-tender stage. Where the subcontract works are not critical, the subcontract period may be open to negotiation. For example, if activity B–G in Figure 5.2 (*supra*) represents work to be done by a subcontractor, the options for the subcontract period may be

- commence at the beginning of the fourth week and complete in six weeks (earliest start);
- commence at the beginning of the tenth week and complete by the end of the fifteenth week (latest start);
- commence at the beginning of the fourth week and complete by the end of the fifteenth week (earliest start and latest finish);
- commence at any period between the beginning of the fourth week and the end of the fifteenth week (which may be more or less than six weeks' duration).

These options may have a bearing on the subcontractor's price for executing the subcontract works and should therefore be discussed before submission of the subcontractor's tender (whether the subcontractor is domestic or nominated). They may also have a bearing on the contractor's attendance (for example, the period required for scaffolding). In the case of a domestic subcontractor, the contractor can use the optimum solution to arrive at the best tender for the main works or (if arising after award of the principal contract) to obtain a saving on its original estimate for the works. In the case of a nominated subcontractor, the employer may enjoy the benefit of the optimum solution.

Another difficulty arises where the subcontract is executed on or about the date of commencement of the main works, but the subcontract works are due to commence several months later. Delays to the main works which occur prior to the date of commencement of the subcontract works may qualify for an extension of time (for completion of the main works). However, the progress of the subcontract works has not been delayed (since the subcontractor has not yet commenced work) and there may be no provision to adjust the completion date of the subcontract works. It is therefore important to make provision in the subcontract for the commencement and completion dates of the subcontract works to be adjusted in such circumstances. This may be overcome by stating a period for completion of the subcontract works and providing for the subcontractor to commence work within a specified period of the contractor's written notice. This may be ideal for contractors, but subcontractors may require provisions to enable them to recover any additional costs which may arise from delayed commencement.

6.5 Extensions of time for completion of subcontract works

Most standard forms of subcontract contain provisions for extensions of time to be made for the following reasons:

- delay for which the contractor is entitled to an extension of time for completion of the works pursuant to the principal contract;
- delay or default on the part of the contractor, or persons for whom the contractor is responsible (such as other subcontractors).

If the subcontract work is on the critical path, a qualifying delay which affects the subcontract works will have equal effect to the completion periods for the subcontract works and the main works. If the subcontract work is not on the critical path, delays which occur may have different effects on the relevant completion dates. For example, delay on the critical path may give rise to an extension of time for completion of the main works, but no extension of time may be necessary for completion of the subcontract works. Alternatively, a qualifying delay to the progress of subcontract works may justify an extension of time for completion of the subcontract works, but no extension may be necessary for completion of the main works (subject to the contractor subsequently needing an extension – see Chapter 5, *supra*).

The interpretation of the various subcontracts ought to run in parallel with the main contracts (back to back). That is to say, all provisions in the main contract which are in connection with the subcontractor's obligations, rights and remedies are required to work together as if the main contract provisions were set out in the subcontract. For example, clause 4.1 of the FIDIC Subcontract (1994) provides for an unpriced copy of the main contract to be made available for the subcontractor to inspect and the subcontractor is deemed to have full knowledge of it. In addition, clause 12.1 states:

> 'The provisions of Clause 54 of the Conditions of Main Contract in relation to Contractor's Equipment, Temporary Works, or materials brought on to the Site by the Subcontractor are hereby incorporated by reference into the Subcontract as completely as if they were set out in full herein.'

There is a 'back to back' subcontract for use with the NEC 3 forms of main contract and the JCT publish subcontracts for use with the JCT range of main contracts. Remarkably the ICE 7th Edition and the FIDIC 1999 Edition standard forms of main contract do not have standard forms of subcontract to accompany them.

It is important that the subcontractor's obligations to give notice should be consistent with the contractor's rights to give notice in such a way that neither the contractor nor the subcontractor are disadvantaged. The FCEC (Federation of Civil Engineering Contractors – now CECA – Civil Engineering Contractors' Association) blue subcontract form for use with the 6th edition ICE main contract and the FIDIC subcontract (1994) for use with the 1987 fourth edition do not quite achieve this. Subclause 7.2 of the 1994 FIDIC subcontract provides for the subcontractor to be entitled to an extension of time (*inter alia*) for reasons for which the contractor would be entitled to an extension under the main contract. The sub-clause goes on to say:

> 'Provided that the Subcontractor shall not be entitled to such extension unless he has submitted to the Contractor notice of the circumstances which are delaying him within 14 days of such delay first occurring...and in any case to which [the Contractor may obtain an extension under the Main Contract] the extension shall not exceed the extension to which the Contractor is entitled under the Main Contract.'

These provisions may be difficult if not impossible to reconcile in some circumstances.

Firstly, clause 44 of the FIDIC main contract requires the contractor to give notice within twenty-eight days after such event has arisen and the engineer has the discretion to grant an extension if notice is not given within twenty-eight days.

Clause 7.2 of the subcontract provides for the requisite notice to be a *condition precedent* to the subcontractor's rights to an extension, whereas Clause 44 of the main contract is not a condition precedent (see *Bremer* v. *Vanden* in 1.7, *supra*).

If a subcontractor gives notices of a delaying event, which would justify an extension of time, fifteen days after the event had first arisen and the contractor gave notice under the main contract before the expiry of the twenty-eight day period, the contractor would be entitled to an extension and the subcontractor would not be entitled to an extension. This could lead to absurd results. The main contractor and the subcontractor would have different programmes and other subcontractors would not know which programme they had to work around. For example, in Figure 5.3 (*supra*), the contractor could be granted an extension of time of two weeks due to delay (D1) and the subcontractor executing activity B–E (if he could not obtain an extension due to his late notice) may have to accelerate to complete by his original completion date (end of week nine). If activity E–F was to be executed by another subcontractor, would this subcontractor be obliged to commence at the beginning of week ten (the original programme date and immediately following completion of the delayed subcontractor's work) or would he be required to commence work at the beginning of week twelve (consistent with the delayed programme incorporating the extension of time granted to the main contractor)? Can the main contractor programme all activities (after activity B–E) to start and finish on the original dates and build in two weeks' float for himself?

What happens if the subcontractor's work is not on the critical path for the main contract?

Assume that the period for completion of the subcontract works is six weeks commencing in the fourth week (as activity B–G in Figure 5.4, *supra*). There is six weeks' float in the main contractor's original programme. A suspension order issued by the engineer under the main contract causes delay to the subcontract works for a period of two weeks (delay D1). If the subcontractor gave notice to the main contractor within fourteen days and the contractor gave notice under the main contract within twenty-eight days, the contractor may not be entitled to an extension (because the delay will not delay completion of the works) and on the express wording of the subcontract, the subcontractor will not be entitled to an extension.

If there should be provisions in the subcontract for liquidated damages for late completion contemplated by sub-clause 7.4 of the sample Conditions of Particular Application to the FIDIC subcontract, is the contractor entitled to levy liquidated damages if the subcontractor fails to complete within the original subcontract period of six weeks? The subcontractor is blameless. The contractor is blameless (the engineer caused the problem). If no extension can be granted to the subcontractor, does time for completion of the subcontract works become at large?

Under some forms of contract (*infra*), delays by other subcontractors (or by the contractor) may entitle the subcontractor to an extension of time, but the contractor may not be able to obtain an extension of time for completion of the main works. In such circumstances, various claims and counter-claims may arise (see Chapter 7, *infra*).

6.6 Named subcontractors

A named subcontractor is a subcontractor who has been named by an employer as a subcontractor the main contractor must contract with. Through his professional team, the employer will usually have been involved in the tendering process to identify the work the named subcontractor is to carry out and consequently the name of the subcontractor. However, once named, the employer has no liability to the main contractor for the actions of the named subcontractor. An employer may provide a number of names from which the main contractor can choose one.

The JCT Intermediate Building Contract 2011 makes provision for named subcontractors (clause 3.7).

6.7 Design and drawings provided by the subcontractor

The usual contractual arrangement is that the employer is responsible for the design of the works unless the contract is one under which the contractor is responsible for designing part, or the whole, of the works. The principle applies whether or not any part of the works are to be carried out by a subcontractor. In design and build contracts the contractor may require a subcontractor to carry out design work or to provide installation drawings. If a subcontractor is to design an element of the works, he will be responsible for any failure in that design. However, where the subcontractor is required to provide installation drawings, these may not be considered to be design drawings and the subcontractor will be liable to the contractor for any delay caused by late issuance of installation drawings: *H. Fairweather & Co Ltd* v. *London Borough of Wandsworth* (*supra*).

What constitutes a design drawing and what constitutes an installation drawing? There are no reasons why these should not be defined in the principal contract (definitions in the subcontract may be of no consequence since the contractor may argue that such definitions were not part of the principal contract). In the absence of such definitions, it is suggested that the following principles may be applied:

- design drawings include drawings which require calculation and/or co-ordination with other parts of the works (such as works being designed by other subcontractors);
- installation drawings include drawings which merely represent the subcontractor's interpretation of the design having regard to all design information provided by the employer's design team.

In the former case, the design of the subcontract works may depend on design development of other parts of the works, for which the employer assumes responsibility for design. The design team will have to ensure that the design of all installations, and the building, fit together. In the latter case, the subcontractor must be given sufficient information on all other installations to enable him to complete his installation drawings.

Some contracts attempt to place responsibility for co-ordination of design by subcontractors (in addition to co-ordination of the installation) upon the contractor, or on the various subcontractors. This is a recipe for disaster and employers should be advised to avoid this practice. It is likely to cause considerable delay and extra cost which, in spite of careful drafting of the contractual provisions, will almost certainly end up being the responsibility of the employer.

6.8 Variations to the subcontract works

Provided the subcontract is 'back to back' with the main contract a main contractor will not usually have a conflict of interest in circumstances where a subcontractor is entitled to claim for the value and time taken to comply with a variation to the subcontract works. If the subcontract is not 'back to back' with the main contract the main contractor will have a conflict of interest in circumstances where the subcontractor is entitled to claim for the value and time taken to comply with a variation to the subcontract works but the main contractor is not be entitled to recover under the main contract. In such circumstances the subcontractor may have a battle on his hands to obtain his entitlement. It should be irrelevant to a subcontractor whether the main contractor is entitled to time and payment. As stated at the end of this chapter, general legal principles applicable to main contracts also apply to subcontracts and whether the subcontractor is entitled to payment (and extension of time) for a variation will depend on the precise wording of the subcontract, not the wording of the main contract.

A main contractor who is not entitled to the value of a variation (and extension of time) under the main contract may be in some difficulty if he is required to grant an extension of time to a subcontractor which would cause the main contractor to miss the completion date stated in the main contract. It is important for main contractors to ensure that subcontracts do not put them in this unenviable position.

Apart from the problem the main contractor will have if he is liable to a subcontractor for the value (and time) caused by a variation to the subcontract works, a main contractor may also find that there are consequences relating to other subcontractors. If, for example, a ground works subcontractor is entitled to an extension of time to complete ground works operations which has a knock on effect that prevents a bricklaying subcontractor from commencing brickwork, the main contractor may find himself liable to the brickwork subcontractor as well as to the ground works subcontractor.

As a consequence of the problems identified above, it is common for subcontracts to state that the subcontractor shall work in accordance with the main contractor's progress. While such a provision is beneficial to the main contractor, subcontractors should seek to avoid it as it can be tantamount to providing an obligation on the part of the subcontractor to keep resources available for an open ended period of time at no extra cost.

6.9 Delay and disruption claims

To a large extent the comments referred to at section 6.8 are applicable to section 6.9.

Subcontractors are likely to be delayed by various causes. Subcontractor's claims for delay or disruption to the progress of the subcontract works for reasons which give rise to a claim against the employer are likely to receive the contractor's co-operation to ensure that the full effects are reflected in extensions of time and additional payment made under the principal contract. The sooner the contractor and subcontractor can recognise the merits of co-operating on the keeping of records, giving notices and the means of formulating a claim, the greater the chance of maximising the remedy and reimbursement of additional payment. A joint approach which is consistent is a powerful tool, provided that the claim has merit and substance.

However, claims for delay or disruption to the progress of the subcontract works by the contractor, or other subcontractors, are likely to be resisted by the contractor for various reasons:

- if the delay is concurrent with a delay which is the employer's responsibility, the contractor's claim against the employer may be prejudiced;
- the contractor may have difficulty in disentangling the causes and effects of delays caused by himself and/or various other subcontractors, thereby increasing the likelihood that the cost will have to be borne by the contractor.

If the contractor can clearly identify the culprit(s) to whom a subcontractor's claim may directed, he may be less resistant to the claim. Much will depend on the chance of recovering the costs from the defaulting subcontractor(s). Where the contractor is to blame for the delay or disruption, settlement will depend on the contractor's and subcontractor's records and the subcontractor's ability to present his claim with clarity. Onerous subcontract conditions and counterclaims will often feature in negotiations and it may be in the contractor's interest to do a deal in order to conceal the nature of the dispute from the employer's professional advisers (particularly if the delay is one which is concurrent with delays which may give rise to additional payment under the principal contract). Subcontractors who recognise a vulnerable contractor can often achieve a prompt and satisfactory settlement.

Subcontractors who cause delay and/or disruption may find themselves liable for claims from three directions (all usually recoverable through the provisions of the subcontract):

- claims for the contractor's own costs caused by the subcontractor's default;
- liquidated damages levied against the contractor by the employer;
- claims from other subcontractors against the contractor resulting from the subcontractor's default.

However, if the subcontractor is delayed by matters for which the employer is responsible, the contractor may be able to obtain a remedy under the main contract subject to the subcontractor complying with the relevant provisions in the subcontract. Provisions for giving notice and particulars vary with subcontracts. As stated in 6.5, subcontracts ought to be 'back to back' with the main contract. In the case of the older FCEC/CECA and FIDIC subcontracts (*supra*), the provisions are somewhat better than those for extensions of time. Sub-clause 11.1 of the 1994 FIDIC subcontract states:

> '...whenever the Contractor is required by the terms of the Main Contract to give any notice or other information to the Engineer or to the Employer, or to keep contemporary records, the Subcontractor shall in relation to the Subcontract Works give a similar notice or such other information in writing to the Contractor and keep contemporary records as will enable the Contractor to comply with the terms of the Main Contract.'

From a practical point of view, it is vital that contractors and subcontractors maintain good relationships and co-operate with each other in order to obtain the fullest benefit under the main contract. Contractors who are at odds with both the employer and their subcontractors are likely to be at a disadvantage whichever way they turn to obtain payment of claims.

6.10 Liquidated damages

Most subcontracts do not have an express provision for liquidated damages (one of the exceptions being the 1994 FIDIC subcontract, Part II). If a subcontractor causes delay which results in the main contractor being liable for liquidated damages, then the main contractor will seek to recover the amount of liquidated damages from the subcontractor, usually through an express term in the subcontract or as damages.

Problems arise if the subcontract works are small in comparison to the main contract but vital to the timely completion of the works. It could be the case that a few days' delay by a subcontractor may bring with it a liability for liquidated damages far in excess of the value of the subcontract works. The fact that the amount of liquidated damages is high compared to the subcontract sum does not make the amount a penalty and there is little the subcontractor can do to avoid liability after the event. If the subcontractor can show that the liquidated damages provisions in the main contract were never properly communicated to it, then that may be a defence. The best advice to any small subcontractor is to obtain a limit on its liability for any claims for delay before submitting a tender for the work.

If liquidated damages for late completion are stipulated in a subcontract, then it is possible that the courts may construe this head of damages as including the main contractor's and other subcontractors' claims for delay: *M.J. Gleeson Plc* v. *Taylor Woodrow Plc* (1990) 49 BLR 95. The example sub-clause in Part II of the FIDIC subcontract states that the amounts stipulated for liquidated damages are the only damages for delay. However, there is no reason why there could not be provisions for both liquidated damages and other claims for delay, provided that the definition of liquidated damages clearly sets out what is included in the stipulated sum. Any head of claim which did not fall within the definition may (subject to it being a direct consequence of the delay) be claimed in addition under a suitable express term in the subcontract.

If more than one subcontractor is guilty of delay to completion (and possibly the main contractor is also at fault), the allocation of liability to each subcontractor may be fraught with difficulty. The contractor is required to link cause and effect and to illustrate how and why each subcontractor is liable and for what amount: *Mid Glamorgan County Council* v. *J. Devonald Williams* (see 1.6, *supra*).

6.11 The law applicable to the subcontract

As a general rule, the law applicable to main contracts is just as applicable to subcontracts. However, it is not uncommon for contractors to choose the law applicable to the subcontract which is at odds to the law applicable to the main contract. This usually happens by error rather than by a conscious decision and it can create significant problems. If, for example, the main contract was subject to the Laws of South Africa and a subcontract was subject to English Law, how would penalties work if the subcontractor caused delay? Even the laws of Scotland and England differ to some extent and it is prudent to ensure that subcontracts operate under the same law as the main contract for the sake of consistency.

What would be the situation if the subcontract work was on the critical path and the subcontractor was the only culprit that caused delay, as a result of which the main contractor finished late and the employer deducted penalties? The subcontractor would no doubt argue that the penalties could not be levied against it. The main contractor would argue that the deduction of penalties by the employer was part of the damages suffered as a result of the subcontractor's delay, which were contemplated and of which the subcontractor was aware by having notice of the main contract provisions.

Response to Claims: Counter-Claims

7.1 General policy

No one likes to be on the receiving end of a claim. From the employer's point of view it will mean additional cost by way of loss of revenue and/or additional payments to be made to the contractor. From the point of view of the professional advisers to the employers, it may reflect on the firms' competence in preparing contract documents and on their skills in contracts administration. They may also be faced with additional costs of administration which cannot be recovered from the employer. When contractors receive claims from subcontractors, they will be mindful of the fact that the claim may arise out of their poor organisational skills, in which case they will not be able to obtain reimbursement from the employer or other subcontractors.

Nevertheless, valid claims are a fact of life in modern construction projects. They are an essential feature of small and large contracts and the machinery to deal with them should be regarded as an important element of control. Prompt submission of notices and particulars, followed by a considered response from the recipient as soon as possible will usually facilitate early remedial action and settlement.

The employer's professional advisers will normally be required to act as independent valuer or certifier under the contract and/or advise the employer on the contractor's rights and entitlements. In *Pacific Associates Inc and Another* v. *Baxter and Others* (*supra* – Chapter 1), it was held that the contractor had no recourse against the engineer if he should fail to certify properly and act fairly. The contractor would, however, be able to recover from the employer. Consultants should therefore be aware that they are likely to be the target for negligence claims from the employer if the contractor's claims arise out of their failure to value or certify in accordance with the conditions of contract. Employers should also be aware that their interference with the impartial certifying function of their consultants will be self-defeating (*Morrison-Knudsen* v. *B.C. Hydro & Power* and *Nash Dredging Ltd* v. *Kestrell Marine Ltd*, Chapter 1 – *supra*).

Consultants who fend off claims to avoid criticism of their own performance may only be compounding the problem and laying themselves, and the employer, open to greater claims from contractors. Delay in recognising a claim and responding to it may cause any hope of effective remedial action to be lost. Poor advice given by consultants to the employer upon which the employer relies to embark upon the road to litigation or arbitration which could otherwise have been avoided may lay the consultants open to claims from the employer.

If claims are to be dealt with effectively, employers and their professional team should decide on policy at the outset. There should be a system of referral to experienced staff who are not responsible for the day-to-day administration of the project. Advice from an independent consultant may be appropriate from time to time. A policy statement should include the following:

- consultation as soon as the first notice from the contractor is received (or as soon as any member of the professional team recognises a potential claim);

- delegation of responsibilities to verify facts;
- consultation to determine the validity, merits and substance of the claim;
- consultation to analyse the causes and effects of the matters which are the subject of the claim;
- recommendations on the quantum of the claim;
- content of written response and necessary certificates to be issued.

Whatever policy is adopted, the timing and content of the first response to a claim situation may be critical to its successful conclusion with the minimum exposure to delay and additional cost. It is important that the response should reflect the opinion of the certifier (which may take into account the various matters discussed during consultations with other members of the professional team and the opinions of persons to whom the claim may have been referred).

The content should be sufficiently detailed to show that the matter has been properly considered and the door should be left open to allow the contractor to submit further arguments or facts in support of the claim.

7.2 Extensions of time

Prompt response to any situation which may jeopardise progress and completion of the works by the due date is necessary for practical and contractual reasons. From a practical point of view, it is essential to have a valid programme which is consistent with progress and the latest extended completion date. Without continual review which takes account of actual delay and entitlement to extensions of time, there is no means to plan future issuance of details and instructions and there is no yardstick by which to measure future delays. Extensions of time granted several months after the event (or even several months after completion of the project) are of no practical use and any opportunity which may have existed to reduce the delay may have been lost.

From a contractual point of view, time to exercise the powers to grant an extension may be critical to the employer's rights to levy liquidated damages (*Miller* v. *London County Council*, Chapter 1 – *supra*). Some doubt has been expressed on the validity of the argument that if extensions of time are not granted within the time contemplated by the contract, the employer's rights to liquidated damages are extinguished. In *Temloc Ltd* v. *Erril Properties Ltd* (Chapter 1 – *supra*), the *employer* argued that since the architect had failed to grant an extension of time within the twelve-week period provided in clause 25.3.3 of JCT80, the employer could not recover liquidated damages but he could recover general damages in lieu of liquidated damages (which in this case had been £nil in the appendix to the contract). The judge took the view that the twelve-week period was *directory only* and not mandatory. The JCT Standard Building Contract 2011 includes similar provisions (clause 2.28.5). This view has been highly criticised by distinguished authors on construction contracts. However, since it was the *employer* who was seeking to rely on this provision in order to recover damages which it could not otherwise claim under the liquidated damages provision in the contract, it is not surprising that the judge did not see fit to allow the employer to benefit from his own architect's failure to grant an extension within the time limits laid down in the contract. If this practice was condoned by the courts, nothing would prevent employers from encouraging architects to delay granting an extension of time if the general damages were found to be greater than the liquidated damages specified in the contract. It is submitted that the contractor would still be able to succeed in arguing that the employer could not rely on the liquidated damages provisions in the contract, if the architect did not grant an extension of time within the twelve-week period, notwithstanding the judge's view in *Temloc* v. *Erril Properties.*

In an Australian case it was held that the employer had the option to levy liquidated damages (if the architect issued the necessary non-completion certificate) or, if no certificate was issued, the employer may levy general damages which may exceed the amount stipulated for liquidated damages: *Baese Pty Ltd* v. *R.A. Bracken Building Pty Ltd* (1989) 52 BLR 130. The commentary to the case (at pp. 131 and 132) suggests that the judgement is of limited application and should not be regarded as creating a precedent giving rise to a general right to opt for liquidated damages or general damages.

The requirement to grant an extension of time within the periods contemplated by the contract does not mean that the architect's or engineer's opinion must be the right one. The architect, or engineer, need only consider the delay and grant, or refuse to grant, an extension of time within the requisite period. Provided that there was a genuine attempt to deal with the matter, and the contractor was notified of the extension, or reasons for refusing an extension, within the period, then the contractual provisions will be satisfied and the employer's rights to rely on the liquidated damages provisions will be preserved. A refusal, or insufficient extension, which is not based on a genuine attempt to assess the delay (but merely to preserve the liquidated damages provisions), may not be effective. No response, or protracted exchanges of correspondence with no conclusion, may not preserve the employer's rights to liquidated damages if it should be subsequently held that an extension of time ought to have been granted at the appropriate time.

The case of *Aoki Corp* v. *Lippoland (Singapore) Pte Ltd* [1995] 2 SLR is likely to be regarded as introducing a change to the existing ground rules. This Singapore decision dealt with the peculiar wording of clause 23.2 of the SIA (Singapore Institute of Architects) form of contract in which the architect is required to give an initial intimation of his decision as to whether, or not, a delaying matter deserves an extension of time, in principle within one month of the contractor's notice of delay, without having to give his opinion on the amount of the extension in his initial intimation. The contractor argued that the architect's failure to give his initial decision in principle within one month had the effect of the architect losing his power to grant an extension, that time (for completion) was 'at large' and that the employer lost its rights to levy liquidated damages.

The judge found in favour of the employer. That is to say, the architect's initial intimation was not given too late in the circumstances of this particular case. Certainly, the wording of clause 23.2 of the SIA form does not make it a *condition precedent* to the architect's rights to grant an extension of time that the initial intimation should be given within one month. That much can be gleaned from *Bremer Handelsgesell-Schaft M.B.H.* v. *Vanden Avenne-Izegem P.V.B.A.* (*infra*), in which the judge stated that there must be express wording to bar an entitlement or right if notice was not given within the prescribed time.

However, the Singapore case did not deal with the issue as to when the extension of time itself should ultimately be granted. In the circumstances of this case, the judge took the view that the initial intimation (given three months after completion of the works) was not too late. However, it is evident that an initial intimation given two-and-a-half years after completion quoted in a reference to an earlier case of *Tropicon*

Contractors Pte Ltd v. *Lojan Properties Pte Ltd* [1991] 2 MLJ 70 (CA); (1989) 2 MLJ 215 (dist) was given too late. Notwithstanding the *Aoki* v. *Lippoland* decision, an architect or engineer who delays any decision regarding an extension of time therefore runs the risk of jeopardising the employer's rights to levy liquidated damages.

It would seem at least arguable that the case of *Aoki* v. *Lippoland* has not affected the existing ground rules for most other forms of contract, but it must be said that there may be a shift in policy on the application of extension of time provisions. What appears to be emerging from the Singapore decision is an acceptance, by the courts, that if an extension of time is not granted within the time contemplated by the contract, then the contractor may be entitled to damages (the costs of acceleration), rather than allowing time to be at large.

Clause 2.28.3 of the JCT Standard Building Contract 2011 requires the Architect/ Contract Administrator to state the relevant event which he has taken into account when making an extension of time and if there is more than one applicable relevant event, allocate periods of time against each relevant event. Clause 2.19.1of the JCT Intermediate Building Contract 2011 does not require the Architect/Contract Administrator to allocate periods against each relevant event. Under the JCT Standard Building Contract 2011 the Architect's/Contract Administrator's response to any notice of delay is required within twelve weeks of receipt of the contractor's notice or particulars, or if the notice is given within twelve weeks of the completion date, using his best endeavours, before the completion date (clause 2.28.2). Under the JCT Intermediate Building Contract 2011 the Architect's/Contract Administrator's response is required as soon as he is able to assess the extension (clause 2.19). In both cases there is provision to review the extensions of time within twelve weeks of practical completion.

It will be interesting to see whether the decision in the case of *Steria Limited* v. *Sigma Wireless Communications Limited* [2007] (*supra*), which concerned a heavily revised MF1 standard form contract, will be applied in the case of the JCT standard forms of contract. In the *Steria* case the court stated that it is not necessary for a contract to expressly state that a condition will operate as a condition precedent provided the contract makes clear in ordinary language that the right to an extension of time is conditional on notification being given. The court was referring to the notification of delay but it seems reasonable to expect the same principle to apply to a contractual provision requiring notification of loss or expense.

The ICE 7th Edition requires the Engineer to grant an interim extension of time 'forthwith' (if he considers that one is due) and consider all circumstances within fourteen days of the date for completion and make a final determination within twenty-eight days of substantial completion.

The 1987 fourth edition of FIDIC (clause 44) is almost non-committal as to when the engineer should respond to a claim for extensions of time. FIDIC requires the engineer to respond 'without undue delay' if he considers that an extension is due.

NEC 3 contemplates a considerable amount of co-operation between the contractor and the project manager with respect to notification and assessment of 'compensation events'. Sub-clause 64.3 states:

> 'The *Project Manager* notifies the *Contractor* of his assessment of a compensation event and gives him details of it within the period allowed for the *Contractor's* submission of his quotation for the same event. This period starts when the need for the *Project Manager's* assessment becomes apparent.'

Section 6 of NEC 3 makes provision for Compensation Events, which include extensions of time and financial compensation. Under NEC 2 there was no sanction for a failure to issue a notice of delay or an extension of time. NEC 3 resolves the anomaly and provides time limits for notification of delay, quotations, responses to quotations and consequences in respect of failures to comply with the time limits. An example of an Compensation Event is provided in Appendix C (*infra*).

The 1999 FIDIC Red, Yellow and Silver Books require the engineer or employer (as the case may be) to respond within forty-two days after receiving the contractor's notice and particulars (sub-clause 20.1). A response may be with approval or with disapproval and with detailed comments. Sub-clause 3.5 requires the engineer (within the forty-two days) to consult with each party in an endeavour to reach agreement or, failing agreement, he must make a fair determination. Under the Silver Book, where the employer deals with such matters (as there is no engineer), the contractor must register his dissatisfaction with the employer's assessment of his claims within fourteen days or he must give effect to it. The text of this clause could have been clearer and there is at least the possibility that the employer's determination could become final and binding if the contractor fails to register his dissatisfaction within fourteen days. If the contractor registers dissatisfaction [within fourteen days], the dispute may be referred to adjudication.

Under the 1999 FIDIC Green Book, no time limits are laid down within which the employer must respond.

The contents of a response to a notice or claim for an extension of time are important. For the following reasons it is good practice to give periods of extension for each separate cause of delay:

- it enables the contractor to be fully aware of the delays which have been considered (within the time limits for granting an extension);
- it facilitates agreement on some of the delays and extensions of time granted therefor, and enables both sides to concentrate on resolving the contentious delays;
- if facilitates agreement on delays which may, in any event, have to be quantified in order to establish the amount of additional payment;
- it enables the contractor to identify which delays apply to which subcontractors so that consistent extensions of time can be granted under each subcontract.

Some common problems which arise are

Late information

Information may be issued late (having regard to the programme) but not actually cause delay to the progress of the works because the contractor is not ready to commence the work which is affected by the late information. Is the contractor entitled to an extension of time? Factors to be considered include the following:

- Is there a lead time? That is to say, does the contractor have to order materials or arrange for the work to be done by a subcontractor? The architect, or engineer, may be already in delay prior to any delay by the contractor and would therefore not have been in a position to anticipate the site progress. It may well be that the information was required before the contractor commenced the affected work

and the contractor had no need to commence prior to receiving the information (see Figure 7.1).
- Is the contractor in delay for matters which would justify an extension, or is he being dilatory?

It may be that even if no extension was justified, the employer could not in any event have been in a position to give the information earlier and could not therefore have obtained use of the project any earlier than the time required to complete the remaining work affected by the late information. The best advice is not to rely on the contractor's delays to put off issuance of information for construction. If it is unavoidable, the contractor may be entitled to the benefit of the doubt and the employer may have no claim against the contractor.

Information and variations issued after the completion date

If the contractor is in culpable delay and liable to liquidated damages, further delay caused by information and instructions issued after the completion date has passed may be difficult to deal with within the contractual machinery. In such circumstances, contractors will seize the opportunity to establish extensions of time for the full period up to the date when the delay ceased to affect the progress of the works, plus an allowance to complete the remaining works. Much will depend on the reasons for the late information or variation (see Chapter 5 – *supra*) and the terms of the contract.

If the contract does not provide for extensions of time after the completion date has passed, or if the provisions allow for extensions of time without preservation of the employer's rights to liquidated damages, the employer and his professional advisers will need to give careful consideration to the need for giving any instructions at all, and if they cannot be avoided, what should be done to protect the employer's interests?

If the architect, or engineer, is of the opinion that an extension of time can, and ought to be made, then an extension should be made having regard to the facts and circumstances. If the architect, or engineer, is of the opinion that no extension can be made, then the contractor should be advised accordingly.

Except in the most straightforward of cases, these circumstances may require expert advice on the meaning of the contractual provisions and the period of extension which may be justified (*see Balfour Beatty Building Ltd* v. *Chestermount Properties Ltd* in 5.3 – *supra*).

Omission of work

The provisions of the JCT Standard Building Contract 2011 contemplate an allowance for any Relevant Omission, which produces a saving in time, when considering the period of any extension of time which may be granted. Clause 2.28.4 requires the Architect/Contract Administrator to:

> 'fix a Completion Date for the Works or that Section earlier than that previously so fixed if in his opinion the fixing of such earlier Completion Date is fair and reasonable, having regard to any Relevant Omissions for which instructions have been issued after the last occasion on which a new Completion Date was fixed' .

The Architect/Contract Administrator may also, after the completion date, fix an earlier completion date than that previously fixed if it should be reasonable to do so

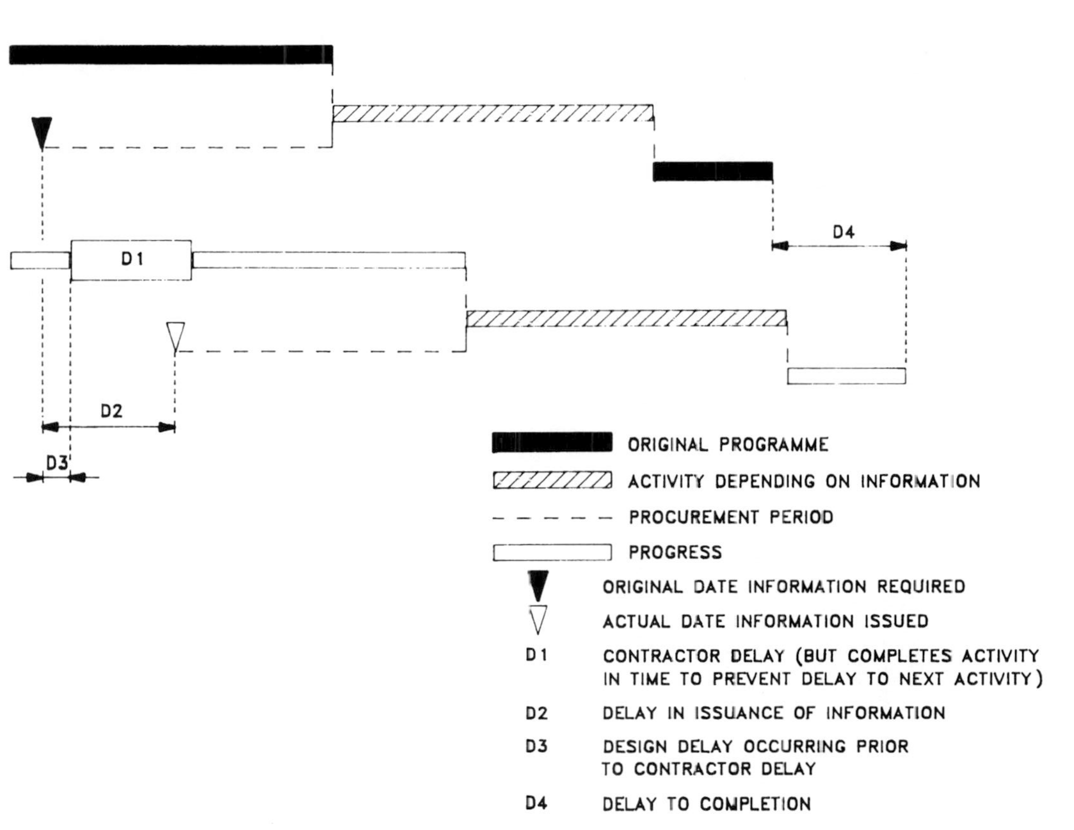

Figure 7.1 *Late information concurrent with contractor's delay*

having regard to omissions ordered after the date of fixing the previous completion date – clause 2.28.5.2.

Whether or not there should be any omissions, the Architect/Contract Administrator is required to grant an extension of time within twelve weeks of the contractor's notice, or before the completion date, whichever is earlier. Even if notices and particulars and extensions of time are given without delay, the contractual provisions may not allow all omissions to be taken into account. There may be a period when omissions occur but which cannot be taken into account (see Figure 7.2). It should also be borne in mind that, where there is delay in granting an extension of time (even if it should be granted within the requisite period), the contractor may issue a programme which is a fair reflection of the extension due with the exception of any omissions. It would be good policy to bring the omissions to the attention of the contractor before work has progressed in accordance with the revised programme to the extent that the benefit of the omission is lost.

In order to prevent these circumstances arising, where the architect is of the opinion that there is a case to make any allowance for omissions, he should address the matter without delay in consultation with the contractor so that there is no doubt as to the reasonableness of any allowance. In any event, an allowance should only be made where the omission is on the critical path, or is of such a nature that resources (previously required to execute the omitted work) can be diverted to execute work on the critical path *and that there will be a benefit in time.* It is insufficient to make a subjective judgement without a proper analysis of the programme and progress to establish that a saving in time was justified.

Neither NEC 3 nor ICE 7th Edition allows the completion date to be brought forward in consequence of work omitted from the contract.

It is important to note that omissions to have the work done by others is a breach of contract and may not qualify to be taken into account (see also Chapter 1 – *supra*).

Concurrent delays

Many architects, and engineers, refuse to grant extensions of time for qualifying delays when the contractor is himself in delay at the same time. Sometimes this is justified, but very often an extension of time is necessary (see Chapter 5 – *supra*).

Once the contractor has given notice of delay, or if the architect, or engineer, is aware of delays on the part of the contractor, it is important that these delays are monitored. The consultants responsible for granting extensions of time and/or certifying additional payment arising out of delay owe a duty of care to the employer to ensure that the contractor is not given any more time or money than is reasonable in all of the circumstances. They will have to consider those matters described in Chapter 5 (*supra*).

In order to ensure that the employer is not exposed to additional costs which should not rightly be borne by the employer, the architect, or engineer, will have to be aware of delays by the contractor at the earliest possible time. Once aware of these delays, it is important to keep contemporary records.

Any response to claims for extensions of time should state which delays (by the contractor) were concurrent with qualifying delays and which (if any) were considered to be delaying completion of the works. This may not necessarily reduce or affect the extension of time to which the contractor is entitled, but the contractor will be aware of the fact that the architect, or engineer, is well informed on the progress of the works.

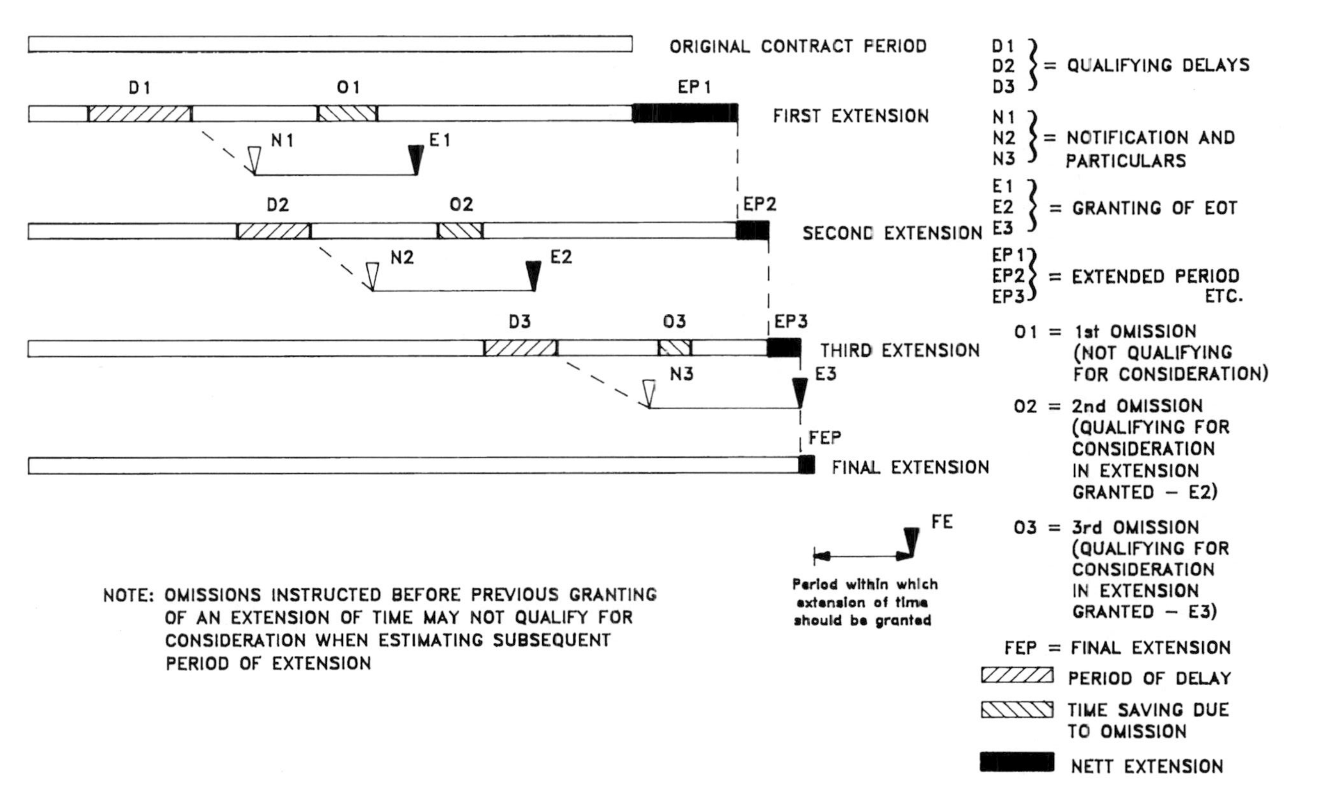

Figure 7.2 Omission of work – clause 2.28.5.2 of JCT 2011

7.3 Claims for additional payment

While a prompt response to claims for extensions of time is essential for practical reasons, and to keep the liquidated damages provisions alive, a response to claims for additional payment is not usually subject to the same urgency. Nevertheless, provided that the contractor gives notice and particulars in accordance with the contractual provisions, assessment of the sums due and certification for payment should be done as soon as possible. It is often in the employer's interests to deal with these claims as early as possible. Agreement of claims and settlement from time-to-time during the course of the project reduces the contractor's ability to collect all outstanding claims into a 'global claim' which may be little more than a statement claiming the difference between the certified value of all completed work and the actual cost.

Many contractors may prefer to wait until the end of the contract before submitting a formal claim. If that is the case, the employer may not be disposed towards any attempt to encourage the contractor to submit his claims as they arise so that they can be settled and set aside. In such circumstances, the employer's professional team should be aware of potential claims and make whatever assessment they can from their own investigations and records. The employer will be interested in knowing the amount of the potential claim, but no action should be taken to effect payment before the contractor has complied with the contractual procedures (unless a deduction in the contract price may be justified). Once the contractor's particulars are received, the assessment can be modified in the light of such particulars and a prompt settlement may be possible.

If the contractor has gone to a great deal of time and trouble to submit a well thought-out claim, with full particulars and sensible calculations, then a written response merits a similar amount of detail, indicating where there is agreement and reasons for any adjustments which, in the opinion of the architect, or quantity surveyor, or engineer are considered to be appropriate. If, on the other hand, the contractor's submission is poorly argued and presented, the temptation to dismiss the claim out of hand should be resisted. A response should explain why the submission is unsatisfactory and it should give the contractor the opportunity to clarify, or amend the claim. Further particulars may be requested, and these should be specified. If it is a frivolous, or unfounded, claim, the contractor should be politely told so. If the claim is justified, and has merit, it is unlikely to go away, in which case it may be appropriate to give the contractor some guidance as to presentation. It may well be that the matter which is the subject of the contractor's claim is one which ought to be dealt with as a variation, thereby giving the engineer, or quantity surveyor, the scope to deal with the matter within the rules for valuation of variations. Provided that the employer is not disadvantaged, this approach may be the most acceptable to all concerned.

The NEC conditions require co-operation and an early response to all compensation events by the project manager within the time provided in sub-clause 64.3 (see above).

The 1999 FIDIC conditions require the engineer to consult with the employer and the contractor and to respond within forty-two days in accordance with sub-clause 20.1. Under the Silver Book, the employer deals with such determination and this may become binding if the contractor fails to register dissatisfaction within fourteen days (see above).

7.4 Counter-claims: liquidated damages: general damages

Many claims which may be levied by the employer against contractors are overlooked or are not considered to be worth pursuing. This may be because employers are fearful that such claims could be the reason for large claims by contractors which may otherwise have been waived.

Claims which may be levied against contractors include those arising out of defective work and failure by the contractor to execute work expressly authorised under the terms of the contract. Some claims may be made under the terms of the contract and the amounts of the claims may be set off against interim or final payments due to the contractor from the employer. Others may be common law claims.

The most common counter-claim against contractors is the deduction of liquidated damages for late completion of the works (or if provided for in the contract, for late completion of sections of the works). In order to be enforceable, a liquidated damages provision must be unambiguous and the sum stated in the contract must be a genuine pre-estimate of the employer's likely loss, estimated at the time of making the contract in the event of delay to completion. If the sum stated is a penalty, the employer cannot rely on the clause (unless the law expressly permits penalties). It will not be deemed to be a penalty merely because the employer's actual loss is less than the liquidated damages (for example, if the liquidated damages were based on realistic anticipated rents at the time of making the contract, and the market had collapsed by the time the works were complete, the contractor could not argue that the sum was a penalty).

The employer's professional team may have to advise the employer on the amount of liquidated damages to be inserted in the contract and on the contractor's potential liability for liquidated damages when the contractor is in delay during the course of the contract. However, consultants should not use the threat of liquidated damages in any response to a contractor's delay claim, even if it is clear that the contractor is in default. Such matters should be for the employer alone, and then only when the consultants have properly considered all delays which may give rise to an extension of time.

The JCT Standard Building Contract 2011 requires the Architect/Contract Administrator to certify that the contractor had failed to complete the works by the completion date (as a fact) before the employer can give notice and deduct liquidated damages – clause 2.32. Many other forms of contract do not require a certificate of any sort as a prerequisite to the employer exercising its rights to deduct liquidated damages.

NEC 3 makes provision for liquidated and ascertained damages as a secondary optional clause (clause X7). The parties are free to choose whether to include the provision. If included in the contract, there is no requirement for the Project Manager to certify non completion before the Employer can deduct liquidated and ascertained damages.

Like the NEC 3 form, the ICE 7th Edition does not require a notice of non completion prior to the employer's entitlement to deduct liquidated and ascertained damages.

It is often argued that the architect cannot certify that the contractor has failed to complete the works by the completion date unless and until he has considered all of the delays for which an extension of time may be granted: *Token Construction Co Ltd* v. *Charlton Estates Ltd* (1976) 1 BLR 48. If, however, a further extension of time is granted after liquidated damages have been deducted, the employer must repay the liquidated damages for the relevant period of further extension (for example, clause 2.32.3 of the JCT Standard Building Contract 2011). It has been held that the contractor is entitled to interest on the liquidated damages withheld, and subsequently repaid: *Department of Environment for Northern Ireland* v. *Farrans* (1981) 19 BLR

1. Clause 47(5) of the seventh edition of the ICE conditions of contract provides for interest on liquidated damages to be repaid to the contractor as a result of further extensions of time.

If there are no provisions in the contract for liquidated damages, the employer may be able to levy a claim for general damages. Where there is a provision for liquidated damages for late completion of the works, but there are no provisions to deduct liquidated damages for late completion of each phase (assuming that the contract contemplates phased completion), the employer may have a claim for general damages for late completion of any phase: *Mathind Ltd* v. *E. Turner & Sons Ltd* (see Chapter 3 – *supra*). Where the employer has lost his rights to liquidated damages, he may be able to claim general damages for late completion (see Chapter 1 – *supra*).

General damages may arise if the employer suffers loss as a result of any breach of contract by the contractor. Provided that the nature and cause of the loss are not identical to those which may be recovered under a liquidated damages provision, then general damages may be recoverable in addition to the liquidated damages for late completion. Some tailor-made conditions of contract provide for liquidated damages *and* general damages for delay. Provided that the nature of the damages are not identical (thereby duplicating the claim for delay), provisions of this kind may be enforceable. For example, if the liquidated damages were a genuine pre-estimate of the loss of revenue and direct costs of supervision during the period of overrun, a separate claim to recover delay costs levied by other contractors (who were delayed by the contractor) would not be a duplication of the same damages and may be recoverable in appropriate circumstances.

The 1999 FIDIC Red, Yellow and Silver Books require the employer to give notice of any claims against the contractor as soon as practicable (sub-clause 2.5) and the engineer (the employer in the case of the Silver Book) is required to determine the claim in accordance with sub-clause 3.5. This procedure is a prerequisite to deduction from sums due to the contractor or payment from the contractor. Under the Silver Book, where the employer deals with such matters (as there is no engineer), the contractor must register his dissatisfaction with the employer's assessment of his claims within fourteen days or it becomes binding. If the contractor registers dissatisfaction within fourteen days, the dispute may be referred to adjudication.

7.5 Claims against subcontractors

There is an increasing incidence of claims made by subcontractors against contractors and by contractors against subcontractors. Some forms of subcontract devised by contractors are aimed at precluding any claim at all from subcontractors and they attempt to provide for claims to be made against subcontractors on dubious grounds with little supporting evidence. Recent cases in the courts have identified the most unreasonable contractors in this regard. Notwithstanding the adverse publicity and understandable indignation expressed by various trade associations, the majority of contractors use recognised standard forms of subcontract and apply the provisions fairly. For those that do not operate the provisions fairly, there are the mandatory provisions of the Housing Grants, Construction and Regeneration Act 1996 (as amended *supra* at 1.10) concerning notice of payment, notice of withholding and adjudication to contend with. Those provisions require clear details of payment and withholding and swift redress in the event that payment is not made in accordance with the contract.

Where a subcontractor is in delay, or is disrupting the progress of the works, the contractor will naturally wish to recover any losses incurred from the defaulting subcontractor. Where there is only one subcontractor in delay, and there are no competing delays, it is possible to establish liability with relative ease. However, it is probable that there will be several delays occurring at the same time, in which case the contractor will be faced with the difficulties which have been mentioned in respect of concurrent delays in Chapter 5 (*supra*). Only the most careful attention to records and regular updating of programme and progress schedules will enable the contractor to establish liability and quantum of damages which may be recoverable from several subcontractors (and possibly from the employer) for what may be substantially the same period of delay.

Where the contractor becomes liable to liquidated damages for late completion of the main works, he will seek to recover some, or all, of the damages from defaulting subcontractors.

Apportionment in the event of delay by several subcontractors is almost bound to cause difficulty. Even where the contractor has been able to calculate the sum which is due from the subcontractor, the provisions for set-off in the subcontract may frustrate the contractor's ability to deduct the amounts due from payments which would otherwise be paid to the subcontractor. The general rule is that the contractor's rights to set-off at common law are not affected by the contractual provisions unless there is clear language in the contract to bar the general right of set-off: *Gilbert Ash (Northern) Ltd* v. *Modern Engineering (Bristol) Ltd* [1974] AC 689. However, where the terms are explicit, and the set-off provisions are exclusively laid down in the subcontract, the contractor's rights to set-off will be determined by the contractual provisions.

The Housing Grants, Construction and Regeneration Act 1996 (as amended *supra*) outlaws 'pay when paid' in construction contracts in the UK. Section 113(1) states:

> 'A provision making payment under a construction contract conditional on the payer receiving payment from a third person is ineffective, unless that third person, or any other person payment by whom is under the contract (directly or indirectly) a condition of payment by that third person, is insolvent.'

In addition to claiming all, or part, of the liquidated damages for late completion of the main works from a defaulting subcontractor, the contractor may also have a claim for other loss and expense, such as prolongation and/or disruption costs incurred by the contractor and by other subcontractors. The quantification of such claims where there are several competing delays is bound to be fraught with problems and, unless a commercial settlement can be reached between the contractor and the subcontractors, the matter may have to be settled by several separate arbitrations or by the same proceedings involving several parties.

Avoidance, Resolution and Settlement of Disputes

8.1 Commercial attitude and policy

Many contractors and subcontractors genuinely wish to avoid claims even when there are good grounds for them. This attitude is usually adopted in the belief that firms with a reputation for claims will not be included on some tender lists, and where they are included, they may be disadvantaged if tenders are very close. In some sectors of the industry, firms may be justified in believing that a history of claims will be a dominant feature in the evaluation of their suitability for new projects. However, provided that the firm submitting the claim follows some simple rules, there is no reason to suppose that the pursuit of valid claims is detrimental in the long term.

It is, of course, very helpful if the contractor has done a good job, finishing as soon as was reasonably possible, and has co-operated with the employer and the design team. However, if the contractor has submitted a poor tender, underestimated the complexity and/or under-resourced the project, his claim may well be seen by the recipient as a means to recover some of the contractor's losses caused by a poor tender and poor management. It is quite natural, in these circumstances, for the employer and his professional advisers to suspect the contractor of employing a pricing policy to obtain work with the intention of using every possible means to recover a much larger sum when the project is complete. It is not surprising if relations between the parties deteriorate almost before the ink on the first interim payment certificate has dried. Very often, this policy will be obvious to the design team if the contractor is complaining of late information at every opportunity even when it is clear that no delay will be caused. Every letter will be an attempt to create evidence for a dubious claim at some future date.

On the other hand, a contractor with a valid claim will be doing himself no favours if he proceeds reasonably well with the project and co-operates with the employer and consultants, but hardly mentions the fact that he intends to submit a claim until the end of the job. Some contractors adopt this policy purely to maintain good relations or in the hope that a favourable opinion on extensions of time and/or borderline compliance with specifications will be forthcoming. It may be expecting too much to believe that the consultant will form a favourable opinion about a substantial claim for additional payment when the consultant has not been given any information to enable the employer to make provision for payment.

The contractor who does a good job and properly manages the project will often stimulate the design team to perform well. If, at the same time, the contractor gives notices and particulars in accordance with the contract, avoiding provocative language and frivolous claims, then he is more likely to be able to resolve his claims painlessly.

Even when contractors have, for commercial reasons, made a policy decision not to submit a valid claim, this policy will be soon reversed if the employer decides to levy a claim for liquidated damages after an insufficient extension of time has been granted. Many consultants and employers have underestimated the potential for the contractor to claim considerable sums of money when he is forced into a corner. For this reason, the employer's professional advisers should monitor all potential claims for extensions of time and additional payment, so that the employer can consider the risks and advantages of levying a claim for liquidated damages. It may be a better decision not to levy a valid claim for liquidated damages if the potential claim from the contractor will far outweigh the claim for liquidated damages. If the contract contains provisions to bar the contractor's claims (failure to give notice and the like), the employer's decision to levy liquidated damages may not be influenced in the same way.

8.2 Claim submissions

Unfortunately, the evaluation of claims is not an exact science. The basis of calculation is dependent on a complex interaction of factors which may be unique to the project. The contractor's method of pricing, allocation of prime cost and overheads in the tender and in the accounting practice, programme, methods of construction, records, monitoring and control systems all have a part to play in the evaluation process. If the contractor has an integrated computerised costing and accounting system with a sensible allocation of cost codes, the evaluation process may be simplified. If the accounting system comprises too many categories it may suffer from a higher incidence of wrongly allocated costs. On the other hand, too few categories may be of no use, thereby necessitating the laborious task of searching through all of the source documents.

Whatever the standard of records and management accounts, even if it is possible to calculate, with precision, the correct amount of the claim, it is a fact of life that the claim is unlikely to be paid in full. For this reason, even the most professionally prepared claim often includes a measure of overvaluation as a negotiating margin.

If the contractor has complied with all of the contractual provisions for claims, the employer's professional advisers may be well advised to settle them during the course of the project, leaving very little to resolve at the end. If this cannot be done, the final claim will probably contain a large negotiating element.

The first submission of a claim requires very careful planning. It must not contain any information, assumptions or calculations which can be used against the party submitting the claim. Several alternative approaches may be necessary in order to establish which is the best and most persuasive presentation. It is important to carry out several crosschecks to ensure that the financial data and assumptions can stand up to scrutiny by the recipient. While there may be justifiable reasons for actual prolongation costs to far exceed those which it may have been possible to derive from the rates for preliminaries in the bills of quantities, it is often an uphill battle to persuade the recipient that the additional costs are a direct result of matters for which the employer is responsible. The contractor may be well advised to anticipate the steps which may be taken by the opposition when scrutinising the claim. Reliance upon the recipient's inexperience and lack of knowledge in the hope of gaining an advantage may be self-defeating. If there is an element in the claim which is found to be dishonest, then the remainder of the claim, no matter how well founded, is likely to be treated with the extra caution which it deserves.

How then, is the contractor to include sufficient margin in his claim to allow for negotiation and at the same time avoid criticism for appearing to be disreputable? Should he include elements which are fairly obvious candidates for rejection so that they can appear to be the basis of the first compromise, leaving the way open for some of the 'grey areas' to be argued vigorously? It is not unusual for some very dubious elements of a claim to succeed merely because they are more palatable to the recipient than other elements which may reflect on the performance of the design team (and which are rejected).

In spite of the fact that a reputable contractor, or his appointed claims adviser, will not deliberately wish to submit a claim which contains dubious elements, they will be aware that it is necessary to include substantial sums in the claim which are expected to be rejected at some stage of negotiations. In some cases, not all of the dubious elements will be rejected, in which case the contractor will recover more than that to which he is entitled. In the long term, the contractor may not be any better off because many claims will be settled below a sum which reflects his full entitlement. Unfortunately, some employers will benefit at the expense of others.

The person, or persons, responsible for preparing the claim will have to establish the basis and quantum of claim which are considered to be correct in all respects. This will take into account all of the facts and particulars which are available and reasonable assumptions where they are necessary. The lowest and highest sums which are likely to be awarded if the matter should proceed to adjudication, litigation or arbitration should be considered, giving each head of claim a rating in order of merit. In cases where there is no evidence of concurrent delay and the contractor has excellent records, it may be possible to quantify prolongation costs with a high degree of certainty. If this is the case, the likely success factor of this head of claim may be as high as one hundred per cent. If there is concurrent delay and incomplete records, the success factor of this head of claim will be reduced accordingly. Claims for disruption will rarely justify a one hundred per cent chance of success.

However, such claims which are based on a logical analysis, where cause and effect are established, will be at the high end of the probability scale. Claims which tend to be based on a global assessment will normally be at the lower end of the probability scale. That is not to say that global claims, in the appropriate circumstances, will not merit a high rating. Some claims for finance charges will be well founded in contract, or in law, while others may be less likely to succeed. The likelihood of recovering the cost of preparing the claim may be zero. In some cases this head of claim may be justified, even if the probability of success is unpredictable.

Having established the likely range of success of the 'real' claim, it will be necessary to decide how, and to what extent, the negotiatng margin can be added. This is not an easy task. If experience has shown that some settlements fall below fifty per cent of the original claim, the contractor is faced with finding plausible methods to double the amount of his first submission. The idealist will view this process with some distaste. The commercial realist will know that it is unavoidable and all of his experience and imagination will be called upon to ensure that the negotiating margin is at least arguable.

Every 'grey area' must be presented as black, or white, depending on the circumstances. Care should be taken to avoid presenting black as white. Under no circumstances should contemporary records be changed, or invented, in order to distort the truth. Dishonesty should be avoided at all costs. The contractor, or subcontractor, submitting the claim should be aware of the probable range of success, the nature and quantum of the negotiating margin, and the strengths and weaknesses of the claim before submission. Any elements which cannot be argued with at least some degree of conviction may have to be discarded.

Most contractors, and subcontractors, will wish to reach an amicable settlement. Some will have decided, before submission of the claim, that under no circumstances will they take the matter to litigation or arbitration if settlement cannot be reached. This attitude is often brought about by the high cost of arbitration, particularly if previous experience has shown that the unrecovered costs of arbitration have not been justified in the light of the award. If this attitude exists, then the negotiating margin is likely to be higher than that which may otherwise have been added. It is, of course, fatal to let the opposition discover that litigation or arbitration has been ruled out. If the case is sound, the contractor may be persuaded to contemplate adjudication at the outset (if the matter cannot be settled). In these circumstances, the negotiating margin may not be excessive. If there are a number of substantial 'grey areas' in the claim, some employers (particularly government bodies) may have no option but to arbitrate, even if there is a willingness to settle. This must be taken into account at the outset.

Many contractors have the resources and capability to prepare their own claims. However, even the best organised contractors (including those who are recognised as being amongst the leading companies in the industry) are often unable to make the most of their case in a written submission. While a poor claim cannot be made into a good one, a good claim can easily fail if it is presented badly. Many good claims fail, at least in part, because the author of the claim is influenced by staff in the company who have vested interests in overlooking any shortcomings in the contractor's case and perhaps by placing too much emphasis on elements of the claim which have caused dispute throughout the contract. If the contractor's staff have been advising management that the claim is well founded and worth several hundred thousand pounds, they will be reluctant to change their view even in the light of valid counter arguments put forward by the other side.

Many final submissions repeat what has already been said, and rejected, in numerous exchanges of correspondence over several months. Even if the contractor is right, it is important to search for alternative arguments and means of persuasion. This is usually difficult to achieve by staff who have lived with the project and have fixed ideas on what happened and who was to blame. In any event, it is good practice to get an independent view of the strengths and weaknesses of the claim, the likely range of settlement, or award, and expert advice on how it should be presented before any submission is finalised for dispatch to the opposition. If there is any potential liability for liquidated or general damages, this should be brought to the attention of management and taken into account in the overall assessment of the likely recovery.

Once the claim is submitted, the contractor will need to ensure that there is a response or some other means of moving forward. The covering letter to the submission should summarise the claim so that any person who is not familiar with the detail, and who may be making important decisions, can appreciate the nature and amount of the claim without reading the detailed submission and appendices. The letter should invite a reply within a reasonable specified period. It may be useful to suggest a meeting to discuss and explain the claim in more detail before a formal reply is expected.

8.3 Negotiation

If the contractor has a valid case, given notices in accordance with the contract, kept accurate contemporary records and presented his case in a logical and professional manner, he will be starting from a position of strength. If a valid claim is not accompanied by these essential ingredients, the recipient will have little difficulty in finding reasons to reject it.

Whatever the merits of the claim, the initial response will usually concede very little. The contents of the response may be positive, giving cause for optimism, or it may be totally negative, rejecting every aspect of the claim. The former will enable both sides to move forward, while the latter will form a barrier to any early progress to resolve the matter. If there is no response at all, or if a negative response cannot be countered by some means of opening a dialogue, the contractor may have little option but to commence proceedings. If he has not already obtained advice before submitting the claim, the contractor should obtain the advice of experts before taking a decision to initiate formal proceedings.

If the response is positive and negotiations commence, then both parties may be able to settle the matters reasonably quickly. The contractor must be wary of employers who are merely going through the motions with no intention to settle at a reasonable figure. Their tactics will be to find out what concessions are on the table and to waste time. A delayed settlement usually means less in real terms, irrespective of any financing element which may ultimately be included (if any). If there are reasonable grounds to suspect that the employer is not genuinely seeking a fair settlement, the decision to commence formal proceedings should be taken sooner rather than later.

Negotiations may be conducted on an open basis (that is to say that the records of the negotiations may be used by the parties in any proceedings), or they may be *without prejudice* (that is to say that they cannot be referred to in any proceedings). In most cases, *without prejudice* negotiations are more satisfactory as they enable the parties to be more frank and they facilitate concessions which can be withdrawn if the other party refuses to make any concession. If there is agreement on any section of the claim, the contractor should endeavour to persuade the employer to make the agreement open and certify any sums which ought to flow from it. The employer will usually resist on the grounds that he will require an overall settlement.

From the employer's point of view, he will be prepared for the contractor's claim if he has been informed by his professional team pursuant to the contractor's previous notices. Even if the contractor has not complied in all respects with the contract to notify the employer's architect, or engineer, the employer ought to have been made aware of potential claims by his consultants. If he is properly advised, he will already have an outline defence to many of the contractor's claims. If the contractual provisions have been followed to the letter, any sums which are, in the opinion of the architect, or engineer, due to the contractor, will have been certified and paid. In practice, in spite of the problems caused by interference by the employer, the architect, or engineer, may be unable to act freely. This is sometimes the case where the architect, or engineer, is an employee of the employer.

Whoever represents the parties at negotiations, it is important to establish at the outset if they have the authority to make an agreement. Negotiations between staff who are not authorised to finalise an agreement may be suitable for initial discussions, but serious negotiations to conclude a settlement must be conducted by staff with full authority to agree on all aspects of the claim. It is particularly important for the contractor to establish whether, or not, the employer's consultants have such authority (they will not normally have this authority as part of their usual agreement with the employer to provide professional services).

If the consultant has such authority, it should be remembered that he stands to be shot at from both sides. If he wrongly certifies, or negotiates a settlement, to the detriment of the employer, he may be sued for negligence by the employer. If he wrongly certifies to the detriment of the contractor, or fails to negotiate a settlement which is satisfactory to the contractor, he may be exposing the employer to unnecessary costs of arbitration or litigation. Finding the right solution may require a careful and critical review of the consultant's own conduct during the contract and possibly acknowledging mistakes which have been made from time-to-time. For this reason, the employer may be well advised to be represented by an experienced negotiator who has not been involved with the day-to-day administration of the project and who is not tied by previous decisions.

Both parties should decide on the team which will be present to advise and support the negotiator. The temptation to field a large team should be resisted. It is important to select a team that is fully conversant with the matters under negotiation. It should be possible to verify or reject allegations, facts, matters of law or contract, principles of evaluation and the like by reference to members of the team. The negotiator should decide whether any difficult points should be discussed in the presence of the other party, or if negotiations should adjourn to enable private discussions to take place. The team should not be changed unless there is a clash of personalities which is hindering a settlement. Sufficient time should be allowed to prepare for each meeting and a common approach should be established so that no divisions between team members will become evident at the meetings.

Concessions should be considered before any meeting and the negotiator should be ready to concede at the appropriate time if it should be necessary to do so. Concessions should not be made too lightly, and then only if the other party is showing a willingness to give ground. It may not be the best policy to concede too many points unconditionally. At the end of negotiations, both parties will seek a satisfactory overall settlement. Too much given away on individual heads of claim may make it impossible to agree on the entire claim.

If one of the parties is not genuinely seeking a fair settlement, they may field a team which does not have authority and who have to report to others to verify facts or decide on important points. Perhaps they will change the negotiating team when it seemed that progress was being made and the other party finds that the entire process has to begin from first base again. If this becomes evident, it may be appropriate to break off negotiations and commence proceedings without delay. It may be worthwhile preparing a notice of arbitration or a letter of claim in accordance with the Pre-Action Protocol for Construction and Engineering Disputes to issue at the meeting, perhaps leaving the door open to serious negotiations in a parting statement to the team leader.

If an agreement can be reached, it is important to have the terms of the agreement recorded and signed by the authorised representatives of the parties before the meeting is concluded. The agreement should make it clear that it covers all matters which were the subject of the negotiations, and if both parties intended the agreement to cover every claim and counter-claim (so that no other claims could be brought against the parties) it should clearly say so. Indemnities may be required with respect to possible claims from subcontractors and/or third parties. The date of payment should be specified and there should be provision for interest to be added in the event of late payment.

8.4 Resolution of disputes by third parties

If, despite all efforts to come to an amicable agreement, no agreement can be reached, it may be possible to resolve the dispute by an independent third party. The most common means of resolution are adjudication, litigation, mediation and to a lesser extent arbitration. The process of each is briefly described below along with other lesser used methods of dispute resolution.

Litigation

Litigation is a process of dispute resolution governed by rules determined by the state. In England and Wales the rules are set out in the Civil Procedure Rules, which are often amended and easily found on the World Wide Web.

Construction disputes of more than £50,000 in value are usually referred to the Technology and Construction Court, which is a specialist sub division of the High Court. For sums in dispute of less than £50,000 the dispute should normally be referred to the County Court.

In order to seek to avoid or narrow construction disputes referred to court, the parties to a dispute are required to comply with the Pre-Action Protocol for Construction and Engineering Disputes. The procedure requires the party intending to commence proceedings (the 'claimant') to issue a claimant's letter setting out the nature of the dispute (factual and contractual/legal principles relied on), the full names of the parties to the dispute and their addresses and the redress sought. The potential defendant is required to acknowledge receipt of the letter of claim in full within fourteen days of receiving it and to answer it in full within a further fourteen days. The parties are then normally required to attend a meeting to seek to narrow or resolve the dispute and to consider alternative means of dispute resolution. The meeting is held on a without-prejudice basis. Failure to comply with the protocol may mean that the defaulting party incurs the cost of the litigation proceedings irrespective of whether the court finds in its favour in respect of the substantive dispute.

Mediation

Mediation is becoming a very popular method of dispute resolution. It is included in the current JCT range of contracts as a possible method of resolving disputes (although it is not mandatory).

Mediation is only successful if the parties in dispute have a genuine intention to resolve their dispute. It has often been thought of as an easy delaying tactic but statistically it has proved to be an efficient and cost effective method of dispute resolution.

Any person can be proposed and appointed as the mediator by the parties and if they cannot agree on an appointment there are a number of bodies, such as the Chartered Institute of Arbitrators and the Centre for Dispute Resolution (CEDR), who will assist the parties to find the right mediator.

The procedural rules for mediation are entirely flexible and are subject to agreement between the parties to the dispute.

The mediator normally meets the parties separately and he may be empowered, if the parties cannot be persuaded to settle their differences, to make a recommendation on the matters in dispute. As a general rule any confidential information which is made available to the mediator at private meetings with one party cannot be divulged to the other party. However, in the case of *Farm Assist Limited (in liquidation)* v. *The Secretary of State for the Environment, Food and Rural Affairs (No. 2)* [2009] BLR 399, The Honourable Mr. Justice Ramsey stated that confidentiality may be waived if it is in the interests of justice. While not usually being conducted in the formal manner normally associated with litigation and arbitration, mediation proceedings may be conducted with lawyers and other experts to present each party's case to the mediator. The mediator will endeavour to find common ground at these separate meetings and he will try to find means of reaching a settlement. A meeting with both parties present will usually be required at some stage. Whoever represents the parties at these discussions, it is essential that they have the authority to agree and settle the dispute. Failing agreement, the mediator may decide on the matters in dispute if the parties have given him authority to do so, but this is a very rare power given to mediators.

Adjudication

Adjudication is a process in which the parties to a dispute submit their cases to a third party (an individual or a panel) for a decision. This decision is not permanently binding unless both parties gave their prior agreement that it should be, or otherwise if the aggrieved party fails to register his dissatisfaction within a stipulated period (such as under NEC 3 and ICE 7th edition standard forms). If disputed, the matter can subsequently be referred to arbitration or litigation. Without a contractual provision in the contract, or without the parties' agreement, adjudication cannot be imposed on any party unless there are provisions in law to enforce adjudication.

With the advent of the Housing Grants, Construction and Regeneration Act 1996 (the 'Construction Act'), mandatory provisions for adjudication were embodied in English Law (England and Wales). Similar laws are applicable in Scotland and Northern Ireland. The Construction Act has recently been amended by Part 8 of the Local Democracy, Economic Development and Construction Act 2009 ('LDEDCA'). A copy of the relevant provisions of the Construction Act as amended by the LDEDCA is provided in Appendix D.

All of the standard forms of contract in the UK have now been revised to give effect to the provisions required by the Construction Act. The procedure for adjudication may be laid down in the contract, or it may be set out in a separate document (referred to in the contract as the procedure to be adopted). For example, ICE stipulates that adjudication shall be conducted under 'The Institution of Civil Engineers' Adjudication Procedure 1997' or any amendment or modification thereof being in force at the time of the said Notice [of Adjudication]' (sub-clause 66(6)(a)). The latest edition of the ICE Civil Engineers' Adjudication Procedure is the 3rd Edition published in 2005. However, in the UK, if the parties enter into a contract which does not contain suitable provisions for adjudication, a party cannot refuse to have a dispute referred for adjudication. Section 108(5) of the Construction Act states:

> 'If the contract does not comply with the requirements of subsections (1) to (4), the adjudication provisions of the Scheme for Construction Contracts apply.'

The Scheme for Construction Contracts (England and Wales) 1998 has recently been amended to take into account amendments to the Construction Act. A copy of the amended version of the Scheme for Construction Contracts is provided in Appendix E. The Scheme contains detailed provisions for adjudication which include a procedure for the appointment of an adjudicator. The adjudicator may be named in the contract or he may be agreed by the parties. If the adjudicator is not named, or if the parties fail to agree on the appointment of an adjudicator, there is provision for an adjudicator to be appointed by a nominating body named in the contract, or if there is no such body named in the contract, by an adjudicator nominating body.

Under the Scheme for Contruction Contracts the decision of an adjudicator is binding on the parties unless and until the dispute is finally determined by some other form of dispute resolution process (by agreement, litigation or arbitration for example).

A considerable amount of case law has built up on the back of adjudication under the Scheme, particularly in the area of adjudicators' jurisdiction. One matter that the courts have made clear however, is that whether an adjudicator's decision is good, bad or indifferent the courts will enforce it unless there is a good jurisdictional reason for not doing so. The effect is that adjudication provides a 'pay now, argue later' resolution to disputes whereby an adjudicator's decision has to be complied with in the short term but which the aggrieved party can seek to redress at a later date. In practice, the adjudicator's decisions often conclude disputes between parties.

The 1999 FIDIC contracts have taken on board the general trend set in the UK and by the World Bank, and now provide for adjudication in clause 20 (Red, Yellow and Silver Books) and clause 15 (Green Book). The Red, Yellow and Silver Books contemplate either a single adjudicator or a panel of three (The Dispute Adjudication Board – DAB). The Green Book requires a single adjudicator. The DAB may be named in the contract, agreed, or appointed by the appointing body named in Part II of the contract. Any dispute must be referred to the DAB as a prerequisite to arbitration. In the case of the Silver Book, a dispute which has become binding (by the contractor failing to register his dissatisfaction to a determination by the employer within fourteen days in accordance with sub-clause 3.5) may not be referred to the DAB or to arbitration.

The DAB must give its reasoned decision within eighty-four days (or such other period as may be agreed) of the reference and, unless a party gives a notice of dissatisfaction within twenty-eight days of the decision, it shall be final and binding on the parties. If a valid notice of dissatisfaction has been lodged, the parties are required to attempt amicable settlement and, unless the parties agree otherwise, arbitration may be commenced on or after the fifty-sixth day after the date of the notice of dissatisfaction.

Sub-clause 20.6 of the FIDIC contracts provide for the DAB's decision to be used as evidence in arbitration. This does not appear to be conducive to open and frank disclosure in an adjudication forum.

Arbitration

Arbitration has lost favour in England and Wales in recent years and it is no longer the default dispute resolution procedure in the current JCT range of contracts. Prior to the 2005 range of JCT standard forms, to avoid arbitration as the mandatory form of dispute resolution, reference to it had to be expressly deleted from the contract. In the current range of JCT contracts, arbitration has to be expressly included if it is to apply. If it is not expressly included, litigation is the default dispute resolution process.

Arbitration in England was governed by the *Arbitration Acts of 1950, 1975 and 1979*. Different provisions apply in Scotland where until recently the *(Scotland) Arbitration Act of 1894* was used for domestic arbitration and the *UNCITRAL Model Law* was used for international arbitration (or for domestic arbitration if the parties agreed). However, the Arbitration (Scotland) Act 2010 received Royal Ascent on 5 January 2010 and overhauled the Scottish arbitration system. In England, the enactment of the *Arbitration Act of 1996* swept up most of the previous Arbitration Acts, however some parts of the 1950 Act are still applicable.

The parties' agreement is essential before any dispute can be settled by arbitration. Agreement can be made at any time, but it is usual practice for the agreement to be made at the time of entering into the contract for the work. Standard forms of contract have express provisions for arbitration in the articles or in the conditions of contract.

In the event of there being valid arbitration provisions in the contract which cover the matters in dispute, the parties will generally be prevented from having the dispute resolved by litigation. However, if one of the parties commences litigation, and the other party does not, before taking any steps in the litigation, apply to the courts for a stay of proceedings under Section 9 of the *Arbitration Act of 1996* which states

> '(1) A party to an arbitration agreement against whom legal proceedings are brought (whether by way of claim or counterclaim) in respect of a matter which under the agreement is to be referred to arbitration may (upon notice to the other parties to the proceedings) apply to the court in which the proceedings have been brought to stay the proceedings so far as they concern that matter.
> (2) An application may be made notwithstanding that the matter is to be referred to arbitration only after the exhaustion of other dispute resolution procedures.
> (3) An application may not be made by a person before taking the appropriate procedural step (if any) to acknowledge the legal proceedings against him or after he has taken any step in those proceedings to answer the substantive claim.
> (4) On an application under this section the court shall grant a stay unless satisfied that the arbitration agreement is null and void, inoperative, or incapable of being performed.'

Then the dispute may be settled by litigation.

If, before taking any steps in the litigation, an application to stay the proceedings is made, then provided that the applicant is ready and willing to have the dispute settled by arbitration, the power to order a stay of proceedings is usually exercised. A stay of proceedings may be refused for the following reasons:

- the arbitration agreement does not contain provisions for immediate arbitration;
- the matters in dispute do not fall within the ambit of the arbitration agreement;
- there would be undue hardship on the plaintiff if the stay were granted;
- the only matter to be decided in the dispute was a question of law;
- fraud is alleged;
- if there would be two separate sets of proceedings requiring resolution based upon the same facts, one of which would be settled in the courts, and the dispute which was the subject of the application for a stay (if no stay were granted) would be settled in arbitration.

It was thought that the courts did not have the same powers as an arbitrator and they could not open up, or recover, an architect's certificate: *North West Regional Health Authority* v. *Derek Crouch* [1984] 2 WLR 676. Some forms of contract do not restrict the power of the courts. The Singapore Institute of Architect's form of contract expressly states that the courts shall have the same powers as an arbitrator – clause 37(4). The *Courts and Legal Services Act 1990* provides that the High Court may, if all parties agree, exercise the same powers as those conferred upon an arbitrator (section 100, giving effect to an additional section 43A in *The Supreme Court Act 1981*). Other important matters to be considered are the facts that arbitration is held in private and the costs are likely to (but not necessarily) be less than litigation.

The decision in *North West Regional Health Authority* v. *Derek Crouch* has since been overtaken by a House of Lords' decision in the case of *Beaufort Developments Ltd* v. *Gilbert Ash NI Ltd and Others* [1998] 2 All ER 778, in which it was held that under a JCT contract, architects' certificates could be reviewed by any tribunal including the courts. The *Crouch* decision was therefore decided wrongly.

When one of the parties has decided to refer a dispute to arbitration, the most important decision is to select the most appropriate arbitrator. If the resolution of the dispute is likely to turn on questions of law, a legally qualified arbitrator may be the best choice. Section 93 of the *Arbitration Act 1996* provides for a judge of the Commercial Court or an official referee to accept an appointment as a sole arbitrator or umpire by virtue of an arbitration agreement, if the Lord Chief Justice is satisfied that the judge or official referee can be made available. If the dispute is mainly to do with technical matters, then a technical arbitrator may be more appropriate. If the parties agree, a legal assessor, or a technical assessor, can be appointed to facilitate resolution of the dispute. However, the arbitrator must make his own decision, whatever the advice given by the assessor.

If the parties cannot agree on the arbitrator, there is provision in most standard forms of contract for an appointing body (stipulated in the contract) to appoint an arbitrator. Failure to agree on an arbitrator is usually caused by the respondent's desire to delay the proceedings. The disadvantage of having an arbitrator appointed by a third party is that the appointed arbitrator may be a person which neither party would have selected. There may, of course, be valid reasons to object to the other party's choice of arbitrator:

- there may be a conflict of interest (this would in any event be brought to the attention of the parties by the arbitrator);
- the arbitrator may have a reputation for deciding the matters in dispute against the interests of the objecting party (in some cases, the arbitrator's views are well known from published works);
- the arbitrator may have a reputation for poor control of arbitration proceedings, thereby permitting delays to occur and costs to increase (a reluctant party may prefer such an arbitrator).

Some forms of contract specify the procedure to be used in the arbitration. The most common procedures in use in the construction industry are the *ICE Arbitration Procedure* (2005) and the *CIMAR* (Construction Industry Model Arbitration Rules) 2011 for use with JCT contracts.

Foreign arbitration (domestic arbitration in foreign countries) is usually subject to local rules set by the local chamber of commerce or arbitration centre. In international contracting, it is common for the arbitration agreement to require arbitration proceedings to be governed by a recognised international set of rules.

Foreign arbitration subject to local rules

International firms operating in foreign countries may find themselves in disputes which will be resolved according to local rules and law.

Kuwait

Until recently, arbitration under ICC (International Chamber of Commerce) Rules was common in Kuwait. However, administrative contracts between contractors and government departments are more likely to contain provisions for disputes to be referred to the local courts. This process is likely to be costly, requiring all documents to be translated into Arabic (even if the language of the contract and/or correspondence and records are in English). The proceedings will usually be conducted in Arabic. Court fees are required for all proceedings. A judge would normally submit technical issues to the Department of Experts to report on their findings. Appeals are possible to the High Court of Appeal or to the Courts of Cassation. Some contracts in Kuwait may be subject to local arbitration.

Bahrain

Settlements in Bahrain are often referred to arbitrators appointed by the Minister of Justice and Islamic Affairs. There is provision for international arbitration in the Bahrain Legislative Decree No. 9 of 1994. However, in most commercial contracts, it is not unusual to have a locally appointed arbitration committee comprising one arbitrator appointed by one party, one by the other party and a third (the chairman) by agreement of the two appointed members. The principal centre for commercial arbitration is the Bahrain Chamber of Commerce and Industry (BCCI).

United Arab Emirates

The principal Emirates of Dubai, Sharjar and Abu Dhabi rely to varying degrees on the *Shari'a* (ancient Islamic law), commercial practice and statutory provisions. Commercial arbitration in Dubai is usually conducted under the auspices of the Dubai Chamber of Commerce and Industry. Western practices are followed in most cases. Clause 67.3 of the Dubai Municipality's conditions of contract provides for each party to appoint a member to the tribunal within forty-two days of the notice to commence arbitration. The third member is to be mutually chosen by the two appointed members. If the parties fail to appoint members to the tribunal, they shall be appointed by the Dubai Chamber of Commerce and Industry.

Hong Kong

Hong Kong arbitration is based mainly on the English Arbitration Acts embodied in the Hong Kong Arbitration Ordinance, Chapter 341 (for domestic arbitration). From April 1990, Hong Kong adopted the *UNCITRAL Model Law* (for international arbitration). A number of changes have taken place since 1 July 1997 to take account of the 'Basic Law' following transfer of sovereignty to China.

International arbitration

International arbitration is the private adjudication of commercial disputes with international aspects and/or internationally diverse parties. It includes both '*ad hoc*' and 'institutional' arbitration.

Ad hoc arbitration is administered and conducted in a manner specifically designed by the parties. Institutional arbitration is administered by organisations such as the International Chamber of Commerce (ICC) and the United Nations Commission on International Trade Law (UNCITRAL) pursuant to their published rules and procedures.

The ICC Rules are perhaps the most commonly used procedure in international construction contracts. The place of arbitration is fixed by the Court unless agreed by the parties (Article 14). However, it is usual (unless the contract provides otherwise) for the arbitration to be held where the chairman of the tribunal resides (or where the single arbitrator resides if only one arbitrator is required). The ICC usually appoints a chairman from a country other than those from which the parties are nationals (unless otherwise agreed by the parties).

However, in recent years, the greater flexibility of the UNCITRAL Arbitration Rules has led to an increasing acceptance of these Rules for *ad hoc* arbitration. In the UK, the London Court of Arbitration Rules are based on the UNCITRAL Model and they also allow for the parties to agree to arbitration under the UNCITRAL Arbitration Rules with the London Court acting as administrator. A number of countries now embody the UNCITRAL Model Law as part of their arbitration machinery, for example, Hong Kong, USA, Canada and Australia. Provision to opt out of the UNCITRAL Model Law is normally available by agreement between the parties.

8.5 Enforcement of Foreign Awards

International disputes across national boundaries

Arbitration enables the parties to settle international disputes across national boundaries without the unnecessarily high costs which may otherwise arise in various jurisdictions. If the arbitration is structured properly in the contract, the results usually prevent recourse to multiple courts, appeals and extended enforcement procedures. It is important that the award is final and binding as well as being enforceable across international boundaries, otherwise the final resolution to the dispute may involve separate courts, lawyers in several countries and repeats of the process in numerous appeal forums. Separate proceedings for seeking enforcement of a judgement can substantially escalate the cost of resolving such disputes. The parties may have to consider the unpredictability of the final results. Many countries do not necessarily recognise judgements issued in another country.

The 'New York Convention' on the Enforcement of International Awards

The United Nations Convention on the Recognition and Enforcement of Foreign Arbitral Awards, 1958 (the 'New York Convention') is recognised by many, but not all, countries. Contracting states include Australia, India, Japan, Korea, Philippines, Thailand, USA, most Western countries and some eastern European countries. Some countries accede to the Convention subject to reservations. For example, Kuwait acceded to the Convention in March 1978, subject to the reservation that it would only be applied to awards made in territories of other contracting states. The UAE did not accede to the Convention, but signed a treaty between members of the Arab League. Bahrain has no formal treaty with respect to the enforcement of foreign awards. However, many states (including Bahrain) subscribe to the general policy that they will enforce an award made in any country which, in turn, enforces any award made in the respective state (reciprocity).

The main exceptions to the obligation to enforce foreign arbitral awards under Article V of the Convention are

Under paragraph 1

(a) If the parties were under some incapacity to contract or the agreement was not valid under the law to which the parties made it subject (the substantive law or law of the contract)
(b) If a party was not given a proper opportunity to present his case
(c) If the award deals with matters not submitted to arbitration
(d) If the composition of the arbitral authority or the procedure was not in accordance with the agreement of the parties or with the law of the country where the arbitration took place
(e) If the award is not binding or has been set aside or suspended by a competent authority.

Under paragraph 2

(a) If the subject matter or difference is not capable of settlement by arbitration under the laws of the country where enforcement is sought
(b) The recognition or enforcement of the award would be contrary to public policy of the country where enforcement is sought.

Most institutional rules of arbitration specially permit either party to apply to a court for interim relief (awards). In the absence of such a provision, the parties may well be able to seek interim relief from the arbitrator or tribunal. These interim awards may, under the New York Convention, be enforceable in the courts, however the delays involved in appointing the arbitrators and then obtaining interim relief may well be sufficient to dissuade the parties to seek interim relief:

> 'Settle matters quickly with your adversary who is taking you to court. Do it while you are still with him on the way, or he may hand you over to the judge, and the judge may hand you over to the officer, and you will be thrown into prison. I tell you the truth, you will not get out until you have paid the last penny.'
>
> (*Matthew* 5: 25, 26 NIV)

Conciliation

The process is similar to mediation. If the parties are willing to settle, but there are genuine obstacles to settlement, it may be possible to close the gap between the parties and facilitate a settlement by the process of conciliation. This method may not be imposed unilaterally and the agreement of the parties is essential. It involves the appointment of an independent third party, mutually agreed by the parties, to hear both parties' points of view. The conciliator will usually be a recognised expert on the matters in dispute and he will look at the evidence and listen to the arguments put forward by each side. He will contribute his own ideas on the merits of the case. He will not meet any party in private and all discussions take place with both parties present. The parties may have legal advisers present at any meetings, and they may, of course, meet each other without the conciliator being present. The conciliator's aim will be to bring the two sides together to discuss all aspects of the matters in dispute and lead them to an amicable settlement. The conciliator will not make decisions, but he may make recommendations. It is up to the parties to agree on an acceptable settlement. They are not obliged to agree, and if settlement cannot be reached, the parties may pursue the matter in adjudication, arbitration or litigation.

The seventh edition of the ICE contract contains provisions for conciliation as well as adjudication (although conciliation is not mandatory).

Dispute Review Board

Internationally, dispute review boards (DRBs) have been common for a number of years. DRBs carry out a similar function to mediation or adjudication. DRBs are specified as a standard amendment by The World Bank in all contracts funded by them or by a subsidiary. DRBs have also found favour with a number of major developers and governments.

A contract which includes a DRB will usually provide for each party to the dispute to name a party who will sit on the board and provide for an independent third board member. The procedure for resolving disputes should be identified in the contract and it usually resembles something between a mediation and an adjudication.

Appendix A: Sample Claim for Extension of Time and Additional Payment under JCT Standard Building Contract with Quantities 2011

Introduction to the example

The sample claim which follows is for an extension of time and reimbursement of loss and/or expense arising out of the delays (D1), (D2), (D3) and (D4) shown in Figure 5.9 in Chapter 5. Sectional completion has been introduced into the example as a result of which additional activities have become critical.

For simplicity, the claim deals with the subject matter in the main narrative. In practice, particularly for complex claims dealing with many issues, more use would be made of appendices (summarising notices of delay and the like). Copies of relevant correspondence (referred to in the claim), supporting documents, particulars and detailed calculations would also normally be given in an appendix. This example does not contain such appendices (except for programmes and illustrations) but it is assumed that they are submitted. Such particulars ought to be included both to prove the amounts claimed to the Architect/Quantity Surveyor and, if the claim is disputed and referred to adjudication, to an adjudicator, or other dispute resolution board. It is advised that the particulars provided are numbered and referenced in the claim in order that the reader can easily find the supporting evidence.

In this example, clauses referred to in the form of contract are often paraphrased. It is sometimes more appropriate to quote the clauses verbatim.

The claim submission

Covering letter from Better Builders Ltd (the contractor) to T. Square (the architect):

Date 4 April 2011

Dear Sir,

Re: ABC Stores and Depot, New Road, Lower Hamstead, Wilton

Further to our letter of 24 August 2010 requesting a review of extensions of time, our letter of 14 September 2010 giving particulars of loss and/or expense and our letter of 14 February 2011 requesting a copy of the draft final account, to which we have had no response, we enclose herewith our claim for extensions of time, reimbursement of loss and/or expense and damages.

Please note that the contents of this submission do not contain any particulars (with the exception of rates for finance charges for the period after 14 September 2010) which have not been submitted to you previously in correspondence referred to therein. It is our understanding that you have all information necessary for the preparation of the final account and we can see no reason why it should not have been issued prior to this letter.

Our claim is for further extensions of time of two weeks for section A and two weeks for section B (up to the dates of practical completion) and for reimbursement of loss and/or expense and/or damages for the amount of £90,637.42 (including finance charges on liquidated damages).

We are also requesting the issuance of a certificate of making good defects pursuant to clause 2.39 of the contract, a statement pursuant to clause 4.5 of the contract (including all adjustments mentioned in the submission), release of retention of £21010.00, release of liquidated damages amounting to £63000.00 and a final certificate pursuant to clause 4.15 of the contract.
Your early response would be appreciated.

Yours faithfully
For and on behalf of Better Builders Ltd

Better Builders Ltd
Scaffold Road
Hamstead Rise, Wilton

Manufacturing plant and associated works at
New Road, Lower Hamstead, Wilton
for
ABC Industries Ltd
Factory Lane, Hamstead Rise, Wilton

Claim for extensions of time and
reimbursement of loss and/or expense
and/or damages and repayment of
liquidated damages

Architect: T. Square of Drawing Board and Associates
Design Avenue, Hamstead Rise, Wilton

4 April 2011

Claim for extensions of time for completion of Section A and Section B, reimbursement of loss and/or expense and/or damages and repayment of liquidated damages.

1.0 **Introduction.**

1.1 **The parties.**

1.1.1 The employer is ABC Industries Ltd of Factory Lane, Hamstead Rise, Wilton.

1.1.2 The architect is T. Square of Drawing Board and Associates, Design Avenue, Hamstead Rise, Wilton.

1.1.3 The quantity surveyor is R.E. Measure of The Manor, Billings- gate Road, Hamstead Rise, Wilton.

1.1.4 The contractor is Better Builders Ltd of Scaffold Road, Hamstead Rise, Wilton.

1.2 **The works.**

1.2.1 The works comprise the alteration of an existing stores building into a manufacturing plant for motor parts including the construction of a new access road, drainage, diversion of services and landscaping at ABC Stores and Depot, New Road, Lower Hamstead, Wilton.

1.3 **The tender and the contract sum.**

1.3.1 The contractor submitted his tender on 11 January 2010 for the sum of £827333.00. It was a condition of the contractor's tender that work would be permitted on weekends and public holidays and that the employer would underl2take to ensure the presence of the architect or his representative on such days where it was necessary for the supervision and administration of the contract.

1.3.2 The employer unconditionally accepted the contractor's tender by letter dated 26 January 2010.

1.3.3 The contract sum in article 2 of the agreement is £827, 333.00.

1.4 **The contract.**

1.4.1 The contract is the Standard Building Contract, With Quantities, 2011, issued by the Joint Contracts Tribunal.

1.4.1.1 As required by the Sixth Recital the sections of the works are described in the Contract Bills and Contract Drawings.
Section A is described as – 'Completion of all alterations in the existing store building to such state as to enable the employer to commence installation of plant and equipment and as shown on drawings AD/1 to AD/15 and described in these Contract Bills as being within Section A.'
Section B is defined in the Contract Bills as 'All work identified in these Contract Bills and in drawings AD/1 to AD/15 other than that stated to be within Section A.'

1.4.1.2 Sub-clause 2.29.9 (relevant event – exceptionally adverse weather conditions) has been deleted.

1.4.2 The relevant information in the Contract Particulars of the contract are as follows:

1.4.2.1 Clause 1.1 Dates for completion
– Section A – Sixteen weeks including the date of possession.
Section B - Twenty-two weeks including the date of possession.

1.4.2.2 Clause 2.9.1.2 Master Programme
Critical paths are not required.

1.4.2.3 Clause 2.38 Rectification Period
– Six months from the date of practical completion of each Section.

1.4.2.4 Clause 6.7 and Schedule 3 Insurance of the works
– Option C applies.

1.4.2.5 Clause 2.4 Date of possession
– Seven days after the architect's written instruction to take possession of the site for both Section A and Section B.

1.4.2.6 Clause 2.5 Deferment of the date of possession
– Does not apply.

1.4.2.7 Clause 2.32.2 Liquidated and ascertained damages
– £1500.00 per day for Section A and £3000 per day for Section B.

1.4.2.8 Clause 4.20.1 Retention percentage
– Five per cent.

1.4.2.9 Clause 4.21 and Schedule 7 Fluctuations
– Clause 4.21 and Option A shall apply.

1.5 **The programme:**

1.5.1 The contractor's master programme for completion of the works is shown in appendix I hereto (see Figure A.1).

1.5.2 The activities forming section A are F–G, B–G and G–H.

2.0 **Summary of facts**

2.1 **Possession of site: commencement and completion of the works.**

2.1.1 On 9 February 2010, the architect gave written notice to the contractor to take possession of the site on 16 February 2010.

2.1.2 The contractor took possession of the site and commenced work on 16 February 2010.

2.1.3 Pursuant to clause 1.1 of the conditions of contract, the Contract Particulars and the architect's written instruction of 9 February 2010, the dates for completion were:

2.1.3.1 Section A – 7 June 2010.

2.1.3.2 Section B – 19 July 2010.

2.1.4 Practical completion occurred on the following dates:

2.1.4.1 Section A – 28 June 2010 (Architect's certificate of practical completion dated 13 August 2010).

2.1.4.2 Section B – 9 August 2010 (Architect's certificate of practical completion dated 13 August 2010).

2.2 **Delay and extensions of time**

2.2.1 The contractor gave the following notices of delay and particulars pursuant to clause 2.27 of the conditions of contract:

2.2.1.1 Letter dated 23 March 2010 [week 6] – Notice of delay as a result of exceptionally adverse weather conditions affecting activity B–E (Delay D1).

2.2.1.2 Letter dated 25 March 2010 [week 6] – Notice of delay as a result of architect's instruction no 1 (issued 22 March 2010) to alter work partially completed to activity B–G (Delay D2).

2.2.1.3 Letter dated 13 April 2010 [week 9] – Particulars of delay caused by architect's instruction no 1.

2.2.1.4 Letter dated 8 April 2010 [week 8] – Notice of delay as a result of revised and additional work to activity B–G shown on drawings AD/14A and AD/15A issued on 6 April 2010 [week 8] (Delay D3).

2.2.1.5 Letter dated 30 June 2010 [week 20] – Particulars of delay caused by the issuance of drawings AD/14A and AD/15A.

2.2.1.6 Letter dated 14 July 2010 [week 22] – Notice of delay as a result of late issuance of instructions for the supply of mechanical equipment for the effluent plant on activity H–K (Delay D4).

2.2.1.7 Letter dated 9 August 2010 – Particulars of delay caused by late issuance of instructions concerning the supply of mechanical equipment (see 2.2.1.6 hereof).

2.2.1.8 Letter dated 26 August 2010 – Letter requesting the architect to review his extensions of time for Section A and Section B pursuant to clause 2.28.5 of the conditions of contract and giving further particulars.

2.2.2 The architect has made the following extension of time for completion of the works pursuant to clause 2.28 of the conditions of contract:

2.2.2.1 Certificate reference EOT 1 dated 17 August 2010 [week 27] Section A – Extension of time of one week as a result of the additional work to activity B–G shown on drawings AD/14A and AD/15A (Delay D3), giving a revised completion date of 14 June 2010.

2.2.2.2 Certificate reference EOT 2 dated 17 August 2010 [week 27] Section B – Extension of time of one week as a result of the late issuance of instructions for the expenditure of PC sums (Delay D4), giving a revised completion date of 26 July 2010 [week 23].

2.2.2.3 At the date of this submission, the architect has not given a written response to the contractor's request of 26 August 2010 (see 2.2.1.8 hereof).

2.3 **Certificates of non-completion.**

2.3.1 Pursuant to clause 2.31 of the conditions of contract, the architect issued certificates of non-completion dated 16 August 2010 certifying that the contractor had not completed:
Section A – by the extended date of completion of 14 June 2010.
Section B – by the extended date of completion of 26 July 2010.

2.4 **Direct loss and/or expense:**

2.4.1 The contractor notified the architect, pursuant to clause 4.23 of the conditions of contract, that the regular progress of the works had been affected and that he had incurred, and was continuing to incur, direct loss and/or expense as follows:

2.4.1.1 Letter dated 1 June 2010 [week 16] – As a result of delays to activity B–G (Delays D2 and D3).

2.4.1.2 Letter dated 30 June 2010 [week 20] – Further disruption of the regular progress of the works as a result of delay to activity B–G (Delay D3) and as a result of late nomination of the subcontractor for activity H–K (Delay D4).

2.4.2 By letter dated 17 August 2010, the Architect requested further particulars from the contractor in support of his application for reimbursement of direct loss and/or expense.

2.4.3 On 14 September 2010, the contractor provided the further particulars requested by the Architect on 17 August 2010.

2.4.4 At the date of this submission, no sums for loss and/or expense have been ascertained and no further requests for particulars have been made by the architect or quantity surveyor.

2.5 **Payment and final account:**

2.5.1 The latest certificate issued prior to the date of this submission is interim payment certificate no 6 dated 13 August 2010 showing the following amounts:

2.5.1.1	Gross value of work at practical completion	£840400.00.
2.5.1.2	Retention	£21010.00.
2.5.1.3	Nett amount due	£819390.00.
2.5.1.4	Previous certificates	£725200.00.
2.5.1.5	Amount due for payment	£94190.00.

2.5.2 On 16 August 2010, the employer notified the contractor pursuant to clauses 2.32 and 4.12.5 that it would withhold the amount of £63,000.00 as liquidated damages from the amount due pursuant to certificate no 6 dated 13 August 2010.

2.5.3 The employer has paid the amount certified as being due for payment in interim payment certificates, less liquidated damages in the sum of £63,000.00. The nett payment made after deduction of damages was £31190.00.

2.5.4 On 14 February 2011, the contractor requested a copy of the final account showing the value of work executed including all adjustments to the contract sum.

2.5.5 At the date of this submission, no final account has been issued to the contractor.

2.6 **Defects:**

2.6.1 On 10 January 2011, the architect issued a schedule of defects pursuant to clause 2.38 of the conditions of contract and instructed the contractor to make good the said defects.

2.6.2 On 14 February 2011, the contractor notified the architect that he had rectified all defects notified by the architect in his schedule of 10 January 2011 and he requested a certificate of making good defects pursuant to clause 2.39 of the conditions of contract.

2.6.3 At the date of this submission, no certificate of making good defects has been issued.

3.0 **Basis of claim**

3.1 The contract contained the following provisions:

3.1.1 Clause 2.12 requires the architect to provide further drawings and details at the time it is reasonably necessary for the contractor to receive them.

3.1.2 Clause 5.9 – If compliance with an instruction substantially changes the conditions under which any other work is executed, then such work shall be treated as if it had been the subject of an instruction of the architect requiring a variation under clause 3.14. Provided that no allowance shall be made under clause 5.9 for any affect on the regular progress of the works or for any other direct loss and/or expense for which the contractor would be reimbursed by payment under any other provisions in the conditions of contract (clause 5.10.2).

3.1.3 Clause 2.39 – When in the opinion of the architect any defects or other faults which he may have required be made good under clause 2.38 (defects occurring in the Rectification Period), he shall issue a certificate to that effect and the said defects shall be deemed to have been made good on the day named in such certificate.

3.1.4 Clause 2.32.3 – If, under clause 2.28.5, the architect fixes a later completion date the employer shall repay to the contractor liquidated damages allowed under clause 2.32.1 for the period up to such later completion date.

3.1.5 Clause 2.27 – The contractor shall give notice and particulars of delay and shall be entitled to an extension of time for completion if completion of the works (and/or section) are delayed by the following relevant events (specified in clause 2.28);

3.1.5.1 – compliance with architect's instructions constituting variations under the contract – clause 2.29.1;

3.1.5.2 – failure of the architect to comply with clause 2.12 (clause 2.29.7);

3.1.6 Clause 4.23 – If the contractor makes written application to the architect stating that he has incurred or is likely to incur direct loss and/or expense for which he would not be reimbursed under any other provision in the contract due to the regular progress of the works or any part thereof being materially affected by:

3.1.6.1 – failure of the architect to comply with clause 2.12 (clause 4.24.5);

3.1.6.2 – architect's instructions requiring a variation (clause 4.24.1);
and provided that his application was made as soon as possible after it has become, or should reasonably have become, apparent to the contractor that the regular progress of the works has been or is likely to be affected,

and the contractor has in support of his application upon the request of the architect submitted such information as should reasonably be necessary to enable the architect to form an opinion, and

the contractor has submitted to the architect or quantity surveyor upon request such details of loss and/or expense as are reasonably necessary for ascertainment,

then the architect or the quantity surveyor shall ascertain the amount of such loss and/or expense and the amount ascertained shall be added to the contract sum (clauses 4.23 and 4.25).

3.1.7 Section 4 – Half of the retention percentage may be deducted from the amount which relates to work which has reached practical completion (clause 4.20.3) and the remaining half shall be released upon issuance of the final certificate, which shall be issued no later than two months after whichever of the following occurs last (clause 4.15):

3.1.7.1 the end of the Rectification Period;

3.1.7.2 the date of the issue of the certificate of making good defects under clause 2.39;

3.1.7.3 the date on which the architect sent a copy to the contractor of any ascertainment of loss and expense (if applicable) under clause 4.5.2.1 and a statement of adjustments to the contract sum under clause 4.5.2.

3.2 The above provisions apply to sections A and B of the works.

3.3 *Without prejudice* to the contractor's rights to claim damages under the general law (clause 4.25), save as provided in 3.3.1 and 3.3.2 hereof, the contractor's claim is made pursuant to the provisions on the contract hereinbefore mentioned.

3.3.1 The contractor may be entitled to interest on liquidated damages which shall become repayable to the contractor pursuant to a revised extension of time made by the architect – *Department of Environment for Northern Ireland v. Farrans* (1981) 19 BLR 1. The contractor may also be entitled to contractual interest (*infra*).

3.3.2 Where the contractor complies with his obligations with respect to information and particulars for the purposes of preparing the final account and all adjustments to be made to the contract sum, if the architect or quantity surveyor fail to prepare such final account or make all necessary adjustments as aforesaid, the contractor is entitled to reimbursement of the cost incurred in preparing such adjustments – *James Longley & Co Ltd v. South West Regional Health Authority* (1985) 25 BLR 56.

4.0 Details of Claim

4.1 Introduction.

4.1.1 The contractor's master programme for completion of the works and section A within the periods for completion is shown in appendix I (A.1) hereto. Activities A–B to J–K are critical for completion of all of the works in twenty-two weeks. Activities A–B to E–F, F–G and G–H are critical for completion of section A in sixteen weeks. Activities B–C to D–H and H–K are not critical, and will not become critical until all of the float shown on the contractor's master programme has been used up by delays to these otherwise non-critical activities.

4.1.2 The causes of delay referred to in this section are delays which entitle the contractor to an extension of time, or, if no extension of time is permitted for delay by such cause (as in the case of exceptionally adverse weather conditions), the contractor would be entitled to an extension of time for other causes of delay which used the float in the programme as a result of which otherwise non-critical activities became critical and caused delay to completion of Section A or Section B or both.

4.2 **Exceptionally adverse weather conditions – Delay (D1).**

4.2.1 Activity B–E is for the construction of a surface water culvert under the new access road.

4.2.2 The contractor completed the preceding activity (A–B) on programme and was proceeding with the construction of activity B–E in accordance with the programme.

4.2.3 During the weekend of 20 and 21 March 2010, continuous rainfall caused the open trench for the construction of the culvert to be flooded. On 22 March 2010, the contractor hired additional pumps to remove the water from the excavations. However, exceptionally adverse weather conditions continued during the period of two weeks (weeks commencing 23 and 30 March 2010). Records of the rainfall during the period taken at Much Hamstead (four miles from the site) were obtained by the architect for record purposes.

4.2.4 Water had been removed from the trenches and the contractor was able to recommence construction of the culvert on 6 April 2010 (a delay of two weeks).

4.2.5 The contractor gave notice of delay pursuant to clause 2.27 of the conditions of contract.

4.2.6 It is common ground that the contractor was delayed by a period of two weeks as a result of the said weather conditions and that no extension of time is permitted for such delay by virtue of the deletion of clause 2.29.9 of the conditions of contract.

4.3 **Architect's instruction no 1 – Delay (D2).**

4.3.1 Activity F–G is for the construction of an effluent drain under the existing stores and constructing new bases for the plant and equipment to be installed by the employer.

4.3.2 On 23 March 2010, the architect issued instruction no 1 which required the contractor to reposition the effluent drain in order to accommodate foundations for future alterations to the stores by the employer.

4.3.3 At the time of issuance of the said instruction, the construction of the new effluent drain was on programme. The contractor had excavated and laid all pipes within the existing stores and was ready to test the pipes prior to backfilling the trench on 22 March 2010. Records of the work executed prior to the issuance of the said instruction were agreed with the quantity surveyor.

4.3.4 The contractor commenced cutting out the existing floor slab at the revised location of the effluent drain on 24 March 2010. On the same day, some of the resources (labour and plant) were diverted from activity B–E (delayed as a result of the inclement weather described in 4.2 hereof) to commence backfilling to the redundant length of effluent drain.

4.3.5 The contractor excavated the trench for the revised effluent drain and laid the pipes and was ready for testing on 6 April 2010. A delay of two weeks had occurred as a result of the said instruction. The time taken to carry out the work prior to testing (2 weeks) was the same time allowed in the contractor's programme for carrying out the same quantity of work in the originally designed location of the effluent drain.

4.3.6 Backfilling and making good the floor slab at the location of the redundant effluent drain was completed on 6 April 2010. Had the contractor not been able to utilise resources from activity B–E (see 4.3.4 hereof), this work could not have been executed until after the contractor had completed the diversion of the effluent drain to the revised location.

4.3.7 As a result of the foregoing, activity B–G had been delayed by two weeks. No direct delay to completion of Section A or Section B was caused by the said instruction – see appendix II (A.2) hereto.

4.3.8 Notices and particulars of the delay and disruption and loss and/or expense caused by the said instruction were given by the contractor pursuant to clauses 2.27 and 4.23 of the conditions of contract (see 2.2 and 2.4 hereof).

4.4 Additional work – Delay (D3):

4.4.1 On 5 April 2010, the contractor notified the architect, in writing (letter ref BB/10), that he intended to divert resources from activity B–G in order to make up the time lost due to exceptionally adverse weather conditions (Delay D1). The contractor's revised programme showing completion by the original completion date was attached to the said letter – see appendix II – (A.3) hereto. The revised programme was made on the basis of using some of the float on activity B–G. The original float of six weeks had been reduced by two weeks (Delay D2) and the contractor envisaged using two weeks of the remaining four weeks' float so that work could cease on activity B–G until such time as activity B–E was on programme. No delay to completion of Section A or Section B would occur as a result of the reprogramming and two weeks' float would remain in activity B–G.

4.4.2 On 6 April 2010, the architect issued drawings AD/14A and 15A showing four additional bases for machinery (to be installed by the employer) and additional effluent branch drains.

4.4.3 On 6 April 2010, the contractor had set out for the new bases and ordered materials for the additional work. On the same day the contractor notified the architect that he estimated a delay of seven to eight weeks to activity B–G as a result of the said instruction (see 2.2.1.4 hereof). In the same letter, the contractor notified the architect that it would not be of any benefit to divert resources from activity B–G to activity B–E (see 4.4.1 hereof) as completion of Section A was dependent upon the timely completion of activity B–G, which had now become critical as a result of the additional work.

4.4.4 The contractor had completed all work to the revised drawings, by 21 June 2010 (a delay of 7 weeks).

4.4.5 On 22 June 2010 [week 19], the contractor issued his revised programme showing the Delays D1 to D3, completion of section A on 28 June 2010 [end of week 19] and completion of the Works on 2 August 2010 [end of week 24] – see appendix II (A.4) hereof.

4.4.6 Notices and particulars of the delay and disruption and loss and/or expense caused by the said additional work were given by the contractor pursuant to clauses 2.27 and 4.23 of the conditions of contract (see 2.2 and 2.4 hereof).

4.5 **Late instruction for expenditure of provisional sum – Delay (D4).**

4.5.1 The contract bills included the provisional sum £45000.00 for the supply and installation of mechanical equipment to the effluent treatment plant. This was shown on the contractor's original programme as activity H–K commencing in week 19 and the period for installation was one week.

4.5.2 The contractor's covering letter submitted with the said programme indicated that approximately two weeks would be necessary for ordering, manufacture and delivery of standard equipment from several well-known firms. The letter went on to request the architect to notify the contractor in the event of any potential subcontractors requiring a longer period for delivery, manufacture or installation. The necessary instructions (for standard equipment) would be required no later than 25 May 2010 (commencement of week 15).

4.5.3 As a result of Delays D2 and D3 (see 4.3 and 4.4 hereof) the revised latest date for receipt of instructions was 14 June 2010 [week 18].

4.5.4 On 7 June 2010, the architect issued instruction no 7 for the supply and installation of the equipment. The equipment required by the instruction was not standard equipment and the supplier who could provide it by the earliest delivery date was Pumps & Co for the sum of £42250.00. The delivery period for the equipment was quoted as seven to eight weeks and one week was required for installation.

4.5.5 On the same day, the contractor notified the architect by fax (ref BB/77) that the delivery period quoted by Pumps & Co was unacceptable, but he would be prepared to place the order with Pumps & Co provided that the architect would make an appropriate extension of time.

4.5.6 On 8 June 2010, the architect notified the contractor by fax (ref TS/12A) that he would take the delivery period of the pumps into account when making his decision on extensions of time.

4.5.7 On 9 June 2010, the contractor placed his order with Pumps & Co. A formal subcontract was signed between the contractor and Pumps & Co on 21 June 2010.

4.5.8 Pumps & Co delivered their equipment to site on 2 August 2010 and completed the installation, including testing, on 9 August 2010 [end of week 25]. Completion of the works had been delayed by three weeks having regard to the fact that the contractor had been denied the opportunity to reduce the delay caused by exceptionally adverse weather conditions (Delay D1 – see 4.2 and 4.4.1 hereof) – see appendix II (A.5) hereto.

4.5.9 Notices and particulars of the delay and disruption and loss and/or expense caused by the said additional work were given by the contractor pursuant to clauses 2.27 and 4.23 of the conditions of contract (see 2.2 and 2.4 hereof).

4.6 Summary

4.6.1 Completion of Section A has been delayed by three weeks as a result of Delays (D2) and (D3) – (see 4.3 and 4.4).

4.6.2 Completion of Section B has been delayed by three weeks as a result of Delays (D2), (D3) and (D4) – (see 4.3, 4.4 and 4.5 hereof).

4.6.3 The delays referred to hereinbefore are shown in appendix II (A.5) hereto.

4.6.4 The contractor contends that the architect has wrongly deducted the period of two weeks (delay caused by exceptionally adverse weather conditions) from the total delay to completion of three weeks for Section A and Section B. (The architect's reasons for making this adjustment are given in minutes of meeting of 16 August 2010, paragraph 2.3.)

4.6.5 Even if the contractor had not contemplated reprogramming the works to mitigate Delay (D1) – (see paragraph 4.4.1 hereof), the contractor maintains that no deduction should be made for Delay (D1) when, in any event, completion of Section A and Section B were delayed by Delays (D2), (D3) and (D4) which were the responsibility of the employer. Accordingly, the employer could not levy liquidated damages for the period of two weeks when the progress of the works was delayed by matters for which the employer was responsible.

4.6.6 Further, or alternatively, the contractor was prevented from mitigating Delay (D1) as a result of the additional work (see 4.4 hereof) and is entitled to a fair and reasonable extension of time of three weeks pursuant to clause 2.28 of the conditions of contract (relevant events described in clauses 2.29.1 and 2.29.7) until the date of practical completion of Section A and Section B and for reimbursement of loss and/or expense pursuant to clause 4.23 of the conditions of contract (matters described in clauses 4.24.1 and 4.24.5).

5.0 Evaluation of loss and/or expense

5.1 For the reasons given in 4.0 hereof, the contractor is entitled to direct loss/ and or expense as follows:

5.1.1 **Prolongation**

The period of prolongation is 3 weeks. The contractor contends that the issuance of drawings AD/14A and 15A (see 4.3 hereof) substantially changed the conditions under which the work on activity B–E would otherwise have been carried out (see 4.4.1 hereof). Therefore, notwithstanding Delay (D1), pursuant to the provisions of clause 5.10.1 and the proviso in clause 5.10.2, the contractor is entitled to reimbursement for the total period of prolongation pursuant to clause 4.23 (matter referred to in clause 4.24.1).

The contractor is entitled to reimbursement of loss and/or expense caused by Delays (D2) and (D3) pursuant to clause 4.23 (matter described in clause 4.24.1).

The contractor is entitled to reimbursement of loss and/or expense caused by Delay (D4) pursuant to clause 4.23 (matter described in clause 4.24.5).

5.1.1.1 **Head office overheads and profit**

As a result of Delays (D2), (D3) and (D4) described in 4.0 hereof, the contractor was required to retain its key staff and resources on site for an additional period of three weeks and was deprived of making a contribution to overheads and profit. Provided the contractor can evidence a lost opportunity to recover overheads and profit from other contracts, the contractor is entitled to recover this loss pursuant to the provisions mentioned in 5.1.1 hereof.

The contractor's auditors have certified that the contractor's overheads and profit (as percentages of revenue) were as follows:

Year ending 30 September 2009 – 12.76%
Year ending 30 September 2010 – 11.98%

The average percentage for overheads and profit for two years was therefore:

(12.76 + 11.98)/2 = 12.37%

Using *Emden's* formula:

Loss of overheads and profit for three weeks =

$$\frac{\text{Overheads \& profit \%}}{100} \times \frac{\text{Contract Sum}}{\text{Contract period}} \times \text{Period of delay}$$

$$\frac{12.37\%}{100} \times \frac{\text{£}827333.00}{22\ 2\text{weeks}} \times 3\text{ weeks} = \text{£}13955$$

5.1.1.2 **Site overheads and establishment (preliminaries)**

As a result of Delays (D1), (D2) and (D3) described in 4.0 hereof, the contractor was required to retain its key staff and resources on site for an additional period of three weeks. The contractor is therefore entitled to recover the expense of his site overheads and establishment costs for the period of delay pursuant to the provisions mentioned in 5.1.1 hereof.
Delays (D2) and (D3) – 2 weeks – see (A.4) in appendix II hereto.

Costs incurred during weeks 11 and 12:

Excludes costs associated with activity B–G:

Project manager	2 weeks @ £1500.00/week	= £3000.00
General foreman	2 weeks @ £1250.00/week	= £2500.00
Engineer	2 weeks @ £1200.00/week	= £2400.00
Quantity surveyor (part)	2 weeks @ £700.00/week	= £1400.00
Administration staff	2 weeks @ £750.00/week	= £1500.00
Hire of offices	2 weeks @ £900.00/week	= £1800.00
Office equipment	2 weeks @ £200.00/week	= £400.00
Plant & equipment	2 weeks @ £1950.00/week	= £3900.00
Scaffolding	2 weeks @ £1600.00/week	= £3200.00
Small tools & equipment	2 weeks @ £650.00/week	= £1300.00
Electricity charges	£1950.00 × 2/13 weeks	= £300.00
Telephone charges	£975.00 × 2/13 weeks	= £150.00
Security	2 weeks @ £500.00/week	= £1000.00
Stationery and sundries	£90.00 × 14/30 days	= £42.00
Total		£22892.00

Delay (D4) – One week – see (A.5) in appendix II hereto.

Costs incurred during week 23;

Project manager	1 week @ £1500.00/week	= £1500.00
General foreman	1 week @ £1250.00/week	= £1250.00
Quantity surveyor (part)	1 week @ £700.00/week	= £700.00
Administration staff	1 week @ £300.00/week	= £300.00
Hire of offices	1 week @ £900.00/week	= £900.00
Office equipment	1 week @ £200.00/week	= £200.00
Plant & equipment	1 week @ £550.00/week	= £550.00
Small tools & equipment	1 week @ £200.00/week	= £200.00
Electricity charges	£650.00 × 1/13 weeks	= £50.00
Telephone charges	£325.00 × 1/13 weeks	= £25.00
Security	1 week @ £500.00/week	= £500.00
Stationery and sundries	£62.00 × 7/31 days	= £14.00
Total		£6189.00

Total site overheads and establishment costs = £22892.00 + £6189.00= **£29081.00**

5.1.1.3 **Finance charges on delayed release of retention**

Pursuant to clause 4.20 of the conditions of contract, two-and-a-half per cent of the contract sum (being one half of the retention percentage stated in the Contract Particulars to the conditions of contract) should be certified and paid after practical completion (of each section A and section B) and the remaining two-and-a-half per cent upon the issuance of a Certificate of Making Good Defects.

As a result of Delays (D2), (D3) and (D4), the dates when the retention ought to have been released were three weeks later than the dates which would have applied if there had been no delay. Accordingly, the contractor has incurred financing charges by virtue of the fact that interest charges on his overdraft have been accruing for an additional period of three weeks on the amount of retention withheld.

The finance charges incurred are calculated at the rate of two per cent above the bank base rate (as charged by the contractor's bank from time to time) as follows:

First half due to be released.

Period of financing (assume release three weeks after practical completion):

	Planned release	Actual release	Rate
Section A	12 July 2010	2 August 2010	2.5%
Section B	9 August 2010	30 August 2010	2.5%

Amount of retention:

Section A: £14000.00

Finance charges = £14000.00 x 2.5% x 21/365 = £21.14

The works: £21010.00 – £14000.00 = £7010.00

Finance charges = £7010.00 x 2.5% x 21/365 = £10.08

Second half due to be released (Rectification Period – six months).

Period of financing (assume planned release 3 weeks after Certificate of Making Good Defects):

	Planned release	Actual release	Rate
Section A	12 Jan 2011	2 Feb 2011	8%
Section B	9 February 2011	2 March 2011	8%

Amount of retention:

Section A: £14000.00

Finance charges = £14000.00 x 2.5%% x 21/365 = £20.14

Section B: £21010.00 - £14000.00 = £7010.00

Finance charges = £7010.00 x 2.5% x 21/365 = £10.08

Total finance charges on retention

= £21.14 + £10.08 + £21.14 + £10.08 = **£62.44**

5.1.1.4 Fluctuations

The contract does not provide for reimbursement of fluctuations of labour or materials (see 1.4.2.9 hereof). The contractor allowed for the anticipated increase in labour in June 2010 in his tender (for the labour required to execute the work in weeks 20–22 on activity J–K). The hours allowed by the contractor in his tender during this period were as follows:

Craft operatives 3170 hours
Labourers 2700 hours

Due to Delays (D2), (D3) and (D4), the contractor's labour resources in weeks 20–25 were as follows:

Craft operatives 5060 hours
Labourers 4365 hours

Due to the fact that the contractor had been prevented from mitigating the delay caused by exceptionally adverse weather conditions (Delay D1) – see 4.4.1 hereof, the additional costs of labour for the additional hours expended after the wage increase on 28 June 2010 (most of which would have been prevented by the measures proposed by the contractor to mitigate the delay) qualify for reimbursement pursuant to clause 4.23 of the conditions of contract.

The additional costs of labour claimed are calculated as follows:

	Tender	28 June 2010	Increase[1]
Craft operatives	£10.30	£10.30	
NI & Employer's Ins (13%)	£1.34	£1.34	
	£11.64	£11.64	£0.00 (hr)
Labourers	£7.75	£7.75	
NI & Employer's Ins (13%)	£1.01	£1.01	
	£8.76	£8.76	£0.00 (hr)

Hours after 28 June 2010:

Craft operatives	5060 – 3170	= 1890hrs
Labourers	4365 – 2700	= 1665hrs

Therefore, the additional costs caused by Delays (D1), (D2) and (D3) are:

Craft operatives	1890hrs @ £0.00	= £0.00	
Labourers	1665hrs @ £0.00	= £0.00	
Total			0.00

The total increased cost of labour fluctuations is **£0.00**

The contractor ordered all materials at the prices applicable at the date of tender and no claim is made for increased costs of materials.

5.1.1.5 **Total prolongation costs**

Head office overheads & profit (5.1.1.1)	= £13955.00
Site overheads & establishment costs (5.1.1.2)	= £29081.00
Finance charges on retention (5.1.1.3)	= £62.44
Fluctuations (5.1.1.4)	= £0.00
TOTAL	**£43098.44**

5.1.2 **Disruption**

Activity B–G was delayed by nine weeks as a result of Delays (D2) and (D3). Site staff and resources allocated to this activity were required on site for this additional period and the contractor is entitled to reimbursement of expense caused thereby.

5.1.2.1 **Cost of resources allocated to activity B–G**

Section foreman	9 weeks @ £1200.00/week	= £10800.00
Engineer	9 weeks @ £1200.00/week	= £10800.00
Plant & equipment	9 weeks @ £550.00/week	= £4950.00
Scaffolding (part only)	9 weeks @ £800.00/week	= £7200.00
Small tools & equipment	9 weeks @ £400.00/week	= £3600.00
Total		**£37350.00**

5.1.3 **Finance charges on loss and expense**

The contractor has incurred financing charges by virtue of the fact that interest charges on his overdraft have been accruing from the date that each head of loss and expense occurred.

In addition, the contractor has incurred finance charges on the liquidated damages and he claims finance charges under the general law until liquidated damages are repaid to the contractor (see 3.3.1 hereof).

For the purposes of calculating finance charges, the dates when the loss and expense occurred are taken as follows:

Head office overheads & profit (5.1.1.1)	£13955.00	5 August 2010
Site overheads & establishment (5.1.1.2)	£22892.00	5 May 2010
	£6189.00	5 August 2010
Finance charges on retention (5.1.1.3)	£21.14	5 August 2010
	£10.08	5 Sept 2010
	£21.14	5 Feb 2011
	£10.08	5 March 2011
Disruption (5.1.2.1)	£37350.00	5 May 2010
Fluctuations (5.1.1.4)	£0.00	5 August 2010
Total	£80448.44	
On liquidated damages	£63000.00	5 Sept 2010

Therefore, finance charges accrued on the following sums from the dates given below:

£60242.00	5 May 2010
£20165.14	5 August 2010
£63010.08	5 September 2010
£21.14	5 February 2011
£10.08	5 March 2011

The finance charges incurred are calculated at the rate of two per cent above the bank base rate (as charged by the contractor's bank from time to time) in appendix III hereto.

The total finance charges up to 2 April 2011 (the date of this submission) are **£2640.68**, see Appendix III.

5.1.4 **Costs of preparing the claim**

5.1.4.1 The contractor has complied in all respects with his obligations to give notice and full particulars pursuant to clause 4.23 of the conditions of contract (see 2.4 hereof) and the architect has failed to comply with his obligations to ascertain the loss and/or expense due to the contractor.

5.1.4.2 Accordingly the contractor claims reimbursement of the fees paid to Contraconsult Ltd for the preparation of this submission in the sum of **£6050.00** (see 3.3.2 hereof).

5.2 **Summary of loss and/or expense and/or damages**

The following sums are due to the contractor:

Prolongation costs (5.1.1.5)	£43098.44
Disruption (5.1.2.1)	£37350.00
Finance charges (5.1.3)	£2640.68

Cost of preparing the claim (5.1.4)	£6050.00
Total	**£89139.12**

6.0 **Statement of Claim**

6.1 **Extensions of time**

6.1.1 The contractor claims an extension of time pursuant to clause 2.28 of the conditions of contract of a further two weeks giving the following extended dates for completion:

Section A – 27 June 2010

Section B – 8 August 2010

6.2 **Loss and expense and/or damages**

6.2.1 The contractor claims reimbursement of loss and/or expense pursuant to clause 4.23 of the conditions of contract and/or damages for breach of contract amounting to **£89139.12**.

6.3 **Retention**

6.3.1 The contractor is entitled to release of retention in the sum of £21010.00.

6.4 **Adjustments to the contract sum**

6.4.1 The contractor has submitted under separate cover (letter of even date) his statement of account for all adjustments to the contract sum (excluding the loss and/or expense and/or damages herein) and claims payment of the sum £6325.78 being the outstanding amount due to be included in the final statement of account pursuant to clause 4.5.2 of the conditions of contract.

6.5 **Liquidated damages**

6.5.1 The contractor claims repayment of liquidated damages in full for the amount of £63000.00.

6.6 **Finance charges accruing**

6.6.1 The contractor claims finance charges on the sums stated in 6.2 to 6.5 hereof after the date of this submission at the rate of two per cent above the bank base rate.

Appendix I

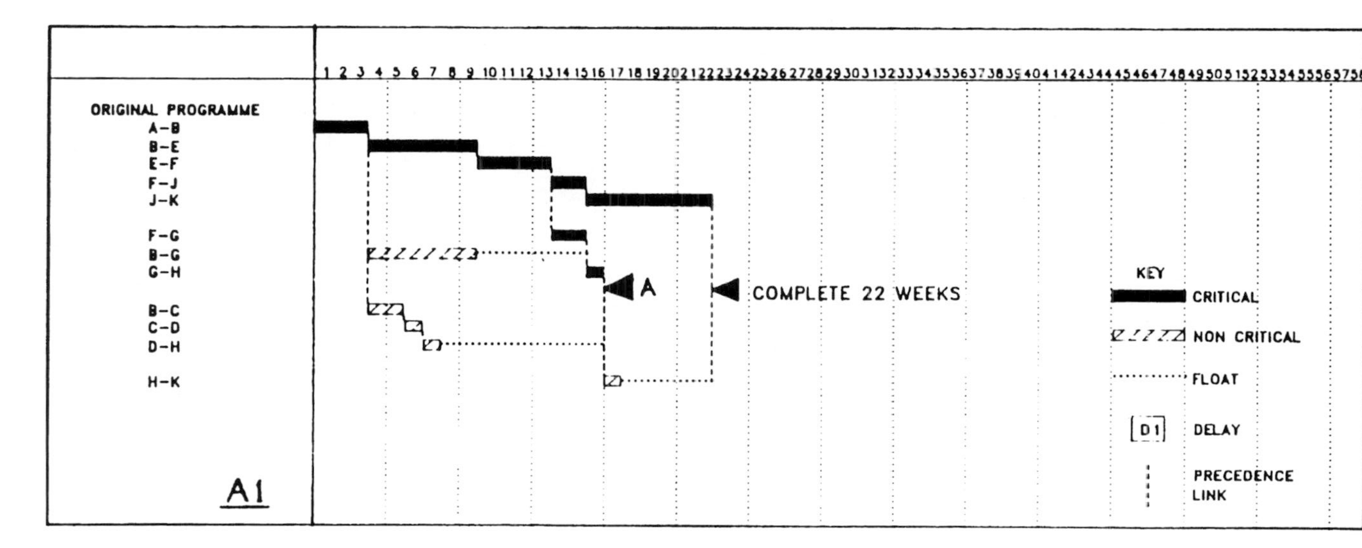

Figure A.1 Precedence (linked) bar chart – original programme

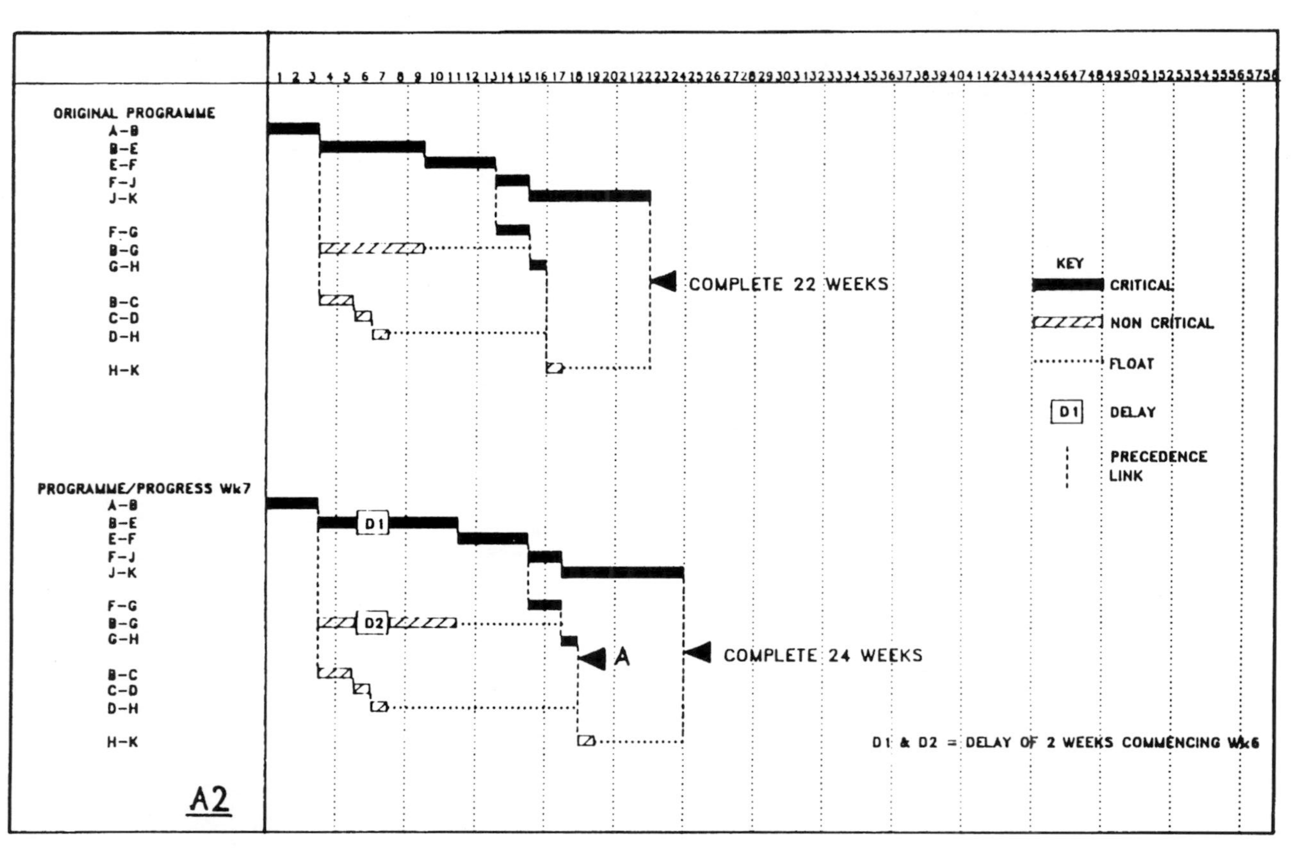

Figure A.2 Effect of delay D2 – Architect's Instruction No. 1

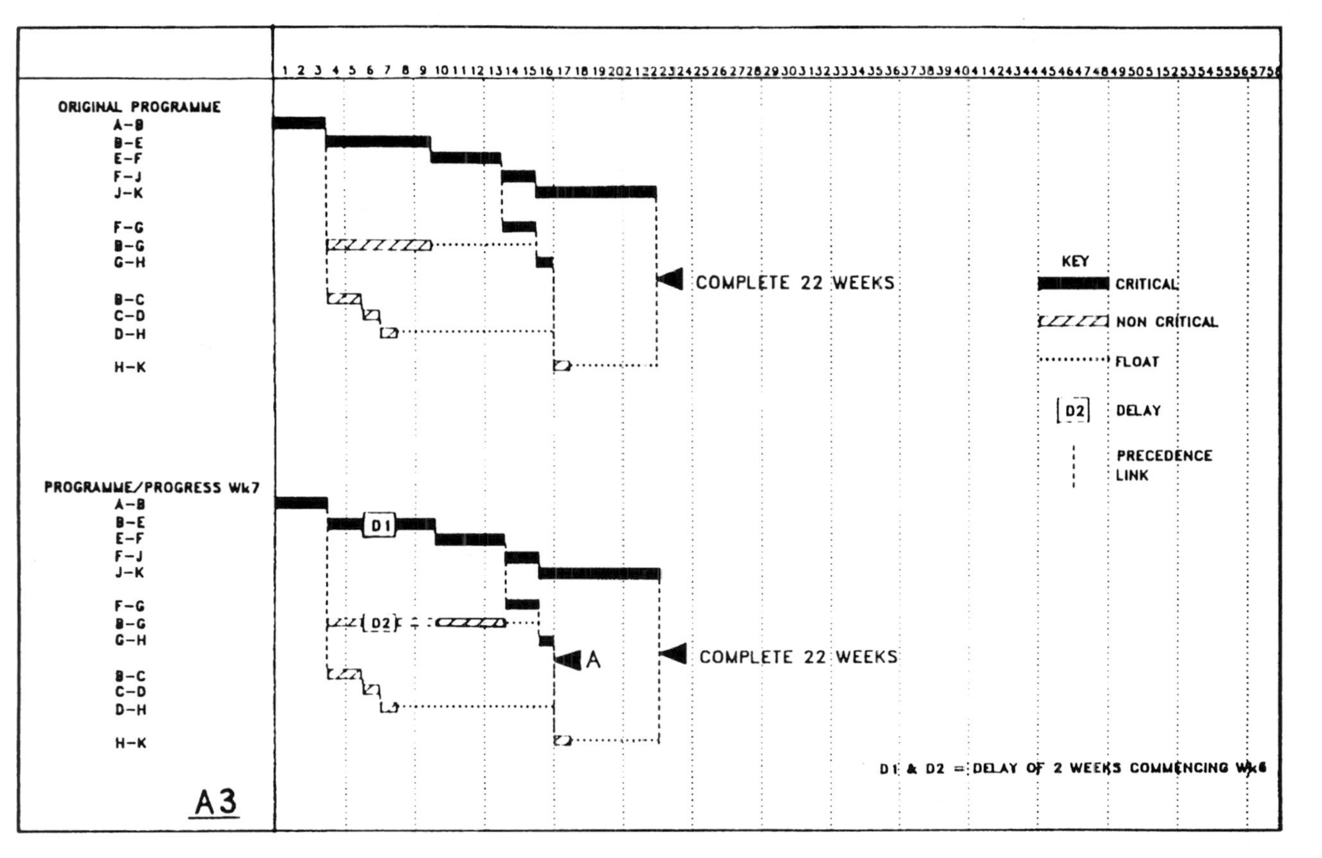

Figure A.3 Contractor's revised programme 5 April 2010

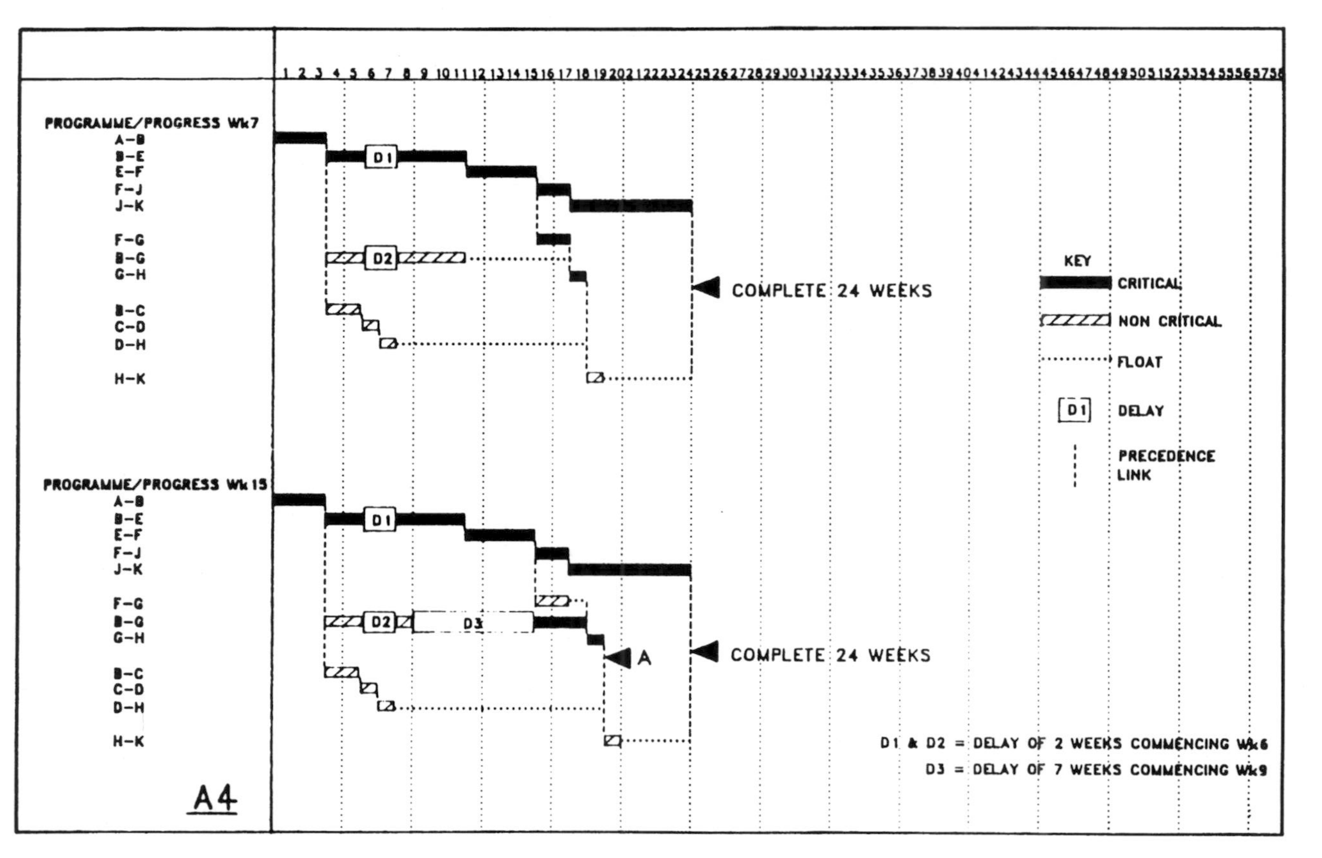

Figure A.4 *Contractor's revised programme 22 June 2010*

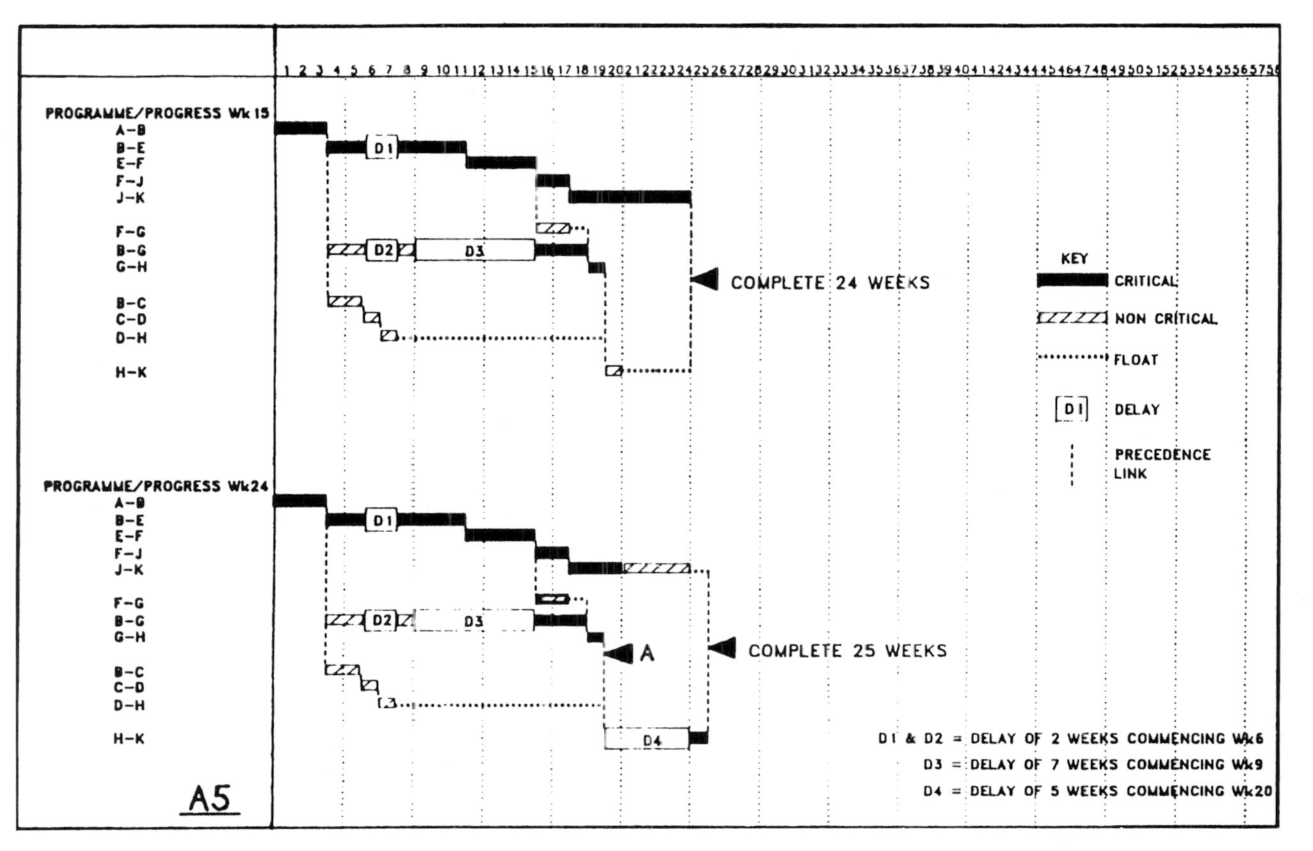

Figure A.5 Effect of delays D2, D3 and D4

Appendix III

Better Builders Ltd finance charges on balance due

Date	*Capital added*	*Capital total*	*Rate*	*Period days*	*Interest*	*Capital plus interest*
	£	£	£	£	£	£
5-May-10	60242.00	60242.00	0.025	57.00	235.19	60477.19
1-Jul-10		60477.19	0.025	35.00	144.98	
5-Aug-10	20165.14	80642.33	0.025	31.00	171.23	
5-Sep-10	63010.08	143652.41	0.025	26.00	255.82	144224.44
1-Oct-10		144224.44	0.025	92.00	908.81	145133.25
1-Jan-11		145133.25	0.025	35.00	347.92	
5-Feb-11	21.14	145154.39	0.025	28.00	278.38	
5-Mar-11	10.08	145164.47	0.025	29.00	288.34	146079.11
1-Apr-11		146079.11	0.025			
2-Apr-11			0.025	1.00	10.01	
	143,448.44				2640.68	146089.12

Dates in bold = rest days for compounding interest. Interest rate after 1 October 2010 assumed at current rates.

Architect's reply to the contractor's letter of 5 April 2011 and the claim submission

Date 10 May 2011

Dear Sirs,

Re: ABC Stores and Depot, New Road, Lower Hamstead, Wilton. I refer to your letter and enclosures of 5 April 2011.

Extensions of time

Having considered the arguments in your submission, I am prepared to fix later completion dates of 28 June 2010 for section A and 9 August 2010 for the works. That is, total extensions of time of three weeks inclusive of the extensions already made in my certificate EOT 1 dated 17 August 2010. I am not empowered to deal with the matter of finance charges on liquidated damages, and I am instructed to inform you that the employer wishes to discuss this with you at a meeting to be arranged next week. In the meantime, I will prepare the necessary certificate and issue it by the end of this week.

Loss and/or expense

I cannot agree that you are entitled to prolongation costs for the period of prolongation caused by Delays (D2) and (D3). The principal cause of delay during this period was exceptionally adverse weather conditions (Delay D1). I have considered your arguments on reprogramming (paragraph 4.4.1 of your submission) and I reject it on the grounds that you would have required additional formwork to make any progress on activity B–E in order to mitigate the delay. No additional formwork was delivered to site for this work.

Further, I cannot agree that your resources were prevented from taking on other work as a result of Delay (D4). According to my records, site offices were removed in week 24 and your resources were decreased commencing the end of week 23. I am prepared to include the part-time cost of your general foreman as part of your claim (subject to substantiation of his time spent on site). I do not accept that you lost any opportunity to make a contribution to overheads and profit as a result of one week delay. Even if I allowed loss of overheads and profit for any part of the prolonged period, I would have to deduct the overheads and profit recovered in the variations and extra work to activity B–G.

I also reject your argument on reimbursement of the costs of preparing the claim.

Although I accept that some additional resources were allocated to activity B–G I do not accept all the resources claimed were required for 9 weeks and therefore both the time and costs claimed will require substantiation.

The quantity surveyor's assessment of loss and/or expense, taking into account the above comments, is £18500.00 inclusive of finance charges up to the date of this letter.

A statement pursuant to clause 4.5.2 of the conditions of contract will be sent to you within the next few weeks.

Yours faithfully
T. Square

Contractor's reply to the architect's letter of 10 May 2011

Date 18 May 2011

Dear Sir,

Re: ABC Stores and Depot, New Road, Lower Hamstead, Wilton.

Thank you for your letter of 10 May 2011.

We cannot agree with your comments on our claim for loss and/or expense and/or damages.

Regarding measures to mitigate the delay caused by exceptionally adverse weather conditions (Delay D1), the work which would have been done in the first week after the delay [week 8] was the excavation of a trench 2.5 metres wide by 2.25 metres deep. No formwork was required until the second week. We enclose herewith the acknowledgement of order for additional formwork which was due to be delivered on 9 April 2010. Accordingly, had we carried out the measures to mitigate the delay, we would have been able to complete activity B–E in accordance with our original programme.

Regarding the removal of site offices and reduction in resources, we had originally planned to remove the site offices before the completion date and our resources would have been reduced commencing week 20 if the project had not been delayed. As a result of Delays (D2), (D3) and (D4) our resources were required for this project for three weeks longer than they would have been if there had been no delay. We reject the argument that we did not lose any opportunity to make a contribution to overheads and profit as a result of the delay. Please find enclosed a copy of the minutes of our board meeting on 6 July 2010 in which it is recorded that we postponed commencement of our own speculative development of twenty-six houses because our labour, staff and plant were retained on this project as a result of the delay.

We also disagree with the proposition that an adjustment should be made for overheads and profit recovered in variations and extra work. This work delayed activity B–G and delayed completion of section A. There was no affect on the period of prolongation (which was a result of late instruction in respect of equipment from Pumps & Co). In other words, the overheads and profit recovered in the additional work to activity B–G would have been earned within the original contract period and no adjustment would have been made.

In the circumstances of this case, we must insist that it is right to reimburse the cost of preparing the claim.

We trust that you will reconsider the matter at your earliest convenience.

Yours faithfully
For and on behalf of Better Builders Ltd.

Footnotes

Some of the arguments in the above example may be persuasive in negotiations. Differences of opinion in the industry on the use of a formula, concurrent delays, adjustment for overheads and profit recovered in variations and the costs of preparing the claim may give rise to real stumbling blocks in the negotiations to settle the sums in dispute.

This example may not cover all that went wrong during the progress of the works. There may have been other delays by the contractor. However, on the facts described in the example, the contractor appears to have reasonable grounds to pursue his claims.

While, in this case, the architect has now granted an extension for the full period of delay, some practitioners may argue that the words used in clause 2.28.1 of JCT 2011:

> 'If, in the Architect/Contract Administrator's opinion, on receiving a notice and particulars under clause 2.27:
>
> 1. any of the events which are stated to be a cause of delay is a Relevant Event; and
> 2. completion of the Works or any Section is likely to be delayed thereby beyond the relevant Completion Date,
>
> then, save where these Conditions expressly provide otherwise, the Architect/Contract Administrator shall give an extension of time by fixing such later date as the Completion Date for the Works or Section as he then estimates to be fair and reasonable.'

do not cover extensions of time in the circumstances of this case. For example, none of Delays (D2), (D3) or (D4) caused completion of the works (or section A or B) to be delayed beyond the completion date. Delay (D1) had already caused the completion of section A and section B to be delayed (or likely to be delayed) beyond the relevant completion date. Unless clause 2.28.5 is intended to allow greater flexibility for granting extensions of time, it would appear to be at least arguable that once the contractor has caused delay which was likely to cause completion of the works to be delayed beyond the relevant Completion Date, the clause does not bite. The editor does not agree with the foregoing proposition. If that was the case, there would be no valid extension of time provision (after the contractor's delay) and all subsequent delays within the control of the employer would put time at large and no liquidated damages could be recovered. This is clearly not the intention of the contract.

Clause 2.32.3 of the JCT Standard Building Contract 2011 provides for repayment of the sum paid as liquidated and ascertained damages in the event that an extension of time is granted after the deduction of liquidated and ascertained damages.

The contractual provisions concerning interest (clauses 4.12.6 and 4.15.7) provide an entitlement to contractual interest '*If the Employer fails to properly pay the amount, or any part of it, due to the Contractor under these Conditions.*'

If the repayment was made promptly there would probably be no entitlement to contractual interest but if it was not then it is arguable that contractual interest should apply.

Note

1. There was no increase in the National Working Rule labour rates in 2010 due to the financial situation at that time. As normally there will be an annual labour rate increase, the calculation is shown as an example.

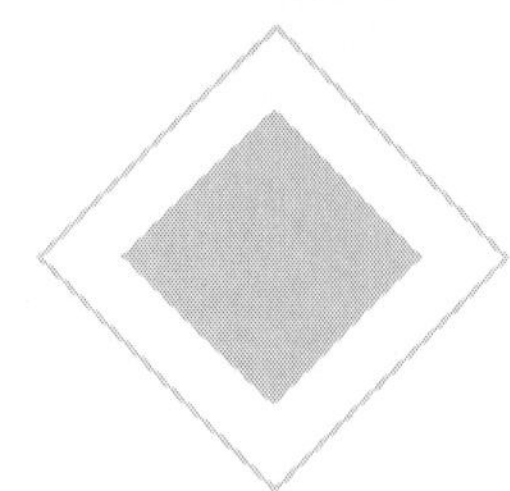

Appendix B: Sample Loss of Productivity Claim (due to disruption)

Introduction and explanation

A contractor for mechanical installations entered into a contract for various pipework systems which were required to be carried out in 13 weeks in accordance with an approved programme.

The contractor's tender was based on estimating norms for productivity; for example, in week 1, the part of the work to be done in accordance with the programme was 35 lineal metres (lin. m.) of 35 mm diameter pipe at 0.6 man-hours per lin. m. (21 man-hours). The total man-hour requirement in week 1 was estimated to be 525 man-hours, to be achieved with a gang of 12 men working 44 hours per week.

The contractor's analysis of total quantities and manpower required to execute the works in 13 weeks is shown in Figure B.1.

The contractor commenced work on time.

From weeks 8 to 17, numerous variation instructions were issued to re-route pipework to avoid conflicts with other installations and to accommodate some changes in layout of the building. Parts of the installation already installed had to be dismantled and re-installed (done on day-work). The contractor had to work out-of-sequence in various parts of the building instead of in an orderly manner as planned. The work actually took 17 weeks to complete. The actual quantities and schedule of work done are shown in Figure B.1. The data from Figure B.1 has been incorporated in Figure B.2, which shows the productivity, incidence of variations and alterations (day-work) on a weekly basis.

It is evident from Figures B.1 and B.2 that during weeks 1–8, the contractor was able to progress the work approximately as planned until the end of week 8, that is 5364 man-hours of work had been achieved compared with the original plan of 5216 man-hours (5216 being the sum of planned man-hours for the first eight weeks in the tender plan). The contractor had achieved this progress using more manpower, because of inefficient working, and for every hour worked an average of 0.936 man-hours of work had been done, that is an earned value or productivity factor (PF) of 0.936 compared with the tender norms of 1.0. Apart from two variations issued during this period, the contractor had not been affected by any adverse factors.

Accordingly, it is reasonable to argue that given no significant external factors to disrupt the contractor's progress, productivity would have been 0.936 man-hour earned for each 1.0 man-hour worked for the duration of the project, that is even without significant disrupting factors, the contractor could not achieve the tender norm of 1.0.

However, from week 9 onwards, it is evident that the number of variations issued and the amount of day-work (dismantling and re-installing work already completed) had an affect on productivity. It is reasonable to conclude that the drop in productivity from week 9 onwards was a direct result of these factors (see Figure B.2).

	Original Quantities and Schedule of Work for 13 Week Programme														Estimated Productivity		
	25 mm			50 mm			100 mm			150 mm			Daywk	Total Planned	Actual No. of	Actual Total	Prod. per hr
	Qty	mh/unit	mh	Qty	mh/unit	mh	Qty	mh/unit	mh	Qty	mh/unit	mh	mh	mh	Men	Mh	PF
w1	35	0.6	21	60	1.2	72	120	2.4	288	40	3.6	144	0	525	12	528	0.994
w2	60	0.6	36	85	1.2	102	140	2.4	336	40	3.6	144	0	618	14	616	1.003
w3	70	0.6	42	80	1.2	96	140	2.4	336	40	3.6	144	0	618	14	616	1.003
w4	60	0.6	36	85	1.2	102	120	2.4	288	60	3.2	192	0	618	14	616	1.003
w5	65	0.6	39	85	1.2	102	170	2.4	408	50	3.2	160	0	709	16	704	1.007
w6	90	0.6	54	80	1.2	96	140	2.4	336	70	3.2	224	0	710	16	704	1.009
w7	90	0.6	54	80	1.2	96	140	2.4	336	70	3.2	224	0	710	16	704	1.009
w8	80	0.6	48	90	1.2	108	170	2.4	408	45	3.2	144	0	708	16	704	1.006
w9	290	0.6	174	100	1.2	120	120	2.4	288	40	3.2	128	0	710	16	704	1.009
w10	210	0.6	126	150	1.2	130	120	2.4	288	35	3.2	112	0	706	16	704	1.003
w11	160	0.6	96	150	1.2	130	110	2.4	264	50	3.2	160	0	700	16	704	0.994
w12	150	0.6	90	100	1.2	120	120	2.4	288	10	3.2	32	0	530	12	528	1.004
w13	30	0.6	18	80	1.2	96	90	2.4	216	5	3.2	16	0	346	8	352	0.983
Total	1390			1225			1700			555				8208	186	8184	1.003

	Actual Quantities and Schedule of Work Done														Actual Productivity			Loss of Productivity		
	25 mm			50 mm			100 mm			150 mm			Daywk	Total Planned	Actual No. of	Actual Total	Prod. per hr	Base Prod.	Loss of Prod.	Loss of Prod.
	Qty	mh/unit	mh	Qty	mh/unit	mh	Qty	mh/unit	mh	Qty	mh/unit	mh	mh	mh**	Men	Mh**	PF		%	mh
w1	35	0.6	21	60	1.2	72	130	2.4	312	55	3.6	198	2	603	14	614	0.982	0.936	0.00	0.00
w2	60	0.6	36	85	1.2	102	120	2.4	288	50	3.6	180	4	606	14	612	0.990	0.936	0.00	0.00
w3	90	0.6	54	90	1.2	108	140	2.4	336	35	3.6	126	6	624	16	698	0.894	0.936	0.00	0.00
w4	80	0.6	48	85	1.2	102	110	2.4	264	65	3.2	208	12	622	16	692	0.899	0.936	0.00	0.00
w5	75	0.6	45	85	1.2	102	155	2.4	372	65	3.2	208	10	727	18	782	0.930	0.936	0.00	0.00
w6	95	0.6	57	85	1.2	102	155	2.4	372	65	3.2	208	12	739	18	780	0.947	0.936	0.00	0.00
w7	95	0.6	57	85	1.2	102	140	2.4	336	70	3.2	224	16	719	18	776	0.927	0.936	0.00	0.00
w8	80	0.6	48	90	1.2	108	170	2.4	408	50	3.2	160	18	724	18	774	0.935	0.936	0.00	0.00
w9	170	0.6	102	90	1.2	108	110	2.4	264	30	3.2	96	80	570	18	712	0.801	0.936	14.51	103.32
w10	120	0.6	72	90	1.2	108	80	2.4	192	35	3.2	112	90	484	18	702	0.689	0.936	26.38	185.16
w11	100	0.6	60	70	1.2	84	80	2.4	192	25	3.2	80	100	416	18	692	0.601	0.936	35.80	247.77
w12	90	0.6	54	80	1.2	96	70	2.4	168	25	3.2	80	110	398	18	682	0.584	0.936	37.68	256.99
w13	80	0.6	48	70	1.2	84	50	2.4	120	10	3.2	32	140	284	16	564	0.504	0.936	46.23	260.73
w14	70	0.6	42	50	1.2	60	30	2.4	72	0	3.2	0	120	174	10	320	0.544	0.936	41.94	134.19
w15	80	0.6	48	50	1.2	60	20	2.4	48	0	3.2	0	130	156	10	310	0.503	0.936	46.26	143.41
w16	70	0.6	42	30	1.2	36	10	2.4	24	0	3.2	0	170	102	10	270	0.378	0.936	59.66	161.08
w17	65	0.6	39	70	1.2	84	30	2.4	72	0	3.2	0	100	195	10	340	0.574	0.936	38.76	131.77
Total	1455			1265			1600			580			1120	8143	Actual	10320	0.789	Total loss prod.		1624.42
Orig.	1390			1225			1700			555			0	8208		8184	1.003	8143 / 0.936		8695.58
Change	65			40			−100			25			1120	−65	Mh** excludes daywork			Check actual		10320.00

Figure B.1 Loss of productivity due to disruption: data

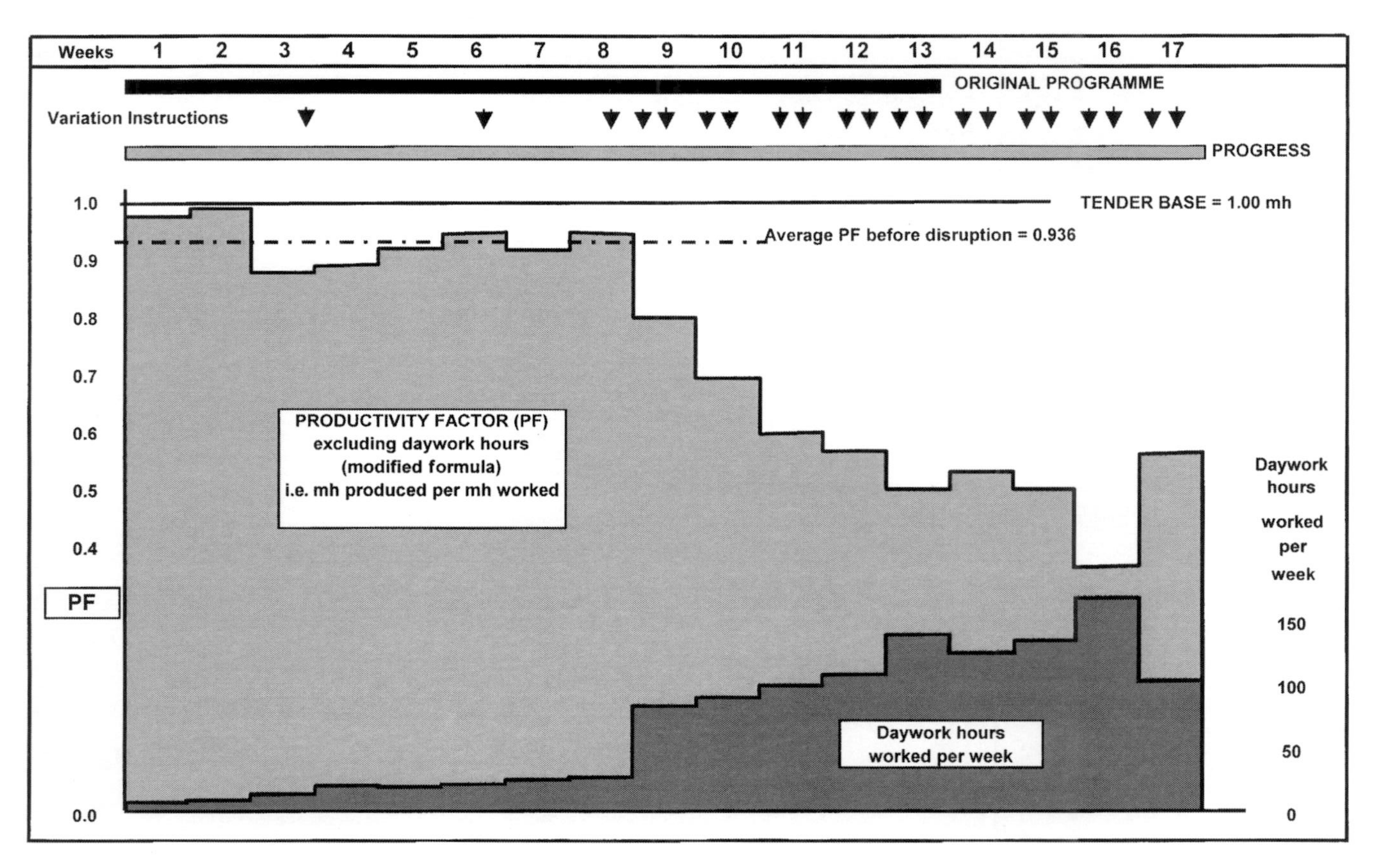

Figure B.2 *Loss of productivity due to disruption: based on data of Figure B.1*

Calculation of loss of productivity

The calculation of loss of productivity is as follows (see Figure B.1):

Productivity factor (PF) in week 1

Man-hours achieved (earned value)

$$= (35 \times 0.6) + (60 \times 1.2) + (130 \times 2.4) + (55 \times 3.6)$$
$$= 21 + 72 + 312 + 198 = 603$$

Man-hours spent (actual) $= 14 \times 44 = 616 - 2$ (day-work) $= 614$

$$\text{Therefore productivity factor (PF)} = \frac{\text{Man-hours work achieved}}{\text{Man-hours spent (actual)}}$$
$$= \frac{603}{614} = 0.982$$

Similar calculations have been done in Figure B.1 for all weeks.

Average productivity factor (PF) for weeks 1–8 (before significant disruption)

$$= \frac{603+606+624+622+727+739+719+724}{614+612+698+692+782+780+776+774} = \frac{5364}{5728} = 0.936$$

That is to say, for every man-hour worked, 0.936 man-hour value of work had been achieved.

Productivity factor (PF) in week 9 (the first week affected by significant disruption)

Man-hours achieved (earned value) $= 102 + 108 + 264 + 96 = 570$

Man-hours spent (actual) $= 18 \times 44 = 792 - 80$ (day-work) $= 712$

$$\text{Therefore productivity factor (PF)} = \frac{570}{712} = 0.801$$

That is to say, for every man-hour worked, 0.801 man-hour value of work had been achieved.

Loss of productivity in week 9

$$= \frac{(0.936 - 0.801)}{0.936} \times 100\% = 14.51\% \text{ (see Note)}$$
$$= 712 \text{ man-hours (spent)} \times 14.51\% = 101.32 \text{ man-hours}$$

Note – All calculations in Figure B.1 are in Excel and are calculated to more than three decimal places. The results in Figure B.1 are therefore more accurate and are given above.

Calculations for weeks 10–17 are also shown in Figure B.1.

The total loss of productivity is 1624.42 man-hours (being the sum of the loss of productivity for weeks 9–17 calculated in the same manner as week 9).

In other words, the contractor's case is that if he had not been disrupted by the numerous changes, instead of spending a total of 10320.00 man-hours to complete the work at a productivity factor (PF) of 0.936, the man-hours spent would have been 8695.58 man-hours calculated as follows:

Total planned man-hours to execute the work done (excluding day-work) = 8143

Man-hours to execute work done at a PF of 0.936 (excluding day-work)

$$= \frac{8143}{0.936} = 8695.58$$

Actual man-hours to execute the work done (excluding day-work) = 10320.00

Loss of productivity = 10320.00 − 8695.58 = 1624.42

The extra cost to the contractor is 1624.42 man-hours at the relevant cost per hour.

Appendix C: Example of the Procedure for the Notification and Assessment of a Compensation Event under the NEC 3 Contract

Introduction to the example

The approach adopted by the NEC in relation to a contractor's entitlement to relief, in terms of both extra time and additional money, is quite different from that of the earlier standard forms such as the ICE, FIDIC and the JCT forms.

The NEC does not contemplate contractor's 'claims', in the traditional (often advanced retrospectively) sense, for extensions of time, additional cost, or loss and/or expense. Under the NEC such matters are dealt with by both parties proactively following the procedures set forth in the contract for the notification and assessment of what are termed 'compensation events'.[1] Any failure by either or both of the parties to adhere to the timetable will be a breach of contract and can, in certain circumstances, lead to a contractor losing his entitlement to relief,[2] or, to an employer paying compensation and/or losing his right to damages for delay in circumstances where he ought not to have done.[3]

Dealing with time

In other standard forms such as the ICE and JCT forms, there are provisions requiring the Contractor to make written applications for extensions of time, and requiring the Engineer, Architect or Contract Administrator to respond, granting (or declining to grant) extensions of time. No such provisions exist in the NEC. Under the NEC, extending time, and thereby revising the Completion Date is dealt with by way of revisions to what is known as the Accepted Programme. This is the Contractor's construction programme, which the Contractor is required to update regularly (at least every 4 weeks) to reflect the actual progress of the work and to submit the revisions to the Employer's Project Manager for his acceptance. The contractual Completion Date is stated in the Contract Data. However, NEC contemplates the Contractor may wish to complete on an earlier date. This date is know as 'planned Completion'. The Contractor shows the date of 'planned Completion' on the first construction programme he submits to the Employer's Project Manager for his acceptance. Once accepted, that programme becomes the Accepted Programme. If, by reason of a compensation event, the Contractor is, or is likely to be delayed beyond the date for planned Completion, the Contractor revises the Accepted Programme showing the impact or expected impact of the compensation event and the delay or expected delay to planned Completion. The Contractor submits the revised programme as part of his quotation for the compensation event. If the Contractor's revised programme is accepted by the Employer's Project Manager, The accepted revision becomes the Accepted Programme and the contractual Completion Date is extended by the period equivalent to the delay to planned Completion shown on the revised Accepted Programme. Thus, the Contractor owns the float, if any, between his planned Completion and the contractual Completion Date.

Financial relief

The philosophy of the NEC is that the Contractor should not lose out, nor indeed should he benefit from a windfall, as a result of a compensation event. The difference in approach from the other standard forms can be explained by reference to the judgement of HH Judge Humphrey Lloyd QC in *Henry Boot Construction Ltd.* V. *Alstom Combined Cycles Ltd. (formerly GEC Alstom Combined Cycles Ltd.)*,[4] with which the Court of Appeal (by a majority) later agreed . The parties had contracted under the ICE 6th Edition, the rules pertaining to the valuation of variations being set out in clause 52 thereof. The Contract contained a price for certain sheet piling works for trench excavation. In the event, those works were the subject of a variation, which, Boot sought to value using the price in the Contract as the basis for its valuation. It was common ground that the price contained mistakes and the price had been arrived at in error. As a result of the error, valuing the variation using the price in the Contract would result in a substantial windfall to Boot. A dispute arose as to the valuation of the variation which was submitted to arbitration. The Arbitrator's award was not based on the price in the Contract but rather on a fair valuation. Boot appealed to the Courts on a point of law. Alstom argued that it was not reasonable to use the price as it would lead to an unjustified 'windfall' for Boot. Judge Lloyd rejected this argument, stating that the rates and prices were 'sacrosanct' and not subject to correction simply because of mistakes. The judge held that the result was not a windfall for Boot but that it was 'all part of the risks of contracting which produces thrills as well as spills'.

Contracting under the NEC produces no such thrills or spills, the outcome, on the facts of *Boot* v. *Alstom* would have been very different. Compensation events are not 'valued' using the Prices in the activity schedule (Main Options A and C) or, where a bill of quantities is applicable, the rates and prices contained therein (Main Options B and D). The Prices are not relevant to the assessment of compensation events.[5] Compensation events are assessed as the effect the compensation event has on the 'Defined Cost' of performing the work. The intention is that assessments will generally be done in advance of the work likely to be affected by the event being carried out, that is, on a forecast basis. The Prices in the activity schedule or bills of quantities as the case may be, are not omitted and substituted by new Prices. To do so would be to correct any errors in the original price. The Prices are simply adjusted to reflect the difference in the 'Defined Cost', due to the compensation event, of performing the work. Accordingly, in the Boot case, the error in the price would not, under NEC, have been compounded in an assessment of the variation, had it been a compensation event under the NEC.

Example

The following example is based on Main Option A, Priced Contract with Activity Schedule. The same procedure would apply in the case of the other Main Options. For simplicity, the example does not include supporting documents in relation to Defined Cost or planned resources but it is assumed that such information would be provided in appendices to the submission. Examples of programmes, notifications, covering letters and activity schedules are provided. Again, for simplicity, the programme periods do not take account of public holidays or other shutdown periods.

The contract

Better Contractors Ltd. has contracted with Waste and Water Ltd. to construct a new Oxidation Ditch, Final Settlement Tank and associated Ancillary Works at an existing Sewage Treatment Works known as Eastlands STW. The main work comprises the placement of new reinforced concrete bases and the construction of new reinforced concrete walls to both the Oxidation Ditch and the Final Settlement Tank. The steel bar reinforcement is free issued by the Employer. The Conditions of Contract are the NEC Edition 3, Main Option A. The starting date is 4 May 2010. The Completion Date is 3 December 2010.

The issue

The Project Manager has issued Better Contractors with revised drawings and bar schedules, changing the design of the steel bar reinforcement to the Oxidation Ditch walls. The original design called for predominantly 20 mm diameter bars. The revised design calls for predominantly 16 mm diameter bars.

The procedure

First letter from Better Contractors Ltd. to the Project Manager.

Our ref: JS/PA/201017/007

A. Brown Esq.,
Project Manager,
Waste and Water Ltd.,
Eastlands STW,
Eastlands
EL5 2XY

7 May 2010

Dear Sirs,

Eastlands STW, Notification of Early Warning.

We refer to your Document Transmittal No. 005 dated 4 May 2010 by which you issued revised drawings ref. EAS/STW/201017/009 Rev B, EAS/STW/201017/010 Rev. B and Reinforcement Bar Schedules Nos. 001B and 002B and, in accordance with clause 16.1 of the Conditions, we enclose our Early Warning Notification EWN 001 relating to same.

We request your attendance at a risk reduction meeting and will hold ourselves available to attend, at your convenience, within the next 7 days.

We look forward to hearing from you.

Yours faithfully
J. Smith.
Better Contractors Ltd.

Project Title: Eastlands STW Project No.: 201016	
NOTIFICATION OF EARLY WARNING	No.001
Early warning is hereby given of a matter, namely; the revised drawings ref. EAS/STW/201017/009 Rev B, EAS/STW/201017/010 Rev. B and Reinforcement Bar Schedules Nos. 001B and 002B, whereby the design of the steel reinforcement to the Oxidation Ditch walls has changed, which could:- Increase the total of the Prices* Delay Completion* ~~Delay the meeting of a Key Date*~~ ~~Impair the performance of the *works* in use*~~ Increase the *Contractor's* total cost* We request your attendance at a risk reduction meeting, time and place to be mutually agreed* * *delete as appropriate*	
Signed: Print name: *Project Manager* Date:	Signed: Print name: J. Smith *Contractor* Date: 7 May 2010
Reply:	
Signed: Print name: *Project Manager* Date:	Signed: Print name: *Contractor* Date:

Second letter from Better Contractors Ltd. to the Project Manager

Our ref: JS/PA/201017/007

A. Brown Esq.,
The Project Manager,
Waste and Water Ltd.,
Eastlands STW,
Eastlands
EL5 2XY

11 May 2010

Dear Sirs,

<u>Eastlands STW, Notification of Compensation Event.</u>

Further to the risk reduction meeting held on 10 May 2010 at which we discussed the matter of the revised drawings ref. EAS/STW/201017/009 Rev B, EAS/STW/201017/010 Rev. B and Reinforcement Bar Schedules Nos. 001B and 002B, in accordance with clause 61.3 of the Conditions, we enclose our Notification of Compensation Event No. 001 relating to same.

We look forward to notification of your decision in accordance with clause 61.4.

Yours faithfully
J. Smith.
Better Contractors Ltd.

<table>
<tr><td colspan="2">Project Title: Eastlands STW
Project No.: 201017</td></tr>
<tr><td>NOTIFICATION OF COMPENSATION EVENT</td><td>No. 001</td></tr>
<tr><td colspan="2">Notice is hereby given that the following event is/is believed to be* a compensation event pursuant to Clause 60.1 (1), an instruction changing the Works Information:-

~~Instruction, namely*~~
~~Change in earlier decision, namely*~~
Event, namely; the issue of revised drawings ref. EAS/STW/201017/009 Rev B, EAS/STW/201017/010 Rev. B and Reinforcement Bar Schedules Nos. 001B and 002B

~~Please submit your quotation in relation to the above compensation event*~~

* *delete as appropriate*</td></tr>
<tr><td>Signed:

Print name:
Project Manager/Supervisor

Date:</td><td>Signed:

Print name: J. Smith
Contractor

Date: 11 May 2010</td></tr>
<tr><td colspan="2">Reply:</td></tr>
<tr><td>Signed:

Print name:
Project Manager/Supervisor

Date:</td><td>Signed:

Print name:
Contractor

Date:</td></tr>
</table>

Letter from the Project Manage to Better Contractors Ltd.

Our ref: JS/PA/201017/007

J. Smith Esq.,
Better Contractors Ltd.,
21, Westlands Road,
Eastlands
EL5 2XY

17 May 2010

Dear Sirs,

Eastlands STW, Notification of Compensation Event Decision.

I attach herewith my decision as to your Notification of Compensation Event No. 001, and look forward to receipt of your quotation in accordance with clause 62.3.

Yours faithfully
A. Brown.
Waste and Water Ltd.

<table>
<tr><td colspan="2">Project Title: Eastlands STW

Project No.: 201017</td></tr>
<tr><td>NOTIFICATION OF COMPENSATION EVENT</td><td>No. 001</td></tr>
<tr><td colspan="2">Notice is hereby given that the following event ~~is~~/is believed to be* a compensation event pursuant to Clause 60.1 (1) an instruction changing the Works Information:-

~~Instruction, namely*~~
~~Change in earlier decision, namely*~~
Event, namely; the issue of revised drawings ref. EAS/STW/201017/009 Rev B, EAS/STW/201017/010 Rev. B and Reinforcement Bar Schedules Nos. 001B and 002B

~~Please submit your quotation in relation to the above compensation event*~~

* *delete as appropriate*</td></tr>
<tr><td>Signed:

Print name:
Project *Manager/Supervisor*

Date:</td><td>Signed:

Print name: J. Smith
Contractor

Date: 11 May 2010</td></tr>
</table>

<table>
<tr><td colspan="2">Reply: I confirm my agreement that the issue of revised drawings ref. EAS/STW/201017/009 Rev B, EAS/STW/201017/010 Rev. B and Reinforcement Bar Schedules Nos. 001B and 002B, constitute a change to the Works Information and as such fall to be a compensation event under clause 60.1 (1). Accordingly, I instruct you to submit your quotation in accordance with clause 62.3</td></tr>
<tr><td>Signed:

Print name: A. Brown

Project Manager~~/Supervisor~~

Date: 17 May 2010</td><td>Signed:

Print name:
Contractor

Date:</td></tr>
</table>

Third letter from Better Contractors Ltd. to the Project Manager.

Our ref: JS/PA/201017/015

A. Brown Esq.,
The Project Manager,
Reliable Water Ltd.,
Eastlands STW,
Eastlands
EL5 2XY

14 June 2010

Dear Sirs,

Eastlands STW, Quotation in relation to Compensation Event No. 001

Further to your decision and instruction dated 17 May 2010 to submit our quotation in respect of the revised drawings ref. EAS/STW/201017/009 Rev B, EAS/STW/201017/010 Rev. B and Reinforcement Bar Schedules Nos. 001B and 002B (Compensation Event No. 001), in accordance with clause 62.3 of the Conditions, we enclose our quotation reference Q/201017/001, comprising our assessment, in accordance with clause 63.1, of the effect of the compensation event on the forecast Defined Cost of the work not yet done, together with our assessment, in accordance with clause 63.3, of the delay to the Completion Date, which is supported by our revised programme for acceptance, Rev.1 dated 14 June 2010.

We look forward to notification of your acceptance of our quotation in accordance with clause 65.1, including notification in accordance with clause 65.4, of the changes to the Prices and the Completion Date.

Yours faithfully
J. Smith.
Better Contractors Ltd.

Better Contractors Ltd.

14 June 2010

Quotation reference Q/201017/001, comprising assessment of the effect of Compensation Event No. 001 on the forecast defined cost of the work not yet done, together with assessment of the delay to the Completion Date

Background

1. The original drawings reference EAS/STW/201017/009 Rev A, EAS/STW/201017/010 Rev. A and Reinforcement Bar Schedules Nos. 001A and 002A dated January 2010 and incorporated in the Works Information, indicate that the steel bar reinforcement to be fixed in the Oxidation Ditch walls was as follows:-

10mm diameter:	2.312 tonnes.
12mm diameter:	2.772 tonnes.
16mm diameter:	15.467 tonnes.
20mm diameter:	120.134 tonnes.
	140.685 tonnes in total

2. The revised drawings and bar schedules (revisions B) of the above indicate the following steel is to be fixed:-

10mm diameter:	1.554 tonnes.
12mm diameter:	4.080 tonnes.
16mm diameter:	135.988 tonnes.
20mm diameter:	42.091 tonnes.
	183.713 tonnes in total

3. The original design was based, predominantly, on the use of 20mm diameter bars. This has changed to, predominantly, 16mm bars.

The effect on time

4. The total tonnage of steel to be fixed has increased by 43.028 tonnes, an increase of about 30%. However, the size of the structure remains the same.

5. The length of steel originally to be fixed in the walls was as follows:-

10mm diameter:	2.312 tonnes @	0.616kg/m =	3,753m
12mm diameter:	2.772 tonnes @	0.888kg/m =	3,121m
16mm diameter:	15.467 tonnes @	1.579kg/m =	9,795m
20mm diameter:	120.134 tonnes @	2.466kg/m =	48,716m
			65,385m in total

6. The length of steel now to be fixed in the walls is as follows:-

10mm diameter:	1.554 tonnes @	0.616kg/m =	2,523m
12mm diameter:	4.080 tonnes @	0.888kg/m =	4,595m
16mm diameter:	135.988 tonnes @	1.579kg/m =	86,123m
20mm diameter:	42.091 tonnes @	2.466kg/m =	17,069m
			110,310m in total

7. The increase in the length of steel to be fixed in the same surface area of wall, in the same confined working area, is 44,925m, an increase of 69%.
8. Further, there is a considerable difference in the rigidity of 16mm diameter steel bars compared with 20mm diameter steel bars, leading to greater difficulty in fixing 6m long bars. Some additional measures are likely to be necessary in order to overcome this problem. These measures may include erecting formwork, out of sequence, to a single side of some of the walls to give greater stability.
9. The current Accepted Programme (copy attached) shows a period of 80 days for steel fixing to the Oxidation Ditch walls. Allowing for the additional length and tonnage, and taking account the lighter, easier to lift, 16mm bars, we estimate that this period will extend by 50% (forty days). Following operations that are dependant upon the steel fixing, i.e., formwork and placing concrete, will of course also be prolonged by the same period.
10. As we consider there is a significant risk of further delay due to the rigidity/ stability of the lighter steel, we have, in accordance with clause 63.6, included a risk allowance of a further 5 days.
11. In accordance with clauses 32.1 and 32.2 and as part of this quotation, we attach a revised programme (Revision 1) for your consideration and acceptance. The revised programme shows that the effect of the compensation event, including the risk allowance is likely delay planned Completion as shown on the Accepted Programme of 16 November 2010, until 18 January 2011, a period of sixty-three days. Accordingly, in accordance with clause 63.3, the Completion Date should be extended by sixty-three days, from 3 December 2010 to 4 February 2011.

The effect on the defined cost

12. In accordance with clause 31.2 we provided you with a statement of how we planned to do the work, including information as to our intended resources. For convenience, we attach a further copy as Appendix 1. We refer you to this information as the basis for our assessment of the Defined Cost of the work, but for the compensation event.
13. Build-ups of the Defined Cost of People in accordance with the Shorter Schedule of Cost Components, together with quotations in relation to the cost of the hire of Equipment and purchase of materials are provided in Appendix 2.

<u>Forecast Defined Cost of the originally intended work to the Oxidation Ditch Walls Steel fixing</u>

14. Steel fixers: 5Nr. @ £122.25 per man per day x 80 days = £48,900.00
    Labourers: 1Nr. @ 84.60 per man per day x 80 days = £6,768.00
    £55,668.00

<u>Forecast defined cost of the changed work to the Oxidation Ditch Walls Steel fixing</u>

15. Steel fixers: 5Nr. @ £122.25 per man per day × 120 days = £73,350.00
    Labourers: 1Nr. @ £ 84.60 per man per day × 120 days = £10,152.00
    £83,502.00
    Risk allowance: 5 days @ £695.85 per day = £3,479.25
    £86,981.25

16. The proposed change to the Prices in relation to item 060 on the *activity schedule*, Oxidation Ditch - Steel fixing to walls is £31,313.35 (£86,981.25 – £55,668.00) plus the Fee (*direct fee percentage* of 8%) equating to £2,505.07, totalling <u>£33,818.42</u>.

<u>Forecast defined cost of the originally intended work to the Oxidation Ditch Walls Formwork</u>

17. Shuttering Joiners: 6Nr. @ £165.85 per man per day x 80 days = £ 79,608.00
    Labourers: 2Nr. @ £ 84.60 per man per day x 80 days = £ 13,536.00
    Hire of Shuttering: @ £268.00 per day × 80 days = £ 21,440.00
    £114,584.00

<u>Forecast defined cost of the work to the Oxidation Ditch Walls Formwork, prolonged due to the changed steel reinforcement</u>

18. Shuttering Joiners: 6Nr. @ £165.85 per man per day x 120 days =119,412.00
    Labourers: 2Nr. @ £ 84.60 per man per day x 120 days =£ 20,304.00
    Hire of Shuttering: @ £268.00 per day × 120 days = £ 32,160.00
    £171,876.00
    Risk allowance: 5 days @ £1,164.30 per day = £ 5,821.50
    £177,697.50

19. The proposed change to the Prices in relation to item 070 on the *activity schedule*, Oxidation Ditch - Formwork to walls is £63,113.50 (£177,697.50 – £114,584.00) plus the Fee (*direct fee percentage* of 8%) equating to £5,059.08, totalling <u>£68,162.58</u>.

<u>Forecast defined cost of the Originally intended work to the Oxidation Ditch Walls Concrete</u>

20. Labourers: 3Nr. @ £ 84.60 per man per day × 80 days = £20,304.00

<u>Forecast defined cost of the work to the Oxidation Ditch Walls Concrete, prolonged due to the changed steel reinforcement</u>

21. Labourers:	3Nr. @ £ 84.60 per man per day × 120 days =	£30,456.00
Risk allowance:	5 days @ £253.80 per day =	£1,269.00
		£31,725.00

22. The proposed change to the Prices in relation to item 080 on the *activity schedule*, Oxidation Ditch – Place concrete to walls is £11,421.00 (£31,725.00 – £20,304.00) plus the Fee (*direct fee percentage* of 8%) equating to £913.68, totalling £12,334.68

Forecast defined cost of the Site establishment and management before the delay due to the compensation event

23. Site establishment and management: @ £ 934.00 per day × 141 days = £131,694.00

Forecast defined cost of the site establishment and management including the delay due to the compensation event

24. Site establishment and management: @ £ 934.00 per day x 181days = £169,054.00

Risk allowance:	5 days @ £934.00 per day =	£ 4,670.00
		£ 173,724.00

25. The proposed change to the Prices in relation to item 000 on the activity schedule, Site establishment and management is £42,030.00 (£173,724.00 - £131,694.00) plus the Fee (direct fee percentage of 8%) equating to £3,362.40 totalling £45,392.40.

26. Copies of the activity schedule and our proposed revised activity schedule (Revision 1) are attached. The proposed revised Total of the Prices is £796,426.00.

Eastlands STW – activity schedule

Item No.	*Description*	*Price* £
000	Site establishment and management	140,000.00
	Oxidation Ditch	
010	Excavate to formation level	6,500.00
020	Mass concrete formation	3,400.00
030	Base slab steel fixing	25,000.00
040	Base slab formwork	5,600.00
050	Base slab place concrete	13,900.00
060	Steel fixing to walls	65,000.00
070	Formwork to walls	120,000.00
080	Place concrete to walls	22,000.00

090	Strike formwork and fair finish	5,000.00
	Final settlement tank	
100	Excavate to formation level	4,000.00
110	Mass concrete formation	2,000.00
120	Base slab steel fixing	11,800.00
130	Base slab formwork	3,200.00
140	Base slab place concrete	8,000.00
150	Steel fixing to walls	25,000.00
160	Formwork to walls	65,000.00
170	Place concrete to walls	9,500.00
180	Strike formwork and fair finish	3,000.00
	Total of the prices	537,900.00

Eastlands STW – activity schedule (revision 1)

Item No.	*Description*	*Price* £
000	Site establishment and management (incl. CE 001)	185,392.00
	Oxidation Ditch	
010	Excavate to formation level	6,500.00
020	Mass concrete formation	3,400.00
030	Base slab steel fixing	25,000.00
040	Base slab formwork	5,600.00
050	Base slab place concrete	13,900.00
060	Steel fixing to walls (incl. CE 001)	98,818.00
070	Formwork to walls (incl. CE 001)	188,163.00
080	Place concrete to walls (incl. CE 001)	34,335.00
090	Strike formwork and fair finish	5,000.00
	Final settlement tank	
100	Excavate to formation level	4,000.00
110	Mass concrete formation	2,000.00
120	Base slab steel fixing	11,800.00
130	Base slab formwork	3,200.00
140	Base slab place concrete	8,000.00
150	Steel fixing to walls	25,000.00
160	Formwork to walls	65,000.00
170	Place concrete to walls	9,500.00
180	Strike formwork and fair finish	3,000.00
	Total of the prices	**796,426.00**

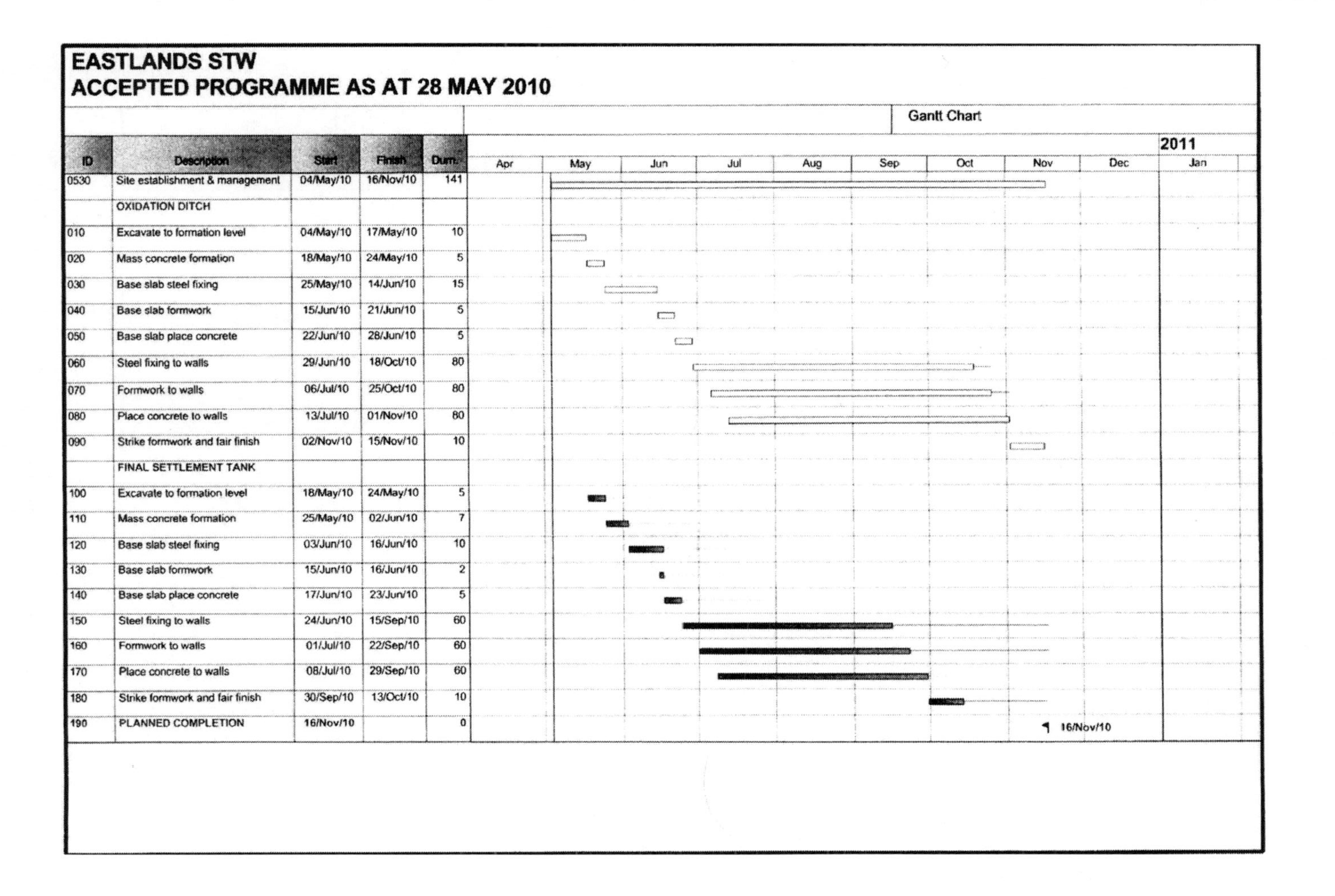

EASTLANDS STW
ACCEPTED PROGRAMME AS AT 28 MAY 2010

Gantt Chart

ID	Description	Start	Finish	Durn.
0530	Site establishment & management	04/May/10	16/Nov/10	141
	OXIDATION DITCH			
010	Excavate to formation level	04/May/10	17/May/10	10
020	Mass concrete formation	18/May/10	24/May/10	5
030	Base slab steel fixing	25/May/10	14/Jun/10	15
040	Base slab formwork	15/Jun/10	21/Jun/10	5
050	Base slab place concrete	22/Jun/10	28/Jun/10	5
060	Steel fixing to walls	29/Jun/10	18/Oct/10	80
070	Formwork to walls	06/Jul/10	25/Oct/10	80
080	Place concrete to walls	13/Jul/10	01/Nov/10	80
090	Strike formwork and fair finish	02/Nov/10	15/Nov/10	10
	FINAL SETTLEMENT TANK			
100	Excavate to formation level	18/May/10	24/May/10	5
110	Mass concrete formation	25/May/10	02/Jun/10	7
120	Base slab steel fixing	03/Jun/10	16/Jun/10	10
130	Base slab formwork	15/Jun/10	16/Jun/10	2
140	Base slab place concrete	17/Jun/10	23/Jun/10	5
150	Steel fixing to walls	24/Jun/10	15/Sep/10	60
160	Formwork to walls	01/Jul/10	22/Sep/10	60
170	Place concrete to walls	08/Jul/10	29/Sep/10	60
180	Strike formwork and fair finish	30/Sep/10	13/Oct/10	10
190	PLANNED COMPLETION	16/Nov/10		0

EASTLANDS STW
REVISED PROGRAMME (Revision 1) Submitted for acceptance 14 JUNE 2010

ID	Description	Start	Finish	Dur.
0530	Site establishment & management	04/May/10	18/Jan/11	186
	OXIDATION DITCH			
010	Excavate to formation level	04/May/10	17/May/10	10
020	Mass concrete formation	18/May/10	24/May/10	5
030	Base slab steel fixing	25/May/10	14/Jun/10	15
040	Base slab formwork	15/Jun/10	21/Jun/10	5
050	Base slab place concrete	22/Jun/10	28/Jun/10	5
060	Steel fixing to walls - Incl. effect of C E 001	29/Jun/10	20/Dec/10	125
070	Formwork to walls - Incl. effect of CE 001	06/Jul/10	27/Dec/10	125
080	Place concrete to walls - Incl effect of CE 001	13/Jul/10	03/Jan/11	125
090	Strike formwork and fair finish	04/Jan/11	17/Jan/11	10
	FINAL SETTLEMENT TANK			
100	Excavate to formation level	18/May/10	24/May/10	5
110	Mass concrete formation	25/May/10	02/Jun/10	7
120	Base slab steel fixing	03/Jun/10	16/Jun/10	10
130	Base slab formwork	15/Jun/10	16/Jun/10	2
140	Base slab place concrete	17/Jun/10	23/Jun/10	5
150	Steel fixing to walls	24/Jun/10	15/Sep/10	60
160	Formwork to walls	01/Jul/10	22/Sep/10	60
170	Place concrete to walls	08/Jul/10	29/Sep/10	60
180	Strike formwork and fair finish	30/Sep/10	13/Oct/10	10
190	PLANNED COMPLETION	18/Jan/11		0

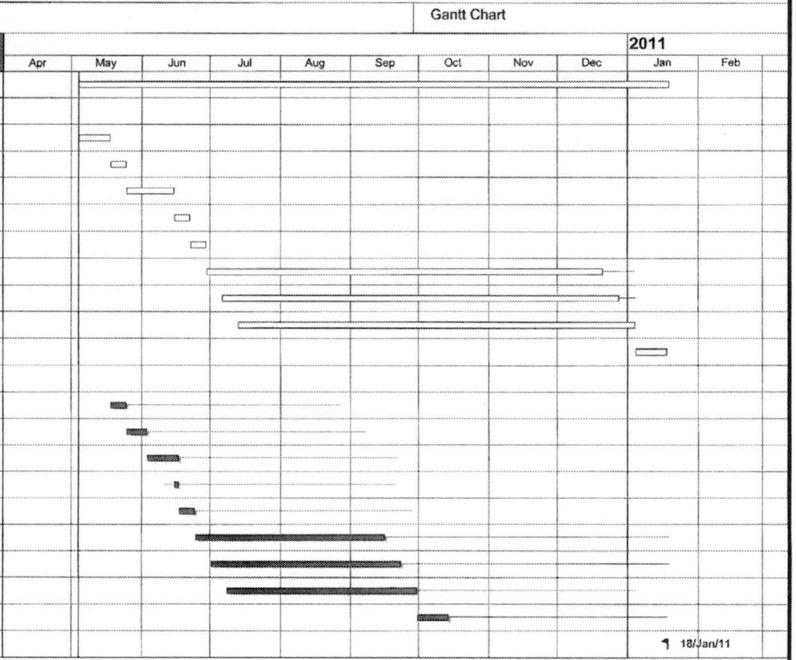

Notes

1. Events, the occurrence of which entitle the Contractor to financial and/or time relief (listed in clause 60.1)
2. See clause 61.3
3. See clauses 61.4, 62.6 and 64.4
4. [1999] EWHC Technology 263 (22nd January, 1999)
5. Except by express agreement of the parties to do so (see clause 63.14 in Option A and clause 63.13 in Options B and D. there is no provision in Option C)

Appendix D: Part II of the Housing Grants, Construction and Regeneration Act 1996 as amended by The Enterprise Act 2002 (Insolvency) Order 2003 (Statutory Instrument 2003 No. 2096), the Communications Act 2003 (Schedule 17, paragraph 137, (Commencement Order No. 1) Order 2003) and Part 8 of the Local Democracy, Economic Development and Construction Act 2009

The following is an amalgamation of the relevant provisions of the foregoing, effective from 1 October 2011 in England.

For construction contracts relating to construction operations in Wales, see the Local Democracy, Economic Development and Construction Act 2009 (Commencement Order No. 2) (Wales) Order 2011, which makes the following provisions effective in Wales from 1 October 2011.

The Construction Contracts Exclusion Orders of 1998 and 2011 should also be referred to if the construction contract concerns sections 38 or 278 of the Highways Act 1980, section 106, 106A or 299A of the Town and Country Planning Act 1990, section 104 of the Water Industry Act 1991 or a private finance initiative arrangement.

Housing Grants, Construction and Regeneration Act 1996

Part II (as amended)

[Section] 104 Construction contracts

(1) In this Part a 'construction contract' means an agreement with a person for any of the following –

(a) the carrying out of construction operations;

(b) arranging for the carrying out of construction operations by others, whether under sub-contract to him or otherwise;
(c) providing his own labour, or the labour of others, for the carrying out of construction operations.

(2) References in this Part to a construction contract include an agreement –
(a) to do architectural, design, or surveying work, or
(b) to provide advice on building, engineering, interior or exterior decoration or on the laying-out of landscape, in relation to construction operations.

(3) References in this Part to a construction contract do not include a contract of employment (within the meaning of the Employment Rights Act 1996).

(4) The Secretary of State may by order add to, amend or repeal any of the provisions of subsection (1), (2) or (3) as to the agreements which are construction contracts for the purposes of this Part or are to be taken or not to be taken as included in references to such contracts.
No such order shall be made unless a draft of it has been laid before and approved by a resolution of each of House of Parliament.

(5) Where an agreement relates to construction operations and other matters, this Part applies to it only so far as it relates to construction operations.
An agreement relates to construction operations so far as it makes provision of any kind within subsection (1) or (2).

(6) This Part applies only to construction contracts which –
(a) are entered into after the commencement of this Part, and
(b) relate to the carrying out of construction operations in England, Wales or Scotland.

(7) This Part applies whether or not the law of England and Wales or Scotland is otherwise the applicable law in relation to the contract.

105 Meaning of 'construction operations'

(1) In this Part 'construction operations' means, subject as follows, operations of any of the following descriptions –
(a) construction, alteration, repair, maintenance, extension, demolition or dismantling of buildings, or structures forming, or to form, part of the land (whether permanent or not);
(b) construction, alteration, repair, maintenance, extension, demolition or dismantling of any works forming, or to form, part of the land, including (without prejudice to the foregoing) walls, roadworks, power-lines, electronic communications apparatus, aircraft runways, docks and harbours, railways, inland waterways, pipe-lines, reservoirs, water-mains, wells, sewers, industrial plant and installations for purposes of land drainage, coast protection or defence;
(c) installation in any building or structure of fittings forming part of the land, including (without prejudice to the foregoing) systems of heating, lighting, air-conditioning, ventilation, power supply, drainage, sanitation, water supply or fire protection, or security or communications systems;
(d) external or internal cleaning of buildings and structures, so far as carried out in the course of their construction, alteration, repair, extension or restoration;
(e) operations which form an integral part of, or are preparatory to, or are for rendering complete, such operations as are previously described in this

subsection, including site clearance, earth–moving, excavation, tunnelling and boring, laying of foundations, erection, maintenance or dismantling of scaffolding, site restoration, landscaping and the provision of roadways and other access works;

(f) painting or decorating the internal or external surfaces of any building or structure.

(2) The following operations are not construction operations within the meaning of this Part –

(a) drilling for, or extraction of, oil or natural gas;

(b) extraction (whether by underground or surface working) of minerals; tunnelling or boring, or construction of underground works, for this purpose;

(c) assembly, installation or demolition of plant or machinery, or erection or demolition of steelwork for the purposes of supporting or providing access to plant or machinery, on a site where the primary activity is –

(i) nuclear processing, power generation, or water or effluent treatment, or

(ii) the production, transmission, processing or bulk storage (other than warehousing) of chemicals, pharmaceuticals, oil, gas, steel or food and drink;

(d) manufacture or delivery to site of –

(i) building or engineering components or equipment,

(ii) materials, plant or machinery, or

(iii) components for systems of heating, lighting, air-conditioning, ventilation, power supply, drainage, sanitation, water supply or fire protection, or for security or communications systems,

except under a contract which also provides for their installation;

(e) the making, installation and repair of artistic works, being sculptures, murals and other works which are wholly artistic in nature.

(3) The Secretary of State may by order add to, amend or repeal any of the provisions of subsection (1) or (2) as to the operations and work to be treated as construction operations for the purposes of this Part.

(4) No such order shall be made unless a draft of it has been laid before and approved by a resolution of each House of Parliament.

106 Provisions not applicable to contract with residential occupier

(1) This Part does not apply –

(a) to a construction contract with a residential occupier (see below).

(2) A construction contract with a residential occupier means a construction contract which principally relates to operations on a dwelling which one of the parties to the contract occupies, or intends to occupy, as his residence.

In this subsection 'dwelling' means a dwelling-house or a flat; and for this purpose –

'dwelling-house' does not include a building containing a flat; and

'flat' means separate and self-contained premises constructed or adapted for use for residential purposes and forming part of a building from some other part of which the premises are divided horizontally.

(3) The Secretary of State may by order amend subsection (2).
(4) No order under this section shall be made unless a draft of it has been laid before and approved by a resolution of each House of Parliament.

106A Power to disapply provisions of this Part

(1) The Secretary of State may by order provide that any or all of the provisions of this Part, so far as extending to England and Wales, shall not apply to any description of construction contract relating to the carrying out of construction operations (not being operations in Wales) which is specified in the order.
(2) The Welsh Ministers may by order provide that any or all of the provisions of this Part, so far as extending to England and Wales, shall not apply to any description of construction contract relating to the carrying out of construction operations in Wales which is specified in the order.
(3) The Scottish Ministers may by order provide that any or all of the provisions of this Part, so far as extending to Scotland, shall not apply to any description of construction contract which is specified in the order.
(4) An order under this section shall not be made unless a draft of it has been laid before and approved by resolution of –
   (a) in the case of an order under subsection (1), each House of Parliament;
   (b) in the case of an order under subsection (2), the National Assembly for Wales;
   (c) in the case of an order under subsection (3), the Scottish Parliament.

108 Right to refer disputes to adjudication

(1) A party to a construction contract has the right to refer a dispute arising under the contract for adjudication under a procedure complying with this section.

For this purpose 'dispute' includes any difference.
(2) The contract shall include provision in writing so as to –
   (a) enable a party to give notice at any time of his intention to refer a dispute to adjudication;
   (b) provide a timetable with the object of securing the appointment of the adjudicator and referral of the dispute to him within 7 days of such notice;
   (c) require the adjudicator to reach a decision within 28 days of referral or such longer period as is agreed by the parties after the dispute has been referred;
   (d) allow the adjudicator to extend the period of 28 days by up to 14 days, with the consent of the party by whom the dispute was referred;
   (e) impose a duty on the adjudicator to act impartially; and
   (f) enable the adjudicator to take the initiative in ascertaining the facts and the law.
(3) The contract shall provide in writing that the decision of the adjudicator is binding until the dispute is finally determined by legal proceedings, by arbitration (if the contract provides for arbitration or the parties otherwise agree to arbitration) or by agreement.

The parties may agree to accept the decision of the adjudicator as finally determining the dispute.

(3A) The contract shall include provision in writing permitting the adjudicator to correct his decision so as to remove a clerical or typographical error arising by accident or omission.
(4) The contract shall also provide in writing that the adjudicator is not liable for anything done or omitted in the discharge or purported discharge of his functions as adjudicator unless the act or omission is in bad faith, and that any employee or agent of the adjudicator is similarly protected from liability.
(5) If the contract does not comply with the requirements of subsections (1) to (4), the adjudication provisions of the Scheme for Construction Contracts apply.
(6) For England and Wales, the Scheme may apply the provisions of the Arbitration Act 1996 with such adaptations and modifications as appear to the Minister making the scheme to be appropriate.

For Scotland, the Scheme may include provision conferring powers on courts in relation to adjudication and provision relating to the enforcement of the adjudicator's decision.

108A Adjudication costs: effectiveness of provision

(1) This section applies in relation to any contractual provision made between the parties to a construction contract which concerns the allocation as between those parties of costs relating to the adjudication of a dispute arising under the construction contract.
(2) The contractual provision referred to in subsection (1) is ineffective unless –
 (a) it is made in writing, is contained in the construction contract and confers power on the adjudicator to allocate his fees and expenses as between the parties, or
 (b) it is made in writing after the giving of notice of intention to refer the dispute to adjudication.

109 Entitlement to stage payments

(1) A party to a construction contract is entitled to payment by instalments, stage payments or other periodic payments for any work under the contract unless –
 (a) it is specified in the contract that the duration of the work is to be less than 45 days, or
 (b) it is agreed between the parties that the duration of the work is estimated to be less than 45 days.
(2) The parties are free to agree the amounts of the payments and the intervals at which, or circumstances in which, they become due.
(3) In the absence of such agreement, the relevant provisions of the Scheme for Construction Contracts apply.
(4) References in the following sections to a payment provided for by the contract include a payment by virtue of this section.

110 Dates for Payment.

(1) Every construction contract shall –
 (a) provide an adequate mechanism for determining what payments become due under the contract, and when, and

(b) provide for a final date for payment in relation to any sum which becomes due.

The parties are free to agree how long the period is to be between the date on which a sum becomes due and the final date for payment.

(1A) The requirement in subsection (1)(a) to provide an adequate mechanism for determining what payments become due under the contract, or when, is not satisfied where a construction contract makes payment conditional on –
  (a) the performance of obligations under another contract, or
  (b) a decision by any person as to whether obligations under another contract have been performed.

(1B) In subsection (1A)(a) and (b) the references to obligations do not include obligations to make payments (but see section 113).

(1C) Subsection (1A) does not apply where –
  (a) the construction contract is an agreement between the parties for the carrying out of construction operations by another person, whether under sub-contract or otherwise, and
  (b) the obligations referred to in that subsection are obligations on that other person to carry out those operations.

(1D) The requirement in subsection (1)(a) to provide an adequate mechanism for determining when payments become due under the contract is not satisfied where a construction contract provides for the date on which a payment becomes due to be determined by reference to the giving to the person to whom the payment is due of a notice which relates to what payments are due under the contract.

(3) If or to the extent that a contract does not contain such provision as is mentioned in subsection (1), the relevant provisions of the Scheme for Construction Contracts apply.

110A Payment notices: contractual requirements

(1) A construction contract shall, in relation to every payment provided for by the contract –
  (a) require the payer or a specified person to give a notice complying with subsection (2) to the payee not later than five days after the payment due date, or
  (b) require the payee to give a notice complying with subsection (3) to the payer or a specified person not later than five days after the payment due date.

(2) A notice complies with this subsection if it specifies
  (a) in a case where the notice is given by the payer
    (i) the sum that the payer considers to be or to have been due at the payment due date in respect of the payment, and
    (ii) the basis on which that sum is calculated;
  (b) in a case where the notice is given by a specified person –
    (i) the sum that the payer or the specified person considers to be or to have been due at the payment due date in respect of the payment, and
    (ii) the basis on which that sum is calculated.

(3) A notice complies with this subsection if it specifies –
  (a) the sum that the payee considers to be or to have been due at the payment due date in respect of the payment, and
  (b) the basis on which that sum is calculated.

(4) For the purposes of this section, it is immaterial that the sum referred to in subsection (2)(a) or (b) or (3)(a) may be zero.
(5) If or to the extent that a contract does not comply with subsection (1), the relevant provisions of the Scheme for Construction Contracts apply.
(6) In this and the following sections, in relation to any payment provided for by a construction contract –

'payee' means the person to whom the payment is due;

'payer' means the person from whom the payment is due;

'payment due date' means the date provided for by the contract as the date on which the payment is due;

'specified person' means a person specified in or determined in accordance with the provisions of the contract.

110B Payment notices: payee's notice in default of payer's notice

(1) This section applies in a case where, in relation to any payment provided for by a construction contract –
   (a) the contract requires the payer or a specified person to give the payee a notice complying with section 110A(2) not later than five days after the payment due date, but
   (b) notice is not given as so required.
(2) Subject to subsection (4), the payee may give to the payer a notice complying with section 110A(3) at any time after the date on which the notice referred to in subsection (1)(a) was required by the contract to be given.
(3) Where pursuant to subsection (2) the payee gives a notice complying with section 110A(3), the final date for payment of the sum specified in the notice shall for all purposes be regarded as postponed by the same number of days as the number of days after the date referred to in subsection (2) that the notice was given.
(4) If
   (a) the contract permits or requires the payee, before the date on which the notice referred to in subsection (1)(a) is required by the contract to be given, to notify the payer or a specified person of –
      (i) the sum that the payee considers will become due on the payment due date in respect of the payment, and
      (ii) the basis on which that sum is calculated, and
   (b) the payee gives such notification in accordance with the contract,

   that notification is to be regarded as a notice complying with section 110A(3) given pursuant to subsection (2) (and the payee may not give another such notice pursuant to that subsection).

111 Requirement to pay notified sum

(1) Subject as follows, where a payment is provided for by a construction contract, the payer must pay the notified sum (to the extent not already paid) on or before the final date for payment.

(2) For the purposes of this section, the 'notified sum' in relation to any payment provided for by a construction contract means –
  (a) in a case where a notice complying with section 110A(2) has been given pursuant to and in accordance with a requirement of the contract, the amount specified in that notice;
  (b) in a case where a notice complying with section 110A(3) has been given pursuant to and in accordance with a requirement of the contract, the amount specified in that notice;
  (c) in a case where a notice complying with section 110A(3) has been given pursuant to and in accordance with section 110B(2), the amount specified in that notice.
(3) The payer or a specified person may in accordance with this section give to the payee a notice of the payer's intention to pay less than the notified sum.
(4) A notice under subsection (3) must specify –
  (a) the sum that the payer considers to be due on the date the notice is served, and
  (b) the basis on which that sum is calculated.

It is immaterial for the purposes of this subsection that the sum referred to in paragraph (a) or (b) may be zero.
(5) A notice under subsection (3) –
  (a) must be given not later than the prescribed period before the final date for payment, and
  (b) in a case referred to in subsection (2)(b) or (c), may not be given before the notice by reference to which the notified sum is determined.
(6) Where a notice is given under subsection (3), subsection (1) applies only in respect of the sum specified pursuant to subsection (4)(a).
(7) In subsection (5), 'prescribed period' means –
  (a) such period as the parties may agree, or
  (b) in the absence of such agreement, the period provided by the Scheme for Construction Contracts.
(8) Subsection (9) applies where in respect of a payment –
  (a) a notice complying with section 110A(2) has been given pursuant to and in accordance with a requirement of the contract (and no notice under subsection (3) is given), or
  (b) a notice under subsection (3) is given in accordance with this section,

but on the matter being referred to adjudication the adjudicator decides that more than the sum specified in the notice should be paid.
(9) In a case where this subsection applies, the decision of the adjudicator referred to in subsection (8) shall be construed as requiring payment of the additional amount not later than –
  (a) seven days from the date of the decision, or
  (b) the date which apart from the notice would have been the final date for payment,

whichever is the later.
(10) Subsection (1) does not apply in relation to a payment provided for by a construction contract where –
  (a) the contract provides that, if the payee becomes insolvent the payer need not pay any sum due in respect of the payment, and

(b) the payee has become insolvent after the prescribed period referred to in subsection (5)(a).

(11) Subsections (2) to (5) of section 113 apply for the purposes of subsection (10) of this section as they apply for the purposes of that section.

112 Right to suspend performance for non-payment

(1) Where the requirement in section 111(1) applies in relation to any sum but is not complied with, the person to whom the sum is due has the right (without prejudice to any other right or remedy) to suspend performance of any or all of his obligations under the contract to the party by whom payment ought to have been made ('the party in default').

(2) The right may not be exercised without first giving to the party in default at least seven days' notice of intention to suspend performance, stating the ground or grounds on which it is intended to suspend performance.

(3) The right to suspend performance ceases when the party in default makes payment in full of the sum referred to in subsection (1).

(3A) Where the right conferred by this section is exercised, the party in default shall be liable to pay to the party exercising the right a reasonable amount in respect of costs and expenses reasonably incurred by that party as a result of the exercise of the right.

(4) Any period during which performance is suspended in pursuance of, or in consequence of the exercise of, the right conferred by this section shall be disregarded in computing for the purposes of any contractual time limit the time taken, by the party exercising the right or by a third party, to complete any work directly or indirectly affected by the exercise of the right.

Where the contractual time limit is set by reference to a date rather than a period, the date shall be adjusted accordingly.

113 Prohibition of conditional payment provisions

(1) A provision making payment under a construction contract conditional on the payer receiving payment from a third person is ineffective, unless that third person, or any other person payment by whom is under the contract (directly or indirectly) a condition of payment by that third person, is insolvent.

(2) For the purposes of this section a company becomes insolvent –

(a) when it enters administration within the meaning of Schedule B1 to the Insolvency Act 1986,

(b) on the appointment of an administrative receiver or a receiver or manager of its property under Chapter I of Part III of that Act, or the appointment of a receiver under Chapter II of that Part,

(c) on the passing of a resolution for voluntary winding-up without a declaration of solvency under section 89 of that Act, or

(d) on the making of a winding-up order under Part IV or V of that Act.

(3) For the purposes of this section a partnership becomes insolvent –

(a) on the making of a winding-up order against it under any provision of the Insolvency Act 1986 as applied by an order under section 420 of that Act, or

(b) when sequestration is awarded on the estate of the partnership under section 12 of the Bankruptcy (Scotland) Act 1985 or the partnership grants a trust deed for its creditors.

(4) For the purposes of this section an individual becomes insolvent –

(a) on the making of a bankruptcy order against him under Part IX of the Insolvency Act 1986, or

(b) on the sequestration of his estate under the Bankruptcy (Scotland) Act 1985 or when he grants a trust deed for his creditors.

(5) A company, partnership or individual shall also be treated as insolvent on the occurrence of any event corresponding to those specified in subsection (2), (3) or (4) under the law of Northern Ireland or of a country outside the United Kingdom.

(6) Where a provision is rendered ineffective by subsection (1), the parties are free to agree other terms for payment.

In the absence of such agreement, the relevant provisions of the Scheme for Construction Contracts apply.

114 The Scheme for Construction Contracts

(1) The Minister shall by regulations make a scheme ('the Scheme for Construction Contracts') containing provision about the matters referred to in the preceding provisions of this Part.

(2) Before making any regulations under this section the Minister shall consult such persons as he thinks fit.

(3) In this section 'the Minister' means –

(a) for England and Wales, the Secretary of State, and

(b) for Scotland, the Lord Advocate.

(4) Where any provisions of the Scheme for Construction Contracts apply by virtue of this Part in default of contractual provision agreed by the parties, they have effect as implied terms of the contract concerned.

(5) Regulations under this section shall not be made unless a draft of them has been approved by resolution of each House of Parliament.

115 Service of notices, &c

(1) The parties are free to agree on the manner of service of any notice or other document required or authorised to be served in pursuance of the construction contract or for any of the purposes of this Part.

(2) If or to the extent that there is no such agreement the following provisions apply.

(3) A notice or other document may be served on a person by any effective means.

(4) If a notice or other document is addressed, pre-paid and delivered by post –

(a) to the addressee's last known principal residence or, if he is or has been carrying on a trade, profession or business, his last known principal business address, or

(b) where the addressee is a body corporate, to the body's registered or principal office,

it shall be treated as effectively served.

(5) This section does not apply to the service of documents for the purposes of legal proceedings, for which provision is made by rules of court.
(6) References in this Part to a notice or other document include any form of communication in writing and references to service shall be construed accordingly.

116 Reckoning periods of time

(1) For the purposes of this Part periods of time shall be reckoned as follows.
(2) Where an act is required to be done within a specified period after or from a specified date, the period begins immediately after that date.
(3) Where the period would include Christmas Day, Good Friday or a day under the Banking and Financial Dealings Act 1971 is a bank holiday in England and Wales or, as the case may be, in Scotland, that day shall be excluded.

117 Crown application

(1) This Part applies to a construction contract entered into by or on behalf of the Crown otherwise than by or on behalf of Her Majesty in her private capacity.
(2) This Part applies to a construction contract entered into on behalf of the Duchy of Cornwall notwithstanding any Crown interest.
(3) Where a construction contract is entered into by or on behalf of Her Majesty in right of the Duchy of Lancaster, Her Majesty shall be represented, for the purposes of any adjudication or other proceedings arising out of the contract by virtue of this Part, by the Chancellor of the Duchy or such person as he may appoint.
(4) Where a construction contract is entered into on behalf of the Duchy of Cornwall, the Duke of Cornwall or the possessor for the time being of the Duchy shall be represented, for the purposes of any adjudication or other proceedings arising out of the contract by virtue of this Part, by such person as he may appoint.

Appendix E: The Scheme for Construction Contracts (England and Wales) Regulations 1998 as amended by the Scheme for Construction Contracts (England and Wales) Regulations 1998 (Amendment) (England) Regulations 2011

Citation, commencement, application and interpretation

1. –(1) These Regulations may be cited as the Scheme for Construction Contracts (England and Wales) Regulations 1998 (Amendment) (England) Regulations 2011 and come into force on 1st October 2011.

(2) These Regulations only apply to construction contracts entered into after the coming into force of these Regulations.

(3) These Regulations do not apply to construction contracts to the extent that they relate to the carrying out of construction operations in Wales.

(4) In these Regulations –

"the Act" means the Housing Grants, Construction and Regeneration Act 1996; and

"the Principal Regulations" means the Scheme for Construction Contracts (England and Wales) Regulations 1998.

The Scheme for Construction Contracts

2. – Where a construction contract does not comply with the requirements of section 108(1) to (4) of the Act, the adjudication provisions in Part I of the Schedule to these Regulations shall apply.

3. – Where –

(a) the parties to a construction contract are unable to reach agreement for the purposes mentioned respectively in sections 109, 111 and 113 of the Act, or

(b) a construction contract does not make provision as required by section 110 or by section 110A of the Act,

the relevant provisions in Part II of the Schedule to these Regulations shall apply.

4. – The provisions in the Schedule to these Regulations shall be the Scheme for Construction Contracts for the purposes of section 114 of the Act.

SCHEDULE

Regulations 2, 3 and 4

THE SCHEME FOR CONSTRUCTION CONTRACTS

PART I

ADJUDICATION

Notice of Intention to seek Adjudication

1. – (1) Any party to a construction contract (the "referring party") may give written notice (the "notice of adjudication") at any time of his intention to refer any dispute arising under the contract, to adjudication.

(2) The notice of adjudication shall be given to every other party to the contract.

(3) The notice of adjudication shall set out briefly –

(a) the nature and a brief description of the dispute and of the parties involved,

(b) details of where and when the dispute has arisen,

(c) the nature of the redress which is sought, and

(d) the names and addresses of the parties to the contract (including, where appropriate, the addresses which the parties have specified for the giving of notices).

2. – (1) Following the giving of a notice of adjudication and subject to any agreement between the parties to the dispute as to who shall act as adjudicator –

(a) the referring party shall request the person (if any) specified in the contract to act as adjudicator, or

(b) if no person is named in the contract or the person named has already indicated that he is unwilling or unable to act, and the contract provides for a specified nominating body to select a person, the referring party shall request the nominating body named in the contract to select a person to act as adjudicator, or

(c) where neither paragraph (a) nor (b) above applies, or where the person referred to in (a) has already indicated that he is unwilling or unable to act and (b) does not apply, the referring party shall request an adjudicator nominating body to select a person to act as adjudicator.

(2) A person requested to act as adjudicator in accordance with the provisions of paragraph (1) shall indicate whether or not he is willing to act within two days of receiving the request.

(3) In this paragraph, and in paragraphs 5 and 6 below, an 'adjudicator nominating body' shall mean a body (not being a natural person and not being a party to the dispute) which holds itself out publicly as a body which will select an adjudicator when requested to do so by a referring party.

3. – The request referred to in paragraphs 2, 5 and 6 shall be accompanied by a copy of the notice of adjudication.

4. – Any person requested or selected to act as adjudicator in accordance with paragraphs 2, 5 or 6 shall be a natural person acting in his personal capacity. A person requested or selected to act as an adjudicator shall not be an employee of any of the parties to the dispute and shall declare any interest, financial or otherwise, in any matter relating to the dispute.

5. – (1) The nominating body referred to in paragraphs 2(1)(b) and 6(1)(b) or the adjudicator nominating body referred to in paragraphs 2(1)(c), 5(2)(b) and 6(1)(c) must communicate the selection of an adjudicator to the referring party within five days of receiving a request to do so.

(2) Where the nominating body or the adjudicator nominating body fails to comply with paragraph (1), the referring party may –

(a) agree with the other party to the dispute to request a specified person to act as adjudicator, or

(b) request any other adjudicator nominating body to select a person to act as adjudicator.

(3) The person requested to act as adjudicator in accordance with the provisions of paragraphs (1) or (2) shall indicate whether or not he is willing to act within two days of receiving the request.

6. – (1) Where an adjudicator who is named in the contract indicates to the parties that he is unable or unwilling to act, or where he fails to respond in accordance with paragraph 2(2), the referring party may –

(a) request another person (if any) specified in the contract to act as adjudicator, or

(b) request the nominating body (if any) referred to in the contract to select a person to act as adjudicator, or

(c) request any other adjudicator nominating body to select a person to act as adjudicator.

(2) The person requested to act in accordance with the provisions of paragraph (1) shall indicate whether or not he is willing to act within two days of receiving the request.

7. – (1) Where an adjudicator has been selected in accordance with paragraphs 2, 5 or 6, the referring party shall, not later than seven days from the date of the notice of adjudication, refer the dispute in writing (the 'referral notice') to the adjudicator.

(2) A referral notice shall be accompanied by copies of, or relevant extracts from, the construction contract and such other documents as the referring party intends to rely upon.

(3) The referring party shall, at the same time as he sends to the adjudicator the documents referred to in paragraphs (1) and (2), send copies of those documents to every other party to the dispute. Upon receipt of the referral notice, the adjudicator must inform every party to the dispute of the date that it was received.

8. – (1) The adjudicator may, with the consent of all the parties to those disputes, adjudicate at the same time on more than one dispute under the same contract.

(2) The adjudicator may, with the consent of all the parties to those disputes, adjudicate at the same time on related disputes under different contracts, whether or not one or more of those parties is a party to those disputes.

(3) All the parties in paragraphs (1) and (2) respectively may agree to extend the period within which the adjudicator may reach a decision in relation to all or any of these disputes.

(4) Where an adjudicator ceases to act because a dispute is to be adjudicated on by another person in terms of this paragraph, that adjudicator's fees and expenses shall be determined in accordance with paragraph 25.

9. – (1) An adjudicator may resign at any time on giving notice in writing to the parties to the dispute.

(2) An adjudicator must resign where the dispute is the same or substantially the same as one which has previously been referred to adjudication, and a decision has been taken in that adjudication.

(3) Where an adjudicator ceases to act under paragraph 9(1) –

(a) the referring party may serve a fresh notice under paragraph 1 and shall request an adjudicator to act in accordance with paragraphs 2 to 7; and

(b) if requested by the new adjudicator and insofar as it is reasonably practicable, the parties shall supply him with copies of all documents which they made available to the previous adjudicator.

(4) Where an adjudicator resigns in the circumstances referred to in paragraph (2), or where a dispute varies significantly from the dispute referred to him in the referral notice and for that reason he is not competent to decide it, the adjudicator shall be entitled to the payment of such reasonable amount as he may determine by way of fees and expenses reasonably incurred by him. Subject to any contractual provision pursuant to section 108A(2) of the Act, the adjudicator may determine how the payment is to be apportioned and the parties are jointly and severally liable for any sum which remains outstanding following the making of any such determination.

10. –Where any party to the dispute objects to the appointment of a particular person as adjudicator, that objection shall not invalidate the adjudicator's appointment nor any decision he may reach in accordance with paragraph 20.

11. –(1) The parties to a dispute may at any time agree to revoke the appointment of the adjudicator. The adjudicator shall be entitled to the payment of such reasonable amount as he may determine by way of fees and expenses incurred by him. Subject to any contractual provision pursuant to section 108A(2) of the Act, the adjudicator may determine how the payment is to be apportioned and the parties are jointly and severally liable for any sum which remains outstanding following the making of any such determination.

(2) Where the revocation of the appointment of the adjudicator is due to the default or misconduct of the adjudicator, the parties shall not be liable to pay the adjudicator's fees and expenses.

Powers of the adjudicator

12. – The adjudicator shall –

(a) act impartially in carrying out his duties and shall do so in accordance with any relevant terms of the contract and shall reach his decision in accordance with the applicable law in relation to the contract; and

(b) avoid incurring unnecessary expense.

13. –The adjudicator may take the initiative in ascertaining the facts and the law necessary to determine the dispute, and shall decide on the procedure to be followed in the adjudication. In particular he may –

(a) request any party to the contract to supply him with such documents as he may reasonably require including, if he so directs, any written statement from any party to the contract supporting or supplementing the referral notice and any other documents given under paragraph 7(2),
(b) decide the language or languages to be used in the adjudication and whether a translation of any document is to be provided and if so by whom,
(c) meet and question any of the parties to the contract and their representatives,
(d) subject to obtaining any necessary consent from a third party or parties, make such site visits and inspections as he considers appropriate, whether accompanied by the parties or not,
(e) subject to obtaining any necessary consent from a third party or parties, carry out any tests or experiments,
(f) obtain and consider such representations and submissions as he requires, and, provided he has notified the parties of his intention, appoint experts, assessors or legal advisers,
(g) give directions as to the timetable for the adjudication, any deadlines, or limits as to the length of written documents or oral representations to be complied with, and
(h) issue other directions relating to the conduct of the adjudication.

14. – The parties shall comply with any request or direction of the adjudicator in relation to the adjudication.

15. – If, without showing sufficient cause, a party fails to comply with any request, direction or timetable of the adjudicator made in accordance with his powers, fails to produce any document or written statement requested by the adjudicator, or in any other way fails to comply with a requirement under these provisions relating to the adjudication, the adjudicator may –
(a) continue the adjudication in the absence of that party or of the document or written statement requested,
(b) draw such inferences from that failure to comply as the circumstances may, in the adjudicator's opinion, justify, and
(c) make a decision on the basis of the information before him attaching such weight as he thinks fit to any evidence submitted to him outside any period he may have requested or directed.

16. – (1) Subject to any agreement between the parties to the contrary, and to the terms of paragraph (2) below, any party to the dispute may be assisted by, or represented by, such advisers or representatives (whether legally qualified or not) as he considers appropriate.
(2) Where the adjudicator is considering oral evidence or representations, a party to the dispute may not be represented by more than one person, unless the adjudicator gives directions to the contrary.

17. – The adjudicator shall consider any relevant information submitted to him by any of the parties to the dispute and shall make available to them any information to be taken into account in reaching his decision.

18. – The adjudicator and any party to the dispute shall not disclose to any other person any information or document provided to him in connection with the adjudication which the party supplying it has indicated is to be treated as confidential, except to the extent that it is necessary for the purposes of, or in connection with, the adjudication.

19. – (1) The adjudicator shall reach his decision not later than –

(a) twenty eight days after receipt of the referral notice mentioned in paragraph 7(1), or

(b) forty two days after receipt of the referral notice if the referring party so consents, or

(c) such period exceeding twenty eight days after receipt of the referral notice as the parties to the dispute may, after the giving of that notice, agree.

(2) Where the adjudicator fails, for any reason, to reach his decision in accordance with paragraph (1)

(a) any of the parties to the dispute may serve a fresh notice under paragraph 1 and shall request an adjudicator to act in accordance with paragraphs 2 to 7; and

(b) if requested by the new adjudicator and insofar as it is reasonably practicable, the parties shall supply him with copies of all documents which they had made available to the previous adjudicator.

(3) As soon as possible after he has reached a decision, the adjudicator shall deliver a copy of that decision to each of the parties to the contract.

Adjudicator's decision

20. – The adjudicator shall decide the matters in dispute. He may take into account any other matters which the parties to the dispute agree should be within the scope of the adjudication or which are matters under the contract which he considers are necessarily connected with the dispute. In particular, he may –

(a) open up, revise and review any decision taken or any certificate given by any person referred to in the contract unless the contract states that the decision or certificate is final and conclusive,

(b) decide that any of the parties to the dispute is liable to make a payment under the contract (whether in sterling or some other currency) and, subject to section 111(9) of the Act, when that payment is due and the final date for payment,

(c) having regard to any term of the contract relating to the payment of interest decide the circumstances in which, and the rates at which, and the periods for which simple or compound rates of interest shall be paid.

21. – In the absence of any directions by the adjudicator relating to the time for performance of his decision, the parties shall be required to comply with any decision of the adjudicator immediately on delivery of the decision to the parties.

22. – If requested by one of the parties to the dispute, the adjudicator shall provide reasons for his decision.

22A. – (1) The adjudicator may on his own initiative or on the application of a party correct his decision so as to remove a clerical or typographical error arising by accident or omission.

(2) Any correction of a decision shall be made within five days of the delivery of the decision to the parties.

(3) As soon as possible after correcting a decision in accordance with this paragraph, the adjudicator must deliver a copy of the corrected decision to each of the parties to the contract.

(4) Any correction of a decision forms part of the decision.

Effects of the decision

[Editor's note: Paragraph 23(1) is not used]

23. –(2) The decision of the adjudicator shall be binding on the parties, and they shall comply with it until the dispute is finally determined by legal proceedings, by arbitration (if the contract provides for arbitration or the parties otherwise agree to arbitration) or by agreement between the parties.

[Editor's note: Paragraph 24 is not used]

25. – The adjudicator shall be entitled to the payment of such reasonable amount as he may determine by way of fees and expenses reasonably incurred by him. Subject to any contractual provision pursuant to section 108A(2) of the Act, the adjudicator may determine how the payment is to be apportioned and the parties are jointly and severally liable for any sum which remains outstanding following the making of any such determination.

26. – The adjudicator shall not be liable for anything done or omitted in the discharge or purported discharge of his functions as adjudicator unless the act or omission is in bad faith, and any employee or agent of the adjudicator shall be similarly protected from liability.

PART II

PAYMENT

Entitlement to and amount of stage payments

1. – Where the parties to a relevant construction contract fail to agree –

(a) the amount of any instalment or stage or periodic payment for any work under the contract, or

(b) the intervals at which, or circumstances in which, such payments become due under that contract, or

(c) both of the matters mentioned in sub-paragraphs (a) and (b) above,

the relevant provisions of paragraphs 2 to 4 below shall apply.

2. –(1) The amount of any payment by way of instalments or stage or periodic payments in respect of a relevant period shall be the difference between the amount determined in accordance with sub-paragraph (2) and the amount determined in accordance with sub-paragraph (3).

(2) The aggregate of the following amounts –

(a) an amount equal to the value of any work performed in accordance with the relevant construction contract during the period from the commencement of the contract to the end of the relevant period (excluding any amount calculated in accordance with sub-paragraph (b)),

(b) where the contract provides for payment for materials, an amount equal to the value of any materials manufactured on site or brought onto site for the purposes of the works during the period from the commencement of the contract to the end of the relevant period, and

(c) any other amount or sum which the contract specifies shall be payable during or in respect of the period from the commencement of the contract to the end of the relevant period.

(3) The aggregate of any sums which have been paid or are due for payment by way of instalments, stage or periodic payments during the period from the commencement of the contract to the end of the relevant period.

(4) An amount calculated in accordance with this paragraph shall not exceed the difference between –

(a) the contract price, and

(b) the aggregate of the instalments or stage or periodic payments which have become due.

Dates for payment

3. – Where the parties to a construction contract fail to provide an adequate mechanism for determining either what payments become due under the contract, or when they become due for payment, or both, the relevant provisions of paragraphs 4 to 7 shall apply.

4. – Any payment of a kind mentioned in paragraph 2 above shall become due on whichever of the following dates occurs later –

(a) the expiry of 7 days following the relevant period mentioned in paragraph 2(1) above, or

(b) the making of a claim by the payee.

5. – The final payment payable under a relevant construction contract, namely the payment of an amount equal to the difference (if any) between –

(a) the contract price, and

(b) the aggregate of any instalment or stage or periodic payments which have become due under the contract,

shall become due on –

(a) the expiry of 30 days following completion of the work, or

(b) the making of a claim by the payee,

whichever is the later.

6. – Payment of the contract price under a construction contract (not being a relevant construction contract) shall become due on

(a) the expiry of 30 days following the completion of the work, or

(b) the making of a claim by the payee,

whichever is the later.

7. – Any other payment under a construction contract shall become due

(a) on the expiry of 7 days following the completion of the work to which the payment relates, or

(b) the making of a claim by the payee,

whichever is the later.

Final date for payment

8. –(1) Where the parties to a construction contract fail to provide a final date for payment in relation to any sum which becomes due under a construction contract, the provisions of this paragraph shall apply.

(2) The final date for the making of any payment of a kind mentioned in paragraphs 2, 5, 6 or 7, shall be 17 days from the date that payment becomes due.

Payment Notice

9. –(1) Where the parties to a construction contract fail, in relation to a payment provided for by the contract, provide for the issue of a payment notice pursuant to section 110A(1) of the Act, the provisions of this paragraph apply.

(2) The payer must, not later than five days after the payment due date, give a notice to the payee complying with sub-paragraph (3).

(3) A notice complies with this sub-paragraph if it specifies the sum that the payer considers to be due or to have been due at the payment due date and the basis on which that sum is calculated.

(4) For the purposes of this paragraph, it is immaterial that the sum referred to in sub-paragraph (3) may be zero.

(5) A payment provided for by the contract includes any payment of the kind mentioned in paragraph 2, 5, 6, or 7 above.

Notice of intention to pay less than the notified sum

10. – Where, in relation to a notice of intention to pay less than the notified sum mentioned in section 111(3) of the Act, the parties fail to agree the prescribed period mentioned in section 111(5), that notice must be given not later than seven days before the final date for payment determined either in accordance with the construction contract, or where no such provision is made in the contract, in accordance with paragraph 8 above.

Prohibition of conditional payment provisions

11. – Where a provision making payment under a construction contract conditional on the payer receiving payment from a third person is ineffective as mentioned in section 113 of the Act, and the parties have not agreed other terms for payment, the relevant provisions of –

(a) paragraphs 2, 4, 5, 7, 8, 9 and 10 shall apply in the case of a relevant construction contract, and

(b) paragraphs 6, 7, 8, 9 and 10 shall apply in the case of any other construction contract.

Interpretation

12. – In this Part of the Scheme for Construction Contracts –

'claim by the payee' means a written notice given by the party carrying out work under a construction contract to the other party specifying the amount of any payment or payments which he considers to be due and the basis on which it is, or they are calculated;

'contract price' means the entire sum payable under the construction contract in respect of the work;

'relevant construction contract' means any construction contract other than one –
(a) which specifies that the duration of the work is to be less than 45 days, or
(b) in respect of which the parties agree that the duration of the work is estimated to be less than 45 days;

'relevant period' means a period which is specified in, or is calculated by reference to the construction contract or where no such period is so specified or is so calculable, a period of 28 days;

'value of work' means an amount determined in accordance with the construction contract under which the work is performed or where the contract contains no such provision, the cost of any work performed in accordance with that contract together with an amount equal to any overhead or profit included in the contract price;

"work" means any of the work or services mentioned in section 104 of the Act.

Bibliography

Books and publications

Books

Abrahamson, M. W., *Engineering Law and the I.C.E. Contracts* (Elsevier Applied Science Publishers, 1979).

Bickford-Smith, S., *Emden's Building Contracts and Practice, 8th edition, Volume 2* (Butterworths, 1990).

Bramble, B. B. and Callahan, M. T., *Construction Delay Claims* (John Wiley & Sons, 1987) (Note: *Third edition*, by the same authors, published by Aspen Law and Business, 2000).

Callahan, M.T., Bramble, B.B., and Lurie, P. M., *Arbitration of Construction Disputes* (John Wiley & Sons, 1990).

Chambers, A., *Hudson's Building and Engineering Contracts, 12th edition* (Sweet & Maxwell Ltd, 2010).

Chapman, M. J., *Commercial & Consumer Arbitration* (Blackstone Press, 1997).

Coleman, T., *The Railway Navvies* (Penguin Books, reprinted 1981).

Demetriou, A., Savvides, L., Tsirides, C., Longworth, N. and Thomas, R., *International Arbitration* (James R. Knowles (Middle East) Limited).

Furst QC, S. and Ramsey, Justice V. QC, *Keating on Construction Contracts, 8th edition* (Sweet & Maxwell, 2008).

Glacki, A., *Building Law Information Subscriber Service (BLISS)* (Knowles Publications, 1993, 1994, 1995, 1996, 1997, 1998, 2000).

Glacki, A., *Building Case Law Digest* (Lloyd's, 1999).

Hall, M., *Construction Adjudication* (James R. Knowles Seminars Division).

Hall, M., *The Construction Act 1996* (James R. Knowles Seminars Division).

International Chamber of Commerce, *ICC: International Chamber of Commerce Rules of Arbitration, 1998* (International Chamber of Commerce, 1998).

Joint Contracts Tribunal, *JCT Practice Note 20* (RIBA Publications Ltd, 1993).

Joint Contracts Tribunal, *JCT Amendment Number 18, issued by the Joint Contracts Tribunal, 1998* (RIBA Publications Ltd, 1998).

Keating, D., *Keating on Building Contracts, First Supplement to the fourth edition* (Sweet & Maxwell Ltd, 1982).

Knowles, R., *Recent Legal Cases* (Knowles Publications, 1991).

Knowles, R., *Thirty Crucial Contractual Issues and Their Solutions* (Knowles Publications, 1992).

United Nations Commission on International Trade Law, *UNCITRAL Arbitration Rules, 1976* (United Nations Commission on International Trade Law).

Knowles, R. and Carrick, D., *Claims: Their Mysteries Unravelled, first* and *second editions* (Knowles Publications, 1989).

Knowles, R., *50 Contractual Nightmares and Their Antidotes* (James R. Knowles Seminars Division).

Knowles, R. and Entwhistle, M. (ed. Ann Glacki), *Building Law Information Subscriber Service (BLISS) Weekly Bulletins: BLISS Annuals* (Knowles Publications, 1988, 1989, 1990, 1991).

Knowles, R., Price, D. and Thomas, R., *Private Finance Initiative* (James R. Knowles Seminars Division).

Latham, Sir M., *Constructing the Team (The Latham Report)* (HMSO, 1994).

Langdon, H. W. (Chapter II), *House and Cottage Construction, Volume IV* (Caxton Publishing Company Limited, estimated 1927).

May, Hon Sir A., *Keating on Building Contracts, 5th edition* (Sweet & Maxwell Ltd, 1991).

May, Hon Sir A., *Keating on Building Contracts, 6th edition* (Sweet & Maxwell Ltd, 1995).

Mead, L., *Hudson on Building and Engineering Contracts, 6th edition* (1933).

Milne, M. *The Arbitration Act 1996* (James R. Knowles Seminars Division).

Morgan, R., *The Arbitration Ordinance of Hong Kong: A Commentary* (Butterworths, 1997).

O'Connor, P., *Alternative Dispute Resolution* (Knowles Publications, 1991).

Pickavance, K., *Delay and Disruption in Construction Contracts* (LLP Limited, 1997).

Price, D. and Thomas, R., *Privately Funded Infrastructure Projects (BOT)* (James R. Knowles International Division, 1999).

Powell-Smith, V., *Problems in Construction Claims* (BSP Professional, 1990).

Powell-Smith, V. and Furmston, M., *A Building Contract Casebook, 2nd edition* (BSP Professional, 1990).

Powell-Smith, V. and Sims, J., *Building Contract Claims, 2nd edition* (BSP Professional, 1988) (Note: *3rd edition*, updated by David Chappell, published by BSP Professional, 2000).

Price, D. and Thomas, R. *Practical Explanation of the New FIDIC Contracts* (James R. Knowles International Division, 2000).

The Institution of Civil Engineers, *ICE: The Institution of Civil Engineers' Adjudication Procedure, 1997* (Thomas Telford, 1997).

The Joint Contracts Tribunal, *CIMAR: Construction Industry Model Arbitration Rules, 1998* (RIBA Publications, 1998).

The World Bank, *Standard Bidding Documents: Procurement of Works* (The World Bank, 1995).

Thomas, R. and Binnington, C., *All You Need to Know About Construction Claims (South Africa)* (James R. Knowles and Binnington, Copeland and Associates, 1995).

Trickey, G. and Hacket, M., *The Presentation and Settlement of Contractors' Claims* (E. & F.N. Spon Ltd, 1983).

Wallace, I. N. D., *Construction Contracts: Principles and Policies in Tort and Contract* (Sweet & Maxwell Ltd, 1986).

Wallace, I. N. D., *Hudson's Building and Engineering Contracts, 10th edition* (Sweet & Maxwell Ltd, 1970).

Wallace, I.N. D., *Hudson's Building and Engineering Contracts, tenth edition, First Supplement* (Sweet & Maxwell Ltd, 1978).

Wallace, I. N. D., *Hudson's Building and Engineering Contracts, 11th edition* (Sweet & Maxwell Ltd, 1994).

United Nations Commission on International Trade Law, *UNCITRAL Model Law on International Commercial Arbitration, 1985* (United Nations Commission on International Trade Law).

Reports and Directives

The Banwell Report – The Placing and Management of Contracts for Building and Civil Engineering Work (HMSO), 1964.
The Excluded Sectors Directive, 90/531 OJ L2971/29 October 1990 (EC), amended by Council Directive 98/4/EC dated 16 February 1998 (EC).
The Joint Contracts Committee of Hong Kong.
The Public Supplies Directive, 77/62/EEC, amended 22 March 1988 88/295/EEC (EC), amended by Council Directive 93/36/EEC dated 14 June 1992 (EC).
The Public Works Directive, 77/305/EEC, amended 18 July 1989 89/440/EEC (EC), amended by Council Directive 93/37/EEC dated 14 June 1993 (EC).

Journals and conference papers

Binnington, C. and Thomas, R., 'Some Reminiscences on South African Legal Cases', *South African Builder*, December 1995/January 1996.
Falkner, R., 'Public Procurement Directives', Conference paper at seminar '*The Construction Industry, Europe and 1992*', organised by Legal Studies & Services Limited, 10 December 1990.
Nielson, K., 'Anatomy of a construction project', *International Construction*, November 1980.
Thomas, R., 'The Use and Abuse of FIDIC in South East Asia, Parts I, II and III', *International Construction*, May 1996, July 1996, September 1996.
The Society of Construction Law Delay and Disruption Protocol October 2002, October 2004 reprint.
Webb, A., 'The Origin and Use of Cost Performance Measurement: Part 1', *The Project Manager Today*, November/December 1991.

Acts of Parliament; Statutory Instruments; Laws and legal codes (and pages of this book on which referred to)

Arbitration (Scotland) Act 2010 – p. 193.
Article 81 of the European Treaty – p. 49.
Article 82 of the European Treaty – p. 49.
Bahrain Legislative Decree No. 9 of 1994 – p. 195.
Civil Procedure Rules for England and Wales – p. 190.
Communications Act (Commencement Order No. 1) 2003 – Appendix D
Competition Act 1998 – p. 49, Appendix D.
Enterprise Act 2002 – p. 49, Appendix D.
Housing Grants, Construction and Regeneration Act 1996 – pp. 24, 148, 182, 183, 191, Appendix D.
Late Payment of Commercial Debt (Interest) Act of 1998 – p. 148.
Local Democracy, Economic Development and Construction Act 2009 – p. 191, Appendix D.
Local Democracy, Economic Development and Construction Act 2009 (Commencement Order No. 2)(Wales) Order 2011 – Appendix D.
Pre-Action Protocol for Construction and Engineering Disputes – pp. 189, 190.
Public Contracts Regulations 2006 (SI 2006 No 5) – p. 50.
Scheme for Construction Contracts (England and Wales) Regulations 1998 – p. 191, Appendix E.

Scheme for Construction Contracts (England and Wales) Regulations 1998 (Amendment)(England) Regulations 2011 – p. 191, Appendix E.
The Arbitration Acts of 1950, 1975, 1979 and 1996 – p. 192.
The Commercial and Civil Code of Kuwait – p. 4.
The Commercial and Civil Code of Thailand – p. 98.
The Contracts Act of Malaysia – p. 15.
The Construction Contracts (England) Exclusion Order 2011 – Appendix D.
The Conventional Penalties Act of South Africa, Act 15 of 1962 – p. 15.
The Courts and Legal Services Act 1990 – p. 194.
The Enterprise Act 2002 (Insolvency) Order 2003 (Statutory Instrument 2003 No. 2096) – p. 49.
The Hong Kong Arbitration Ordinance; Chapter 341 – p. 195.
The Public Contracts (Scotland) Regulations 2006 (SSI 2006 No 1) – p. 49.
The Public Contracts (Amendment) Regulations 2009 (SI 2009 No. 2992) – p. 49.
The Unfair Contract Terms Act 1977 – p. 98.
The United Nations Convention on the Recognition and Enforcement of Foreign Arbitral Awards 1958 – p. 196.
The Utilities Contracts (Scotland) Regulations 2006 (SSI 2006 No 2) – p. 49.
Utilities Contracts Regulations 2006 (SI 2006 No 6) – p. 50.

List of cases (and pages of this book on which referred to)

J. and J. C. Abrahams v. *Ancliffe* [1938] 2 NZLR 420 – p. 96.
Alghussein Establishment v. *Eton College* [1988] 1 WLR 587 – p. 14.
Amalgamated Building Contractors v. *Waltham Holy Cross UDC* [1952] 2 All ER 452 – p. 12.
Amec Building Ltd v. *Cadmus Investment Co Ltd* [1996] 51 ConLR 105 – pp. 9, 20, 106, 147.
Aoki Corp v. *Lippoland (Singapore) Pte Ltd* [1995] 2 SLR – p. 173.
Appeal of Eichleay Corporation, ASBCA 5183, 60–2 BCA (CCH) 2688 (1960) – p. 108.
Appeal of Pathman Construction Co, ASBCA 14285, 71–1 BCA (CCH) 8905 (1971) – p. 128.
Baese Pty Ltd v. *R.A. Bracken Building Pty Ltd* (1989) 52 BLR 130 – p. 173.
Balfour Beatty Construction Limited v. *The Mayor and Burgesses of the London Borough of Lambeth* [2002] BLR 288 – p. 85.
Balfour Beatty Building Ltd v. *Chestermount Properties Ltd* (1993) 62 BLR 1 – pp. 91, 92, 94, 95.
Barry D Trentham Ltd v. *McNeil* (1996) SLT 202 – p. 46.
Beaufort Developments Ltd v. *Gilbert Ash NI Ltd and Others* [1998] 2 All ER 778 – p. 194.
Beechwood Development Company (Scotland) Limited v. *Stuart Mitchell t/a Discovery Land Surveys* [2001] (unreported) – p. 107.
Bolt v. *Thomas* (1859): (*Hudson's Building and Engineering Contracts, tenth edition* at page 196) – p. 6.
Borough of Kingston-upon-Thames v. *Amec Civil Engineering* [1993] 35 ConLR 39 – p. 147.
Bouissevan v. *Weil* (1948) 1KB 482 – p. 6.
Boyd & Forrest v. *Glasgow S.W. Railway Company* [1914] SC 472 – p. 16.

Bramall and Ogden v. *Sheffield City Council* (1983) 29 BLR 73 – p. 15.
Bremer Handelsgesell-Schaft M.B.H. v. *Vanden Avenne-Izigem P.V.B.A.* [1978] 2 Lloyds LR 109 – pp. 22, 173.
British Airways Pension Trustees Ltd v. *Sir Robert McAlpine and Son* (1995) 72 BLR 26 – p. 20.
British Steel Corporation v. *Cleveland Bridge Engineering Co Ltd* (1981) 24 BLR 94 – p. 59.
Bryant and Sons Ltd v. *Birmingham Saturday Hospital Fund* [1938] 1 All ER 503 – p. 7.
Bush v. *Whitehaven Port and Town Trustees* (1888) 52 JP 392 – p. 9.
Capital Electric Company v. *United States* (Appeal No 88/965, 7.2.84) 729 F.2d 143 (1984) – p. 109.
Carslogie S.S. Co. v. *Norwegian Government* [1952] AC 292 – p. 119.
CFW Architects (a firm) v. *Cowlin Construction Limited* [2006] (unreported) – p. 108.
City Inn Ltd v. *Shepherd Construction Ltd* [2010] Scot CSIH 68 – p. 85.
Commission of the European Communities v. *Ireland* (1988) 44 BLR 1 – p. 58.
J. Crosby & Sons Ltd v. *Portland Urban District Council* (1967) 5 BLR 121 – p. 19.
Davis Contractors Limited v. *Fareham U.D.C.* [1956] AC 696 – p. 56.
Department of Environment for Northern Ireland v. *Farrans* (1981) 19 BLR 1 – p. 181.
Ellis-Don v. *Parking Authority of Toronto* (1978) 28 BLR 98 – p. 107.
E.C. Ernst, Inc v. *Koppers Co* 476 F.Supp. 729 (W.D.Pa 1979) – p. 133.
Fairclough Building Ltd v. *Rhuddlan Borough Council* (1985) 30 BLR 26 – p. 160.
H. Fairweather & Co Ltd v. *London Borough of Wandsworth* (1987) 39 BLR 106 – p. 60.
Farm Assist Limited (in liquidation) v. *The Secretary of State for the Environment, Food and Rural Affairs (No. 2)* [2009] BLR 399 – p. 190.
A. E. Farr Ltd v. *Ministry of Transport* [1960] 1 WLR 956; (1965) 5 BLR 94 – p. 8.
Fernbrook Trading Co. Ltd v. *Taggart* [1979] 1 NZLR 556 – p. 12.
J.F. Finnegan v. *Sheffield City Council* (1989) 43 BLR 124 – p. 108.
Fratelli Constanzo SpA v. *Comune di Milano* [1990] 3 CMLR 239 – p. 58.
General Insurance Co of America v. *Hercules Construction* 385 F.2d 13 (8th Cir 1967) – pp. 130, 131.
George Mitchell (Chester Hall) Ltd v. *Finney Lock Seeds Ltd* (1983) 1-CLD-05-18 – p. 97.
Gilbert Ash (Northern) Ltd v. *Modern Engineering (Bristol) Ltd* [1974] AC 689 – p. 183.
M.J. Gleeson (Contractors) Ltd v. *London Borough of Hillingdon* (1970) 215 EG 165 – p. 60.
M.J. Gleeson Plc v. *Taylor Woodrow Plc* (1990) 49 BLR 95 – p. 170
Glenlion Construction Ltd v. *The Guinness Trust* (1987) 39 BLR 89 – p. 56.
GMTC Tools & Equipment Ltd v. *Yuasa Warwick Machinery Ltd* (1995) 73 BLR 102 – p. 20.
Government of Ceylon v. *Chandris* [1965] 3 All ER 48 – p. 119.
Group 5 Building Limited v. *The Minister of Community Development* 1993(3) SA 629 (A) – p. 13.
Henry Boot Construction (UK) Limited v. *Malmaison Hotel (Manchester) Limited* (1999) 70 ConLR 32 – p. 85.
Holme v. *Guppy* (1838) 3 M & W 378 – p. 9.

Housing Authority v. *E W Johnson Construction Co* 573 S W 2d at 323. – p. 125.
Howard Marine & Dredging v. *Ogden* (1978) 9 BLR 34 – p. 17.
Imperial Chemical Industries v. *Bovis Construction Ltd and Others* (1993) 32 ConLR 90 – p. 20.
James Longley & Co Ltd v. *South West Regional Health Authority* (1985) 25 BLR 56 – p. 149.
John F Burke Engineering and Construction, ASBCA 8182, 1963 BCA – p. 125.
Kelly & Hingles Trustees v. *Union Government (Minister of Public Works)* 1928 TPD 272 – p. 13.
Larut Matang Supermarket Sdn Bhd v. *Liew Fook Yung* [1995] 1 MLJ 375 – p. 15.
Leyland Shipping Company v. *Norwich Union Fire Insurance Society* [1918] AC 350 – p. 117.
London Borough of Merton v. *Stanley Hugh Leach Ltd* (1985) 32 BLR 51 – pp. 19, 127.
Marsden Construction Co Ltd v. *Kigass Ltd* (1989) 15 ConLR 116 – p. 58.
Martin Grant & Co Ltd v. *Sir Lindsay Parkinson & Co Ltd* (1984) 29 BLR 31 – p. 154.
Mathind Ltd v. *E. Turner & Sons Ltd* [1986] 23 ConLR 16 – p. 60.
McMaster University v. *Wilchar Construction Ltd* (1971) 22 DLR (3d) 9 – p. 58.
Michael Salliss & Co Ltd v. *E.C.A. Calil and William F. Newman & Associates* [1989] 13 ConLR 68 – p. 24.
Mid-Glamorgan County Council v. *J. Devonald Williams & Partner* [1992] 29 ConLR 129 – pp. 19, 20.
Miller v. *London County Council* (1934) 151 LT 425 – pp. 12, 97.
Mitsui Construction Co Ltd v. *Attorney General of Hong Kong* (1986) 33 BLR 1 – p. 142.
Moon v. *Witney Union* (1837): (*Hudson's Building and Engineering Contracts, tenth edition* at page 113) – p. 2
Morgan Grenfell Ltd v. *Sunderland Borough Council and Seven Seas Dredging Ltd* (1991) 51 BLR 85 – pp. 23, 145.
Morrison-Knudsen v. *B.C. Hydro & Power* (1975) 85 DLR 3d 186 – p. 23.
Morrison-Knudsen International Co Inc and Another v. *Commonwealth of Australia* (1980) 13 BLR 114 – p. 17.
Mowlem Plc v. *Stena Line Ports Limited* [2004] (unreported) – p. 60.
Multiplex Constructions (UK) Ltd v. *Honeywell Systems* [2007] BLR 195 – pp. 77, 95, 98.
Nash Dredging Ltd v. *Kestrell Marine Ltd* (1986) SLT 62 – p. 23.
Natkin & Co v. *George A. Fuller Co* 347 F.Supp.17 (WD Mo 1972), reconsidered 626 F.2d 324 (8th Cir 1980) – pp. 131, 133.
North West Regional Health Authority v. *Derek Crouch* [1984] 2 WLR 676 – p. 194.
Ovcon (Pty) Ltd v. *Administrator Natal* 1991 (4) SA 71 – p. 124.
Owen L Schwam Construction Co, ASBCA 22407, 79–2 BCA (CCH) – p. 126.
Pacific Associates Inc and Another v. *Baxter and Others* (1988) 44 BLR 33 – p. 24.
Peak Construction (Liverpool) Ltd v. *Mckinney Foundations Ltd* (1970) 1 BLR 111 – p. 11.
Penvidic Contracting Co. Ltd v. *International Nickel Co. of Canada Ltd* (1975) 53 DLR (3d) 748 – p. 127.

Percy Bilton Ltd v. *The Greater London Council* (1981) 17 BLR 1 (CA); (1982) 20 BLR 1 (HL) – p. 157.
Perini Pacific Ltd v. *Greater Vancouver Sewerage and Drainage District Council* [1967] SCR 189 – p. 10.
Philips Hong Kong Ltd v. *The Attorney General of Hong Kong* (1990) 50 BLR 122 – p. 15.
Philips Hong Kong v. *The Attorney General of Hong Kong* (1993) 61 BLR 41 (PC) – p. 15.
Property and Land Contractors Ltd v. *Alfred McAlpine Homes North Ltd* (1996) 76 BLR 59 – p. 107.
Rapid Building Group Ltd v. *Ealing Family Housing Association Ltd* (1984) 29 BLR 5 – pp. 11, 14.
Rees and Kirby Ltd v. *Swansea City Council* (1985) 30 BLR 1 – p. 147.
Royal Brompton Hospital NHS Trust v. *Frederick Hammond & Others* (No. 7) [2001] 76 ConLR 148 – p. 85.
Secretary of State for Transport v. *Birse–Farr Joint Venture* (1993) 62 BLR 36 – pp. 29, 146, 147.
C.J. Sims v. *Shaftesbury Plc* (1991) QBD; 8-CLD-03-10 – p. 60.
Song Toh Chu v. *Chan Kiat Neo* [1973] 2 MLJ 206 – p. 15.
St Modwen Development Ltd v. *Bowmer and Kirkland* [1996] 38 BLISS 4 – p. 105.
Steria Ltd v. *Sigma Wireless Communications Ltd* [2007] 2008 BLR 79 – pp. 22, 70, 77, 98, 99, 174.
Strachan & Henshaw Limited v. *Stein Industrie (UK) Limited* (1998) 87 BLR 52 – p. 97.
Sun Shipbuilding & Dry Dock Co v. *United States Lines Inc.* 76 US C.Cls 154 (1932) – p. 125.
Sutcliffe v. *Thackrah and Others* (1974) 4 BLR 16 – p. 23.
Sydney Constructions Co No 21377, 77–2 BCA (CCH) – p. 126.
Tate & Lyle Food Distribution Ltd and Another v. *Greater London Council* [1982] 1 WLR 149 – p. 104.
Temloc Ltd v. *Erril Properties Ltd* (1987) 39 BLR 31 – p. 14.
Tersons Ltd v. *Stevenage Development Corporation* (1963) 5 BLR 54 – p. 16.
Token Construction Co Ltd v. *Charlton Estates Ltd* (1976) 1 BLR 48 – p. 181.
Tropicon Contractors Pte Ltd v. *Lojan Properties Pte Ltd* [1991] 2 MLJ 70 (CA); (1989) 2 MLJ 215 (dist) – pp. 173, 174.
Victoria Falls and Transvaal Power Co Ltd v. *Consolidated Langlaagke Mines Ltd* (1915) AD – p. 124.
Waghorn v. *Wimbledon Local Board* (1877): (*Hudson's Building and Engineering Contracts, tenth edition* at page 114) – p. 2.
Wates Construction (London) Ltd v. *Franthom Properties Ltd* (1991) 53 BLR 23 – p. 19.
Wegan Construction Company Pty. Ltd v. *Wodonga Sewerage Authority* [1978] VR 67 – pp. 9, 145.
Wells v. *Army and Navy Co-operative Society Ltd* (1902) 86 LT 764 – pp. 10, 124.
Wharf Properties Ltd and Another v. *Eric Cumine Associates and Others* (1988) 45 BLR 72; (1991) 52 BLR 1 PC – p. 19.
Whittall Builders Company Ltd v. *Chester-le-Street District Council* (1985) unreported – pp. 107, 127, 130.
William Lacey (Hounslow) Ltd v. *Davis* [1987] 2 All ER 712 – p. 58.

Wood v. *Grand Valley Railway Co* (1916) 51 SCR 283 – pp. 127.
Woon Hoe Kan & Sons Sdn Bhd v. *Bandar Raya Development Bhd* [1973] 1 MLJ 60 – p. 15.
Yorkshire Water Authority v. *Sir Alfred McAlpine and Son (Northern) Ltd* (1985) 32 BLR 114 – p. 66.

Forms of contract

JCT: *Joint Contracts Tribunal* (2011 standard form contracts published by Sweet & Maxwell – part of Thompson Reuters (Professional) UK Limited).
Design and Build Contract 2011 ***JCT63***: *Standard Form of Building Contract*, 1963.
JCT80: *Standard Form of Building Contract (**JCT98**)*, 1998.
JCT Standard Building Contract 2011.
IFC84: *Intermediate Form of Contract (**IFC98**)*, 1998.
Intermediate Building Contract 2011.
MW80: *Minor Works Form of Contract (**MW98**)*, 1998.
Minor Works Form of Building Contract 2011.
CDPS: *Contractor's Designed Portion Supplement.*
Standard Form of Contract with Approximate Quantities.
Fixed Fee Form of Contract.
JCT87: *Standard Form of Management Contract*, 1998.
Management Building Contract 2011.
*Prime Cost Contract, **PCC98***, 1998.
Prime Cost Building Contract 2011.
*Standard Form of Building Contract with Contractor's Design, **CD98***, 1998.
NSC/1: *The Standard Form of Nominated Subcontract Tender and Agreement.*
NSC/4A: *The Standard Form of Nominated Subcontract.*
ICE: *NEC Standard Forms of Contract*, published by Thomas Telford Limited (a wholly owned subsidiary of The Institution of Civil Engineers).
***ICE** Standard Forms of Contract*, published by Thomas Telford Limited, sponsored by The Institution of Civil Engineers, The Association of Consulting Engineers and The Civil Engineers Contracting Association.
***ICE** Conditions of Contract, 5th edition*, 1986.
***ICE** Conditions of Contract, 6th edition*, 1995.
***ICE** Conditions of Contract, 7th edition*, 1999.
***ICE** Design and Build Construct Conditions of Contract*, 1992.
***ICE** Conditions of Contract for Minor Works, 2nd edition*, 1995.
NEC: *The New Engineering Contract; The Engineering and Construction Contract, 2nd edition.*
NEC 3 Engineering and Construction Contract 2005.
FIDIC: *Fédération International des Ingénieurs-Conseils.*
Conditions of Contract for Works of Civil Engineering Construction, fourth edition (**The Red Book**), 1987, reprinted 1992.
Conditions of Contract for Electrical and Mechanical Works, third edition (**The Yellow Book**), 1987, reprinted 1999.
Conditions of Contract for Design–Build and Turnkey, first edition (**The Orange Book**), 1995.
Conditions of Contract for Construction, first edition (**The Red Book**), 1999.

Conditions of Contract for Plant and Design–Build, first edition (**The Yellow Book**), 1999.
Conditions of Contract for EPC Turnkey Projects, first edition (**The Silver Book**), 1999.
Short Form of Contract, first edition (**The Green Form**), 1999.
Conditions of Subcontract for Works of Civil Engineering Construction, first edition, 1994 The Institution of Electrical Engineers, The Institution of Mechanical Engineers and The Association of Consulting Engineers, *Model Form of General Conditions of Contract for Use in Connection with Home or Overseas Contracts for the Supply of Electrical, Electronic or Mechanical Plant – with Erection, 1988 (MF/1), Revision 3*, 1995 (The Joint IMechE/IEE Committee).
The Federation of Civil Engineering Contractors (The Civil Engineering Contractors' Association), ***FCEC:*** *Form of Sub-Contract* (**The Blue Form**) 1991; ***CECE*** *Form of Subcontract*, 1998 (The Federation of Civil Engineering Contractors (The Civil Engineering Contractors' Association)).
GC/Works/1 – Edition 3: *General Conditions of Government Contracts for Building and Civil Engineering Works*, prepared by The Department of the Environment (HMSO).
GC/Works/1 *with Quantities (1998): General Conditions of Government Contracts for Building and Civil Engineering Works*, prepared by The Property Advisers to the Civil Estate (PACE) (HMSO).
Standard Form of Contract, issued by the Dubai Municipality.
Malaysian Standard Form of Building Contract, issued by Purtubohan Arkitec Malaysia (PAM) [The Malaysian Association of Architects].
SIA Conditions of Contract: The Conditions of Contract, issued by The Singapore Architects' Association.
RIBA Model Form of Contract, issued by The Royal Institute of Architects.

Miscellaneous abbreviations

BAS: Building Automation System.
BLISS: Building Law Information Subscriber Service.
BQ: Bill of Quantities.
CCTV: Closed Circuit Television.
CESMM: *Civil Engineering Standard Method of Measurement*, issued by the Institution of Civil Engineers.
EC: European Commission.
ECU: European Currency Unit.
EEC: European Economic Community.
EOT: Extension of Time.
FIDIC: Fédération Internationale des Ingénieurs-Conseils.
HMSO: Her Majesty's Stationery Office.
HVAC: Heating, Ventilating and Air Conditioning.
ICE: Institution of Civil Engineers.
ISO: International Standards Organisation.
JCT: Joint Contracts Tribunal.
OJ: *The Official Journal of The European Commission.*
P & Gs: Preliminary and General Costs.
PC: Prime Cost.

RIBA: Royal Institute of British Architects.
SIA: Singapore Institute of Architects.
SMM: *Standard Method of Measurement*: published by the Royal Institution of Chartered Surveyors and The National Federation of Building Trades Employers (Building Employers Confederation).
SMM7: *Standard Method of Measurement (seventh edition)* issued by the Royal Institution of Chartered Surveyors and the National Federation of Building Trades Employers (Building Employers Confederation).

Index

Printed and bound by CPI Group (UK) Ltd, Croydon, CR0 4YY